Risk, Systems and Decisions

Series Editors

Igor Linkov, U.S. Army ERDC, Vicksburg, MS, USA

Jeffrey Keisler, College of Management, University of Massachusetts, Boston, MA, USA

James H. Lambert, University of Virginia, Charlottesville, VA, USA

Jose Rui Figueira, CEG-IST Instituto Superior Técnico, University of Lisbon, LISBOA, Portugal

Health, environment, security, energy, technology are problem areas where man-made and natural systems face increasing demands, giving rise to concerns which touch on a range of firms, industries and government agencies. Although a body of powerful background theory about risk, decision, and systems has been generated over the last several decades, the exploitation of this theory in the service of tackling these systemic problems presents a substantial intellectual challenge. This book series includes works dealing with integrated design and solutions for social, technological, environmental, and economic goals. It features research and innovation in cross-disciplinary and transdisciplinary methods of decision analysis, systems analysis, risk assessment, risk management, risk communication, policy analysis, economic analysis, engineering, and the social sciences. The series explores topics at the intersection of the professional and scholarly communities of risk analysis, systems engineering, and decision analysis. Contributions include methodological developments that are well-suited to application for decision makers and managers.

More information about this series at https://link.springer.com/bookseries/13439

Adam Izdebski · John Haldon · Piotr Filipkowski
Editors

Perspectives on Public Policy in Societal-Environmental Crises

What the Future Needs from History

 Springer

Editors
Adam Izdebski
MPI for the Science of Human History
Jena, Thüringen, Germany

John Haldon
Princeton University
Princeton, NJ, USA

Piotr Filipkowski
Polish Academy of Sciences
Berlin, Germany

ISSN 2626-6717   ISSN 2626-6725   (electronic)
Risk, Systems and Decisions
ISBN 978-3-030-94136-9   ISBN 978-3-030-94137-6   (eBook)
https://doi.org/10.1007/978-3-030-94137-6

This Springer imprint is published by the registered company Springer Nature Switzerland AG
The registered company address is: Gewerbestrasse 11, 6330 Cham, Switzerland

Acknowledgements

This book is an outcome of intensive collaboration and cooperation among many people and between several institutions. Most of the contributions in the volume come from two interconnected academic events—both organised virtually under COVID-19 pandemic restrictions. The first was a conference at Princeton University in the fall semester of 2020, organised by the Climate Change and History Research Initiative (CCHRI), entitled "Past answers to current concerns: approaches to understanding historical societal resilience". The second was a series of video lectures finalised with a workshop entitled "What sort of the Past does the Future need?", which was organised by the Center for Historical Research in Berlin of the Polish Academy of Sciences together with the Max Planck Institute for the Science of Human History and the Institute of Philosophy and Sociology of the Polish Academy of Sciences. These are our home academic institutions, and we would like to thank their Directors and all our Colleagues who supported us in organising them—and helped us, also financially, to transform the results into this publication.

This work was supported by the National Socio-Environmental Synthesis Center (SESYNC) under funding received from the National Science Foundation DBI-1639145. The Open Access publication of this book has been supported by

- CCHRI, Princeton University,
- Jagiellonian University in Krakow, Excellence Initiative, POB Heritage,
- Max Planck Institute for the Science of Human History, *Palaeo-Science and History* Independent Research Group,
- Georgetown University, Department of History,
- Polish Academy of Science, Institute of Philosophy and Sociology.

Adam Izdebski
John Haldon
Piotr Filipkowski

Contents

Introduction: What Sort of Past Does Our Future Need?

Adam Izdebski, John Haldon, and Piotr Filipkowski

Abstract In this short introduction we set out the aims of the volume, which represents the fruits of two seminars held in the autumn of 2020. The chapters respond to one big thematic issue: how to research and understand historical societal resilience; and one big question: what sort of past does the future need? They attempt to address these through three linked themes: can history be made more relevant to modern policy in respect of environmental and climate challenges? To what extent do our various sources indicate awareness and management of risk and/or the implementation of mitigating strategies in the past? And how can we identify 'resilience' in the social praxis of historical agents?

Keywords Resilience · Agency · Risk · Complexity · Societal perceptions · Mitigation

The histories we tell never emerge in a vacuum, and history as an academic discipline that studies the past is highly sensitive to the concerns of the present and the heated debates that tear apart entire societies. But does the study of the past also have something to teach us about the future? Can history help us in coping—on different levels: philosophical, psychological, scientific, socio-economic, socio-technological...—with the planetary crisis we are now facing? Does history in the Anthropocene have a new task, does it need to change? Can it help us in facing the current pandemic and finding ways out of the post-Covid crisis that is looming on the horizon? By

A. Izdebski
Max Planck Institute for the Science of Human History, Jena, Germany

Jagiellonian University in Krakow, Krakow, Poland

J. Haldon (✉)
Princeton University, Princeton, NJ, USA
e-mail: jhaldon@princeton.edu

P. Filipkowski
Berlin Centre for Historical Research, Polish Academy of Sciences, Berlin, Germany

Institute of Philosophy and Sociology, Polish Academy of Sciences, Warsaw, Poland

© The Author(s) 2022
A. Izdebski et al. (eds.), *Perspectives on Public Policy in Societal-Environmental Crises*, Risk, Systems and Decisions, https://doi.org/10.1007/978-3-030-94137-6_1

analyzing historical societies as complex adaptive systems, we contribute to contemporary thinking about societal-environmental interactions in policy and planning and consider how environmental and climatic changes, whether sudden high impact events or more subtle gradual changes, impacted human responses in the past. We ask how societal perceptions of such changes affect behavioral patterns and explanatory rationalities in premodernity, and whether a better historical understanding of these relationships can inform our response to contemporary problems of similar nature and magnitude, such as adapting to climate change.

This collection of papers is drawn from the presentations delivered in two linked seminars at Princeton University and at the Berlin Centre for Historical Research of the Polish Academy of Science in the autumn of 2020. The first addressed a range of issues connected with societal resilience under the rubric "Past answers to current concerns. Approaches to understanding historical societal resilience"; the second addressed the question: "What sort of past does the future need?" This volume, therefore, addresses both sets of questions and topics by bringing together a team of scholars in the humanities, the social and the natural sciences, based in Germany, Austria, Poland, the UK, the USA and Canada. Together, we reflect on how looking into the past can help us cope with the present and prepare for the future. The question of the extent to which a better understanding of the ways in which past societies dealt with environmental and climatic challenges might inform our response to contemporary problems of a similar nature and magnitude—such as adapting to climate change—remains the focus of a good deal of discussion. Yet it is still the case that the history invoked in this debate often tends towards a simplified and reductionist interpretation of the past. One of the aims of this collection is to look at how societal perceptions of significant change affected behaviors and explanatory rationalities in premodernity. Another is to try to build complexity and multi-causality into our understanding of past examples of how different cultures coped with such stresses, with a view to isolating key structural elements that either facilitate or jeopardise resilience or sustainability.

Our collection of papers sets out to address these issues by approaching them through three key themes. First we ask whether a better historical understanding of past responses to significant threats to people's environment and the world they inhabited can help contemporaries better grapple with comparable risks and challenges today. Can we draw lessons that are not bound to the national, ethnic, geographical, historical etc. context in which our case studies unfold? In short, can history be made more relevant to modern policy with regard to such challenges? Secondly, we wanted to think about 'threat awareness' in the past. To what extent is there any indication or evidence for awareness and management of risk and/or the implementation of mitigating strategies in historical societies? Did people in the society/societies in question perceive or understand major risks or challenges as such, and how did they react/respond? Or rather, were reactions random, contingent, or at times even dysfunctional or socially exclusive, leading to increased conflict? And finally, we ask about how to highlight the differences in resilience or sustainability as perceived by us, as external observers, and as perceived by different groups and agents within the society in question? In other words, how is 'resilience' or its absence to be identified in agents' social behaviors?

The impacts of environmental stress on past societies are still poorly understood, although there is a good deal of information on the eventual results of such impacts in terms of political change and transformation. But what constitutes an existential risk to a given historical society—a risk that could trigger the collapse of a political or cultural system—has to be approached from two angles: that of the external observer; and that of the people who lived through those changes. Past human societies as a whole have been remarkably resilient in the face of severe challenges. They were well able to manage known environmental risks—seasonal challenges resulting from poor weather, for example, occasional flooding or short-term drought. Explanations for such events, and ways of mitigating their impacts, were part of the annual cycle of life. Other, less predictable threats—such as earthquakes or floods—were events that could be mitigated on a limited scale. Major instances of any of these could overburden a society's capacity to absorb the shock, but not necessarily bring about a permanent transformation of breakdown of a system. But just as important were 'internal' factors, the underlying dynamics and systemic constraints and capacities of a given society—both in terms of conflicts between different sets of vested interests as well as in terms of the degree of flexibility in the environmental situation of the society as a whole. Different sets of social and political structures were impacted, and responded, in many different ways, with implications for the developments that ensued—compare the different medium-term outcomes of the Black Death in England and France, illustrative of socio-environmental asymmetries in which different degrees of socio-political complexity and population density precondition the potentials for inherent resilience under stress.

We would argue that with the right questions and appropriate research, historical case studies can offer valuable guidance on present-day issues in designing risk management strategies and sustainable policies. The study of complex historical societies can reveal how past societal and environmental challenges worked to transform structural relationships and daily life. It can also tell us about what happened when the dust settled and how different levels of society re-evaluated their situations. Key terms in the discussion include that of 'resilience'. In historical research it has been invoked most commonly in the context of research on collapse and adaptation, where societies are understood as complex adaptive systems (a concept drawn ultimately from Ecology). Since the basic structural dynamics of a societal system contribute to the types of collapse to which it may be subject, approaches to collapse and resilience that unite structure and process are the best way forward in applying historical examples to contemporary planning initiatives with respect to environmental problems.

Crucial for understanding any society is the role of human agency and belief systems, a facet often ignored in general accounts of historical societal collapse. In his famous presidential address to the American Historical Association from 1931, entitled *Everyman His Own Historian*, Carl Becker tried to reduce history to its core, to what he called its 'lowest terms'. It took him no more than a couple of sentences to formulate an ultra-short definition of history—as 'memory of things said and done'.

However surprising and controversial this simplification might sound, it seems to grasp most of historiographical production of the last century, both academic and non-academic. Let's put aside the notion of memory in Becker's definition, which is very broad and covers what we would rather call historical knowledge. But let's consider "things said and done" for a moment. Indeed, if you take randomly a couple of historical books from any library shelves or a bookstore—in the USA, Poland, Germany or elsewhere, you would most probably immediately realize that they narrate (his)stories of human affairs by reconstructing what people (as individuals or collectives) did or said. Where 'did' also means experience—that is endure, suffer, cope with…. And where 'said' is often understood performatively as a kind of action, too. Thus history is—in Becker's provocation and in common-sense understanding alike—limited to human affairs. Looked at from the outside: no humans-no history.

If we recall our history lessons at school, or maybe also the textbooks we used, we would probably realize that most of what we were taught was not about humans as such, but about groups of people, usually called 'nations', and even more about their ruling elites—kings, queens, dukes, governors, generals, bishops and priests, or even simply 'the people', personified as a simple (and simplistic) monolith. These people were 'doing politics' (often by saying different things), and it was for the most part only within such historical-political frameworks that societies and their particular classes or other social groups were mentioned, including the underprivileged and the underrepresented. Social and cultural history that privileged a bottom-up perspective tried to rescue them from historiographical oblivion—not without success, although varying from one country and educational system to another. Thus history split into 'political' and 'socio-cultural', where the former concentrated on 'facts', while the latter rather preferred talking of human 'experiences'. What mediated between the two was economic history: full of numbers, it looked more scientific and serious, though remained one more story of human affairs. A very particular variety of the 'experience' approach was—and still is—the history of the Holocaust (and other Genocides) which privileged, for good ethical reasons, victims' perspectives. Much of post-Second World War European historiography, for example, was driven more by the moral impulse to 'remember', than by the pure need to know (if the latter can be said to exist at all).

Carl Becker was well aware that his definition of history is very much, if not entirely, focused on one aspect only of an equation that is fundamental for most philosophies of history, namely the relationship between the real world in the past—or past reality as such—and its narrative (or any other) representation. History as 'memory of things said and done' was concentrated on the side of representation. Becker's address (which also dealt with other fundamental historiographical issues) was often interpreted as controversial and 'relativistic' (the term 'postmodernist' was coined some decades later but would fit here perfectly).

From the perspective of environmental history that informs much of the writing in this volume, however, Becker's nutshell definition of history seems to be both a realistic and a very sober description of the practise of the most of field so far. What then changes—on this basic definitional level—with our environmental extension of historiography?

To put it as simply as possible: the relationship between past reality (*res gestae* in the philosophy of history) and the human story about it (respectively: *historia rerum gestarum*) must be conceptualised anew. It is not simply a case of bringing new sources and data to bear (though this is already a great deal and often judged sufficient in historical research). It is much more a radical revision of the 'essence' of historical fact as such. Facts, in environmental history, are no longer just to do with the record of human affairs. They must now be embedded in 'nature', itself no longer merely the stage upon which the theatre of human actions and reactions plays out, but an important and indeed irreplaceable historical actor in its own right.

From now on: No nature—no human history.

History and Public Policy in the Era of Planetary Crisis

What Stories Should Historians Be Telling at the Dawn of the Anthropocene?

Adam Izdebski

Abstract This chapter discusses the ways in which history can contribute to coping with the current planetary crisis. It argues that historians should engage more in interdisciplinary exchange across the humanities-natural sciences divide. Thus they will be able to create historical narratives fitting for the Anthropocene—both in terms of explaining it and shaping our responses to it, in particular to the acute planetary crisis that marks its advent. At the same time, history should not give up its drive to critically dissect and analyse socio-political, economic, cultural and ecological change, contributing to developing balanced and resilient public policy.

Keywords Environmental history · Science for policy · Applied history · Interdisciplinary history · Climate history · Pandemics

Ever since the term 'Anthropocene' was coined, it has become clear that the condition of both the earth and of humankind that it describes presents a disturbing challenge. Even if the term were not to become an official name for a geological epoch, the destabilization of the planet's life-supporting systems—from the heights of the atmosphere to the depths of the oceans—is a recipe for a perfect catastrophe. The intellectual endeavor of writing history—the craft of professional historians—is not spared from the anxieties that accompany the realization of the Anthropocene. There is an ongoing debate on what the Anthropocene means for the field, and answers are many (Chakrabarty 2018, to name just one of the most influential).

This chapter offers an interdisciplinary historian's perspective on what historical storytelling could become in the Anthropocene. It is based on years of, at times, challenging exchanges with natural scientists, which have led to some perceptions of what the strengths of this peculiar academic discipline actually are, and where it could direct its efforts and attention in order to make a difference and contribute to coping with the challenge of humankind's own making. What follows is more of an essay or even a meditation on the role that history can fulfil, rather than a classic

A. Izdebski (✉)
Max Planck Institute for the Science of Human History, Jena, Germany
e-mail: izdebski@shh.mpg.de

Jagiellonian University in Krakow, Krakow, Poland

© The Author(s) 2022   9
A. Izdebski et al. (eds.), *Perspectives on Public Policy in Societal-Environmental Crises*,
Risk, Systems and Decisions, https://doi.org/10.1007/978-3-030-94137-6_2

research or review paper. I begin by providing four different answers to the question that I am asking in the title, and I present two case studies that can help to illustrate the points that I would like to make. While the first three answers are of more interest to historians, even if they do also show the potential role of history in our society today, a role still to fully materialize, the fourth one, which shows history's potential in re-shaping the policy debates that are already ongoing, could prove most useful for the more general reader.

The First Answer: The Anthropocene as a Challenge to Humankind

So what stories should we be telling as historians—and more generally, as people studying the past? This first answer has to do with the Anthropocene understood as a challenge. Thus, if the Anthropocene is a new situation for us, we need new narratives about the past in order to understand how we got here—new stories of the kind that have not been heard or at least promoted before. Consequently, we as historians should start narrating the past in ways that we have hardly tried before. Of course, this does not mean a complete break from or distortion and rejection of the existing tradition. There is a lot in the historiography, especially in the second half of the twentieth century, on which we can build. The transition we need means, first of all, breaking away from old narrative patterns, the way history is usually told. First of all, we need much less of the 19th-century approach to history, in which the past is recounted in order to understand the conflicts between nations that we see around us. These were conflicts having to do with identity, with industrial societies and the massive scale of aggression, power, and military prowess that industrial societies can mobilize. This is one narrative pattern that we might want to break away from in order to be able to tell stories that rise up to the challenge of the Anthropocene.

The other kind of stories we might want to break away from are the progressivist narratives, or, we could say, liberal, neo-liberal, or even Marxist. In other words, these are stories that see history as the progression from less developed to more developed, from childish to mature states. We as human beings who happen to live in the Anthropocene clearly see that history is not that simple. An increase in complexity can lead to crises and, in the long run, the disintegration of empires, which was the case with the Roman Empire in the 5–7th c. CE and with the British Empire between the 1920s and the 1950s, Brexit being potentially another stage in this process of decomposition, this time showing how ideology and belief lag behind reality. Of course, empires and states do rise and fall. So there is change, but change does not necessarily always mean progress, unless we are narrating history from the point of view of the rising empires' elites, as much of history in the 19 and 20th c. has been done (for a fascinating reflection on how expansionistic states have dominated the narrating of history from the dawn of "civilisation", see Scott 2017).

Distancing ourselves from these two dominant narratives—nationalist and progressivist—can help us share with our readers and listeners—with our societies—new stories that we need for the impending Anthropocene. These new stories should, first of all, respond to the fears of our times and not the fears of the early twentieth century, or even the fears of the Cold War, as much of historiography to date has been doing. We should respond to manmade environmental threats such as the pandemic, that we have been experiencing for more than a year now, or climate changes. Historians, along with archaeologists, palaeo-environmental scientists, archaeogeneticists, and other related field specialists need to work on creating a new sense of meaning for what we are experiencing today as societies. We should address these fears and help our contemporaries by sharing stories about the past that contribute to understanding what is going on around us in this decade and what will only become more powerful as the time goes on. The implications of this are that if we are able to provide new meaning to our historical storytelling, if we are able to break away from old patterns of thought and speech, we will be able to empower our communities to act on these very threats. Our stories could actively respond to the challenge of the Anthropocene and provide the motivation and understanding needed to act and adapt to it.

The Second Answer: The Anthropocene as a Viewpoint

The Anthropocene is not only a problem to solve—or rather, a new living context to which we need to adapt and manage—but also a new viewpoint or even a new philosophical stance. Thus, not only can adapting to the Anthropocene on the part of historical storytelling mean changing the contents and ideological angles of the stories, but also their very structure. Historians as writers—who wish to engage the interest of the public with their stories—might want to start looking for new rhetorical devices, that is actors and modes of rhetorical action (for the basic theory of history's rhetorical structure, see White 1978). Humans will still remain at the center of our stories, of course, simply because we humans are telling these stories and the story is meant for other humans. We are not wolves sharing their stories with deer, or pines whispering to birches. History will always remain a means of communicating within a human community. However, this does not mean that our stories should only be about humans. They should incorporate both human and non-human actors on equal terms, actors such as climate, pathogens, ecosystems, specific plants or animals or even non-organic matter—seas, mountains, and so on. We all belong to networks made up of humans and non-humans, and it is precisely because we belong to these networks that we are confronted by the problems that we are currently experiencing, including the covid pandemic. Yet while the historical narratives we create must reflect these entanglements, this poses a major epistemological and methodological challenge (on human-non-human networks, see Latour 2005).

The Third Answer: The Anthropocene as Opportunity for the Research Community

As a challenge to human societies at large, the Anthropocene is also an opportunity for interdisciplinary researchers. The 'terror of the Anthropocene' may prove strong enough to shake the grounds of one of the most traditional institutions of modern society, the university. Developed in the later 19th and early twentieth centuries, the modern university was meant to serve the highly specialized and technical societies of the late industrial age. It is founded on the principle of professional and scientific specialization, through which individuals acquire knowledge and skills in well-defined fields (or disciplines). Given how complex the scientific knowledge is today, we will not avoid specialization in our times either, at least for the foreseeable future: specialization is a way of ensuring efficiency in academic education. It is also crucial for the very process of knowledge production. The formalised and rigid character of academic fields helps to ensure that scientists with relevant expertise end up reviewing their colleagues' research, being able to evaluate specific methods and the evidence it produces. At the same time, however, this compartmentalisation both of knowledge and of research efforts limits our ability to contemplate the world in its complexity and often leads to rejection of innovative ideas, linking different, usually unrelated specialisations into more unified knowledge. Paradoxically, at times, this same compartmentalisation leads to seemingly peer-reviewed publications containing questionable claims. This happens when what qualifies as inter-disciplinary argument is evaluated from the angle of just one of the fields involved, substituting for a more comprehensive review that is often challenging to arrange. For this reason, we urgently need to develop new structures across and within academic fields that help us understand the age we enter in all its facets.

Let us go through this reasoning once again. First, we need new stories to help us understand the planetary crisis we are in. Second, such stories require new actors and new types of connections between them. Third, in order to be able to deliver this, Anthropocene-adequate stories need new evidence (professional history boasts of its evidence-based credibility, after all). It is exactly here that the opportunity for the global research community emerges. If we really want to give voice to Nature, so that we appreciate the fate it has shared with increasingly -powerful humanity, we first need methods to hear this voice, before we make place for this voice in the historical stories we tell.

While this picture may at first seem somewhat pessimistic, in reality the global academic community of the 2020s is very well positioned to achieve this goal. The methods we need already exist, no one needs to be starting from scratch. We are building on more than a hundred years of research in paleo-ecology, paleo-climatology and in the paleo-sciences more broadly, that is in those branches of the natural sciences that work on reconstructing the past of biological (and physical) phenomena other than those specific to the human species. These are the very sciences that study the natural world of the past, its variability and also its agency, the ways in which some natural phenomena influenced each other and in this way also influenced past human communities. This means that we need to incorporate the evidence and ideas coming from the natural sciences into the historical stories that we as humans develop—into stories about human experience and human fate. In the end, this is only possible if we break down the divide between the humanities and the natural sciences and persevere in developing more unity in our understanding of the world. This is not just an intellectual task: it requires a more unitary and flexible organization of academic disciplines, teaching, and research. We should abandon the disciplinary silos we inherited from the nineteenth century and work on such a restructuring of the human, an adventure with research and universities that would be much more appropriate for the times in which we live.

The very idea of breaking up the disciplinary divides can be illustrated with a tree of knowledge (Fig. 1). The roots go into the different archives, that is the different

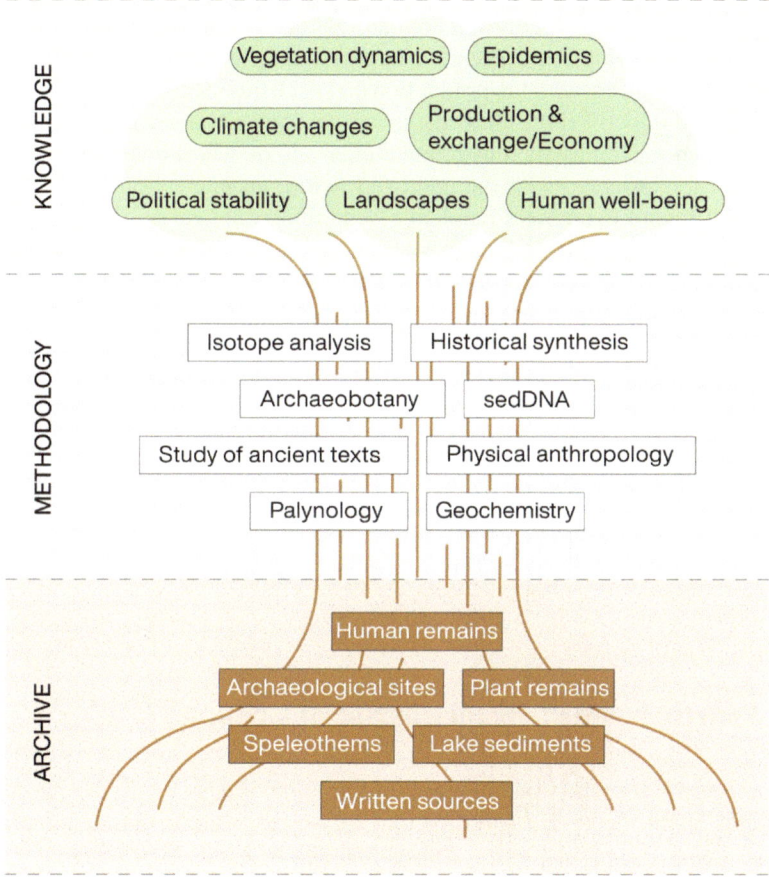

Fig. 1 The interdisciplinary tree of knowledge, uniting archives and methods of different disciplines in order to achieve a holistic knowledge of the past. By Alessia Masi and Adam Izdebski, drawing by Hans Sell, MPI SHH Jena

places or parts of the world around us where we can find material to work on as scholars or scientists of the past. Human remains, archaeological sites, plant remains, speleothems in caves, lake sediments, finally, written sources preserved in human archives and libraries. There is thus a plethora of natural and human archives we can use. What is important is that there is not a single method that we can apply to each of them, on the contrary we can apply several different methods to each archive and in this way create meaning out of these ubiquitous remains of the past. We can analyze the elemental structure of these remains for isotope analysis; we can reconstruct DNA in sediments or in human remains or many other contexts, such as soils in caves formerly occupied by humans. We can read and analyse historical sources (texts) as historians have been doing for generations—but we can also undertake quantitative or digital analysis of such sources. We can apply methods of archaeobotany, palynology, geochemistry and many other paleo-ecological and geological approaches. We can look at human bones through the lens of physical anthropology. No single archive belongs to a single discipline. All these methods are the trunk of the tree that leads us to the leaves and fruits above. Here we learn about different phenomena that have to do with human action, with the production and exchange economy, with the political stability and conflict. Yet the phenomena we can know about are also connected with either exogenous forces such as climate changes, or with human interaction with nature, as, for instance, with the pathogen world, an interaction so obvious to us now at the time of the current pandemic. And thus we have: epidemics, landscapes and ecosystems, as well as human well-being in economic as well as cultural or biological terms. This is how a more unitary, at the same time hybrid and flexible way of approaching the past might appear.

Obviously, this leads to many practical challenges, perhaps the most important of which is the fact that we need new forms of academic writing in order to be able to proceed to new forms of historical narrative, more persuasive and more comprehensive than those currently available. We need to confront different publication cultures, different genres of academic writing that often lead to contradictory perspectives on the same problem. So the new approach to narrating the past in both academic and popular contexts needs to be hybrid, mixed, and most importantly, inclusive of these different perspectives in order to create a more flexible and holistic understanding of the past.

The Fourth Answer: History as Social Critique

Let me begin this section by moving on to my first case study. In fact, it illustrates the points about the hybrid narrative that I have made at the end of the preceding section. We will look at Central Europe in the sixteenth and seventeenth century CE, zooming in on one of the core regions of the Polish-Lithuanian Commonwealth: Greater Poland with its capitals in Poznań and Gniezno, in the west of the country. This province was the most economically developed region of this unusual empire, an aristocratic republic ruled by an elected king and the parliament.

Let me begin with a broad-brush picture. During the 16th and early 17th c., the economy of Poland was booming: every year increasing grain exports were exported to the Netherlands via the Vistula and its tributaries, departing from the port city of Gdańsk. At the same time, there was a steady population growth, largely uninterrupted for more than two centuries (summarised in Haldon et al. 2018). Paleoecological data from many parts of modern-day Poland also show that this period shows a major expansion of cereal cultivation (Izdebski et al. 2016). Yet in contrast the same sources of data indicate a collapsing economy and society in the later 17 and 18th c. A quarter of the population seems to disappear, agricultural expansion stagnates, and grain exports decrease at least by a half and remain highly unstable.

What is the explanation? Much of Polish historical research in the last two hundred years tried to answer this question, and thus explain the inability of the Commonwealth to defend itself against conquest at the hands of its neighbours in the late 18th c. Already in the early 20th c., some historians suggested it was due to natural causes, such as weather extremes or epidemics (Bujak 1938). Today, we can verify or falsify such hypotheses using paleoclimatic data. They show that, while the later 17th c. was indeed particularly cold with respect to summers, similar colder temperatures also prevailed in the mid-16th c, followed by very hot summers at the end of this century (Büntgen et al. 2013). Similarly, whereas there was an increase in droughts in several regions of modern-day Poland in the second half of the 17th c., the core lands of the Commonwealth, such as Greater Poland, Pomerania, Podlasie or Lesser Poland, experienced increased frequency of droughts also in the second half of the 16th c. (Przybylak et al. 2020). There was thus no major difference in climatic conditions during the periods of economic boom and collapse, and in fact there was no difference in general epidemiological conditions either: for instance, waves of plague epidemics were reaching Poland in the 16th c. as they did also in the 17th (Karpiński 2000; Guzowski et al. 2016).

In order to shed more light on this complicated entanglement of natural and social processes, we formed an interdisciplinary team that looked in more detail at the central areas of Greater Poland, specifically between Poznań and Gniezno. We looked first at historical, demographic, and economic data from this area, as well as natural scientific data, which we collected from two peat bogs (Lamentowicz et al. 2019). In terms of the historical data, we see a loss of population also at the local level, but the sources are too scarce to follow this process in more detail, we only know about a general loss between a reconstruction for the later 16th and another one for the later 17th c. (Czerwiński et al. 2019). Secondly, we also looked at crop yields in this region and observed declining agricultural productivity for all major grains already in the early seventeenth century, which can probably be related to climatic instability and cold summers in particular (Kozłowska-Szyc 2019). Still, these declining yields do not explain the scale of the collapse that we see when comparing the late 16th and late 17th c. historical data. In fact, if we consider the data on grain exports from Gdańsk, these declining yields did not impact the booming economy and the positive trend continued.

The answer was in fact to be found in the "voice of nature", a fine-grained reconstruction of the local landscape based on pollen analysis of peat sediments

(Czerwiński et al. 2021). Here, the major watershed very clearly occurred in the middle of the century, in the 1650s, and it was an abrupt process which relieved the landscape of most of the human pressure and allowed for a re-wilding, including a major regrowth of pine forests. After this breakdown point for human activity, there was for several generations no regeneration of the human economy in this part of Poland, thus clarifying the level of demographic and economic loss at the general level.

Of course, it was not a coincidence that the breakthrough point for the collapse of an otherwise booming economy was the 1650s. This was the moment when there were several external invasions of the Polish-Lithuanian Commonwealth: from the north, from Sweden, and from the east, from Russia, together with a massive rebellion in Ukraine, in the south-east of the empire. In this very decade the Commonwealth effectively ceased to exist for a couple of years as a state and an economic system. Its lands were materially devastated and its institutions were severely shaken. It had to rebuild its independence. It seems that it was this external conflict to which the Commonwealth owed the loss of the internal resilience that had allowed for demographic and economic growth in the preceding centuries.

As this example makes clear, the historical and the natural scientific narratives can come together to create a much more meaningful story. First, we see that even a society that is on a strongly expansive trajectory and has at the same time to contain substantial cultural-religious diversity—as the Commonwealth was throughout its entire four hundred year history—can be very resilient to climate change and epidemics that were frequent at the time. Only when the roots of this resilience were destroyed by conflict did the entire social-ecological system falter or break down.

Let me return to the question posed at the beginning: what stories should we be telling at the dawn of the Anthropocene? This case study offers an opportunity to provide a fourth answer to this question, an answer that focuses on the different understandings of the Anthropocene and in particular the criticism of the term itself as too imprecise. According to this critique, 'Anthropocene' does not do justice to what really brought us to the situation in which we find ourselves at the moment, with ongoing climate change and the current pandemic. One way in which to challenge the Anthropocene as a historical and philosophical idea is to emphasise that we actually live in a Capitalocene (Bonneuil and Fressoz 2016a, translated as: b). Indeed, it has been the industrial capitalism of the last two hundred years together with the massive acceleration of consumerism since the Second World War that created the fossil-fuel-based Great Acceleration and which generated the Anthropocene with all of its environmental problems.

So, while rejecting the nationalistic perspective of 19th-c. historiography and the origins of modern historical storytelling, our new stories should not at the same time compromise on one of academic history's strengths, that is, its interest in societies understood as complex entities with internal conflicts and inequalities, as communities ridden by conflicts, both resolving them and endangered by their consequences at one and the same time. The new stories we tell should continue to engage in social criticism. This is something that the natural sciences alone cannot achieve, even if they are formidable in giving voice to nature, simply because they were not developed for this purpose. History, in contrast, has devoted decades reconfiguring itself for the cultural priorities and conflicts of the age—class, gender, ethnic, institutional, etc. This is one of the great assets humanities can bring to this new, interdisciplinary study of the past.

Let me give you a final case study that will illustrate the point I want to make about social critique as the key strength of professional, humanistic history. It is a case study from Byzantium, the proud continuation of the eastern Roman empire in medieval times, and the story I will tell takes place in the 10th c. CE. From the historical records (written sources), we learn that there was a great winter in the years 927/928, and these same records suggest that it was associated with a great famine in the Byzantine Empire (for more details and references for this case study, see Izdebski et al. 2018). The sources claim that, as a consequence, farmers and peasants were forced to sell their lands to the powerful, the elites—and the state was thereby losing tax income from peasant production. Legal measures were undertaken by the state to remedy this situation and stop this process. In other words, the Byzantine texts describe a major societal change and attribute it to a climatic factor, namely the great winter of 927/928. One might say, a crystal-clear and very interesting story of how climate can trigger profound crisis and social transformation.

However, when we look at paleoclimate data available for this part of the world at this time, things become much more complicated, indeed, the above story no longer holds true. Reconstructions of summer temperatures and winter hydroclimate (mostly snowfall) do not show anything unusual for the 920s, and for 927/928 in particular. There was indeed a very cold summer and an increase in snowfall, but it occurred more than a decade later, in 939, and was caused by a major volcanic eruption on Iceland. At that time there was both a cold summer and a cold winter, and several contemporary European texts record these unusual weather conditions and the associated bad yields.

It is also possible to do model simulations of past climate and see to what extent the period 920–030 was different from the general average conditions and from the preceding decades/centuries. Whichever models we employ, there seems to be nothing in particular happening. Just a moderate decade, even slightly warmer than the average as regards the winters.

So what is going on? Why do the Byzantine sources want us to believe in a major famine preceding the state's legal actions against the elites buying up of peasant lands, that is, a major socio-economic crisis? Why should such a crisis happen at the time when—as all the other data, paleoecological and archaeological, shows—the Byzantine economy was in the middle of a major boom cycle? In other words, neither the material nor the natural evidence attest to any major climatic crisis during the tenth century in the Byzantine world.

Are, then, these medieval texts lying? Not necessarily: here is where the careful analysis of texts—the historian's craft—and a social critique—history's strength—come into play. All these texts (except for the legal documents from the 930s, which mention the famine, but not the winter) were most probably composed a generation later. They conflated the social change that was taking place in Byzantium at that time as a result of both economic growth and the associated increasing inequality between lower and upper strata of society, with an unusually heavy winter that did in fact occur, but a decade or so later. So this would mean that in fact a later generation of elites attributed social change to external factors: they were not in a position to acknowledge that there was a major social transformation, one orchestrated by their economic activities and one from which they benefited while others were losing (in fact, in the light of the elites' ideology, the changes which made the lower strata more dependent could have even been seen as a positive development). Such processes of socio-economic change are never simple to observe, neither were they for the educated elites of the culturally most advanced European state of the 10th c. CE. It was in fact not until the nineteenth century, with all the theoretical tools generated by the experience of the Industrial Revolution, including Marxism, that we became capable of fully grasping, comprehending and—last but not least—critically reflecting on social change in its economic context.

* * *

My second case study clearly shows that we need both critical history—such as has been developing over the last century—and we need all the interdisciplinarity that is possible today, in order to really understand what was going on in the past and to show the different levels of human and non-human agency. Only a history that breathes with these two lungs—the humanities and the natural sciences—can help us achieve social justice and the fair transformation that the Anthropocene requires from us to survive as a species. More broadly, only a history which is a dialogue of the old and the new can open our minds to move in the direction of true ecological justice, which might be our hope for survival and flourishing.

References

Bonneuil C, Fressoz J-B (2016a) L'Événement Anthropocène. La Terre, l'histoire et nous. Seuil, Paris

Bonneuil C, Fressoz J-B (2016b) The shock of the Anthropocene: the earth, history and us. Verso, Brooklyn, NY

Bujak F (1938) Czynnik gospodarczy w upadku dawnego Państwa Polskiego. Sekcja Dydaktyczna Oddziału Lwowskiego PTH, Lwów

Büntgen U, Kyncl T, Ginzler C et al (2013) Filling the Eastern European gap in millennium-long temperature reconstructions. PNAS 110:1773–1778. https://doi.org/10.1073/pnas.1211485110

Chakrabarty D (2018) Anthropocene time. Hist Theory 57:5–32. https://doi.org/10.1111/hith.12044

Czerwiński S, Guzowski P, Karpińska-Kołaczek M, et al (2019) Znaczenie wspólnych badań historycznych i paleoekologicznych nad wpływem człowieka na środowisko. Przykład ze stanowiska Kazanie we wschodniej Wielkopolsce. Stud Geohist 7:56–74. https://doi.org/10.12775/SG.2019.04

Czerwiński S, Guzowski P, Lamentowicz M, et al (2021) Environmental implications of past socioeconomic events in Greater Poland during the last 1200 years. Synthesis of paleoecological and historical data. Quaternary Science Reviews in Press:16

Guzowski P, Kuklo C, Poniat R (2016) O metodach pomiaru natężenia epidemii i zaraz w preindustrialnej Europie w demografii historycznej. In: Polek K, Sroka ŁT (eds) Epidemie w dziejach Europy: konsekwencje społeczne, gospodarcze i kulturowe. Wydawnictwo Naukowe Uniwersytetu Pedagogicznego, Kraków, pp 119–144

Haldon J, Mordechai L, Newfield TP et al (2018) History meets palaeoscience: consilience and collaboration in studying past societal responses to environmental change. PNAS 115:3210–3218. https://doi.org/10.1073/pnas.1716912115

Izdebski A, Mordechai L, White S (2018) The social burden of resilience: a historical perspective. Hum Ecol 46:291–303. https://doi.org/10.1007/s10745-018-0002-2

Izdebski A, Pickett J, Roberts N, Waliszewski T (2016) The environmental, archaeological and historical evidence for regional climatic changes and their societal impacts in the Eastern Mediterranean in Late Antiquity. Quatern Sci Rev 136:189–208. https://doi.org/10.1016/j.quascirev.2015.07.022

Karpiński A (2000) W walce z niewidzialnym wrogiem: epidemie chorób zakaźnych w Rzeczypospolitej w XVI-XVIII wieku i ich następstwa demograficzne, społeczno-ekonomiczne i polityczne. Neriton, Warszawa

Kozłowska-Szyc M (2019) Wysokość plonów rolnych w dobrach królewskich dawnej Polski w latach 1564–1665. Stud Geohist 7:17–29

Lamentowicz M, Karpińska-Kołaczek M, Guzowski P et al (2019) Znaczenie wysokorozdzielczych wielowskaźnikowych (multi-proxy) badań paleoekologicznych dla geografii historycznej i historii gospodarczej. SG 56–74. https://doi.org/10.12775/SG.2019.03

Latour B (2005) Reassembling the social: an introduction to actor-network-theory. University Press, Oxford

Przybylak R, Oliński P, Koprowski M et al (2020) Droughts in the area of Poland in recent centuries in the light of multi-proxy data. Clim Past 16:627–661. https://doi.org/10.5194/cp-2019-64

Scott JC (2017) Against the grain. Yale University Press, New Haven, A Deep History of the Earliest States

White HV (1978) Tropics of discourse: essays in cultural criticism. Johns Hopkins University PressBaltimore

The Anthropocene Contract. What Kind of Historian–Reader Agreement Does Environmental Historiography Need?

Jakub Muchowski

Abstract The paper is an attempt to reply to the question on how environmental history can participate in public debates on contemporary world concerns in a reliable and socially relevant way. I argue that the answer to this question lies in environmental history's reading pact, which I call the Anthropocene contract. Its most important element is the principle of equality, which concerns the relationship between historians and their readers. In the first step, I invoke Graeme Wynn's statement to point to important questions about the challenges that the Anthropocene posed to environmental history. Next I critically discuss the answers to these questions provided by historical theory. I then formulate a proposal for a reading pact of environmental history using the theoretical insights of Kalle Pihlainen and the philosophy of Jacques Rancière.

Keywords Environmental history · Reading pact · Anthropocene contract

Introduction

Environmental history intervenes in public debates on contemporary world concerns. For this reason, it asks itself how to participate in them in a reliable and socially relevant way. In this paper, I argue that the answer to this question lies in environmental history's reading pact, which I call the Anthropocene contract. Its most important element is the principle of equality, which concerns the relationship between historians and their readers. In the first step, I invoke Graeme Wynn's statement to point to important questions about the challenges that the Anthropocene posed to environmental history. Next I critically discuss the answers to these questions provided by historical theory. I then formulate a proposal for a reading pact of environmental history using the theoretical insights of Kalle Pihlainen and the philosophy of Jacques Rancière.

J. Muchowski (✉)
Institute of History, Jagiellonian University in Krakow, Krakow, Poland
e-mail: jakub.muchowski@uj.edu.pl

© The Author(s) 2022   21
A. Izdebski et al. (eds.), *Perspectives on Public Policy in Societal-Environmental Crises*,
Risk, Systems and Decisions, https://doi.org/10.1007/978-3-030-94137-6_3

During his 2019 Presidential Address to the American Society for Environmental History, Wynn discussed the challenges facing the modern world and environmental history, and proposed a historiographic approach that could offer a partial response to them. He claimed that historians—both as citizens and as scholars—are currently dealing with global warming, the Anthropocene, the passing of tipping points and planetary boundaries, as well as rising nationalist, populist, nativist sentiment and neoliberalism destroying public institutions and income disparities. In response, historians produce narratives about the fall and the end of the world, which serve both to articulate despair and mobilize action. In Wynn's opinion, however, the dissemination of such stories causes fear, which in turn evokes apathy or fosters a selfish struggle for survival rather than encouraging action for the common good. A more fitting solution would be to build narratives of hope that are critical of the state of our world, that resist the forces that threaten it, and that propose alternative visions of a better future (Wynn 2020: 3, 20–21).

For Wynn, such narratives are the stories of the past endeavours of social change and, particularly, the biographies of thinkers and activists who opposed political and economic power and sought to construct better forms of life in the world. In his speech, Wynn goes straight to two detailed biographical stories: of the ecologist, Pierre Dansereau, who researched the scale of human impact in the biosphere, and of the political theorist C.B. Macpherson, who studied the influence of economic inequalities on the functioning of liberal democracies (Wynn 2020: 5–13, 13–20). According to Wynn, their activities and ideas in the field of environmental protection and creating a democratic society, anticipated our contemporary struggles (Wynn 2020: 3–4).

I have taken note of two important issues here. Wynn strongly emphasizes the social role of environmental history and contributes to the discussion on forms of historiography that would productively carry out this task by indicating one such form (i.e. biography) and then offers examples of its use. The first passages of his speech suggest that Wynn sees the political engagement of environmental history as natural. In the biographical stories, however, he underlines that in the recent past, during the careers of his two protagonists, the academy was convinced that scholars should focus on methods and facts, not on using their knowledge and skills in the political struggle for a better world. It is only in the last few decades that the political commitment of scientists has gained a rank similar to epistemic tasks and has become strongly intertwined with them. Historical approaches that have been linked to social movements outside academia, such as anticolonial, feminist, and environmental history, seem to contribute to this change.

I was surprised at Wynn's choice of biographies as a type of historical writing that would fulfil the political tasks of environmental history in the Anthropocene era. Wynn does not propose a new, inventive form but an established type of writing specific to political or intellectual rather than environmental history. Moreover, within life writing itself, there are many new approaches that question the coherence and agency of the subject, as well as the purposeful order of their life story, such as ecobiographies, autotopography, anti-biographies, or imperial biographies that present life stories as entangled with not only the environment or landscape, but also social, economic, and political conditions (e.g. Glotfelty and Fromm (1996); González 1995; LeGoff 2009; Lambert and Lester 2006). Wynn, however, uses the highly conventional form of the teleological biography of a scholar, presenting his protagonists as consistent, strong and independent subjects, and their whole life as aimed at a goal defined retrospectively by the biographer. This form refers to the ancient model of pragmatic historiography subordinated to the literally understood rule of historia magistra vitae, according to which the story of the ideas and deeds of former great individuals is to be a model for the addressees for how the audiences should manage their own lives. It produces an unequal relationship between the historian and the model biography and the audience, in which the ideal, closed story of the success of a strong individual is offered to—we may imply—lost and helpless readers as an example to imitate. This type of biography enjoys unflagging popularity among many reading groups, but its effectiveness as an instrument of environmental history in dealing with a planetary crisis is questionable.

Wynn's speech raises several questions: which narrative forms does environmental history need to help society cope with the threats generated by the Anthropocene? Does it need new narrative forms for this task, or are those it already has sufficient? To what extent might the environmental history of the Anthropocene be a teacher of life? Can historians help contemporaries better grapple with comparable risks and challenges today? In this paper, I will approach these questions with the resources of the contemporary theory of history, and I propose a new element for the framework of environmental history writing in the form of the Anthropocene contract, an agreement between historians and their audience.

A Response by a Theory of History

Historical theorists participating in the debate on the Anthropocene (I refer primarily to Dipesh Chakrabarty, as well as Zoltán Boldizsár Simon and Marek Tamm, who comment with approval and augment the ideas of the former) point to the need to make changes in how we understand the historical. Instead of reconsidering the ways historians produce stories about the world, they concentrate on refiguring the subject of historical studies by broadening the scope of the historical to include new agents, temporalities and spaces.

Chakrabarty examines the terms 'planet' and 'species' as central categories for the humanities that would cooperate with Earth System Sciences in challenging the Anthropocene. He compares 'planet' with notions of 'earth', 'world' or 'global' to describe the spaces and temporalities of a new historical subject that combines planetary processes with changes taking place in the technosphere and biosphere, and replaces geological time with a more human-centred time (Chakrabarty 2018a, b, 2019). In searching for a term for the human collectivity that experiences the planetary crisis, Chakrabarty measures 'species' with 'human', 'humanity' and 'biological agent'. The latter is a category which operates in environmental history and removes the distinction between natural and human history, but only the term 'species' embraces humans as a geological force. In his analysis, Chakrabarty aims to demonstrate that experience and knowledge produced in the Anthropocene exceed the capacities of contemporary humanities and historical studies (e.g. Chakrabarty 2009).

Zoltán Boldizsár Simon and Marek Tamm share the same goal when reconfiguring the historical by including transhuman and non-human agents, multiscalar history and non-continuous history. They use a distinction between 'more-than-human' and 'better-than-human' to encompass not only the relationships entangling different species and the inanimate environment but also mechanical and digital agents. Similarly to Chakrabarty, Simon and Tamm postulate a multiscalar history that intertwines anthropocentric scales with geological time. However, they underline that it is not about expanding the time scope (as is the case in deep, big or evolutionary history) but entangling different scales (Simon and Tamm 2020).

Simon and Tamm also argue—and I would like to dwell on this here—that approaching the Anthropocene makes it necessary to break with linear, processual or developmental temporalities. The planetary crisis is producing sudden and radical changes with results that cannot be predicted, which requires putting more effort into elaborating on notions of historical disruptions (Simon and Tamm 2020). Simon describes the Anthropocene as an unprecedented event, a rupture between previous experience and expectations for the future, and, as such, it cannot be narrated. He explains that we can construct a continuous story about how it happened but cannot grasp the Anthropocene itself and its consequences, because it is a radically new phenomenon (see Simon 2015, 2017, 2020). I assume that the main consequence of such historical sensitivity is the impossibility of closing the story, which therefore provides neither the full meaning of the representation of the past nor a lesson from the past for the reader.

This way of understanding the historicity of our times is not new. It has been circulating in Western societies since the Second World War, at which time it had already become a challenge for those who attempted to articulate it in narrative form. As a result of the wartime experience, the conviction that Western societies were heading in the right direction was fading, and modern coherent stories of economic and technological progress and emancipation were increasingly subjected to questioning. A sense of confusion and instability dominated, as well as the conviction that no one was able to predict what the future might hold (Gumbrecht 2013).

Nonetheless, treating the Anthropocene as an unprecedented change may have great persuasive value. If we show audiences that the world is in an unprecedented situation, it will be easier to mobilize them and convince them to act quickly. Conversely, if we present the situation as an element of long-term and continuous change, our listeners will believe that it is nothing new and does not require radical decisions. Coherent narratives domesticate the Anthropocene, reconfiguring it into something natural and harmless (e.g. Simon 2020).

Chakrabarty, Simon and Tamm all encounter impasses in their considerations, as they recognize the contradictions in the Anthropocene historical subject. These impasses concern the universality and transdisciplinarity of such a history. The response to the Anthropocene seems to require the production of a universal history, while this way of presenting historical knowledge has been discredited as an instrument of Western imperialism. On the one hand, a universal narrative is justified because the agent in the Anthropocene, as well as the potential victim of the new era, is that of humans as a species. On the other hand, however, the participation of various social and political groups in the exploitation of the planet, which has led to the disturbance of the Earth's system(s) and continues to destabilize it, is very uneven; therefore, it is difficult to use the homogeneous figure of humanity, as this would violate the principle of justice (Chakrabarty 2009, 2018a).

A transdisciplinary approach combining the practice and knowledge of life sciences, Earth system science, social sciences and the humanities is also indispensable to a historical narrative of the environmental crisis. However, as a separate paradigm, Earth system science, within which the concept of the Anthropocene was constructed, cannot be integrated with other disciplines of science. Due to the incommensurable differences, when discussing the phenomenon they label as the Anthropocene, each speaks of something else. Therefore, cooperation between these fields of science encounters problems and is more at risk of errors (e.g. Simon 2019; Chakrabarty 2019).

The Question of Readership

The above-demonstrated important insights and recommendations are focused on reconfiguring the historical, but in doing so they ignore an important issue related to reflection on historical writing: the relationship between historians and their audiences in the Anthropocene era, including the question of the contract that environmental historiography can establish with its readers when it tries to support societies that challenge the planetary crisis. I think it is the process of reception of historiography that may be crucial in answering the question of what sort of environmental history we need. If historians wish to intervene in public debates on environmental crises, they have to consider what their audience makes of their intervention.

Historical writing is, after all, a communicative action which has a sender and receiver. An important element in the process of reading history is the relationship between the historian and the reader. It is inscribed in the text that assigns specific positions to the author and readers. The reading contract established by the text can influence the reception of the knowledge it contains, as well as the relationships that the reader builds with other people, the world, and the planet. Just to be clear, I am focused here on the reading contract that each and every one of us historians inscribes in their writing; I am not focused on the social contract between scientists, researchers, academia and society, states, and power.

Although history theory has for many years been focused on historical writing as a communicative action, it has paid little attention to the question of reader and reception. This question was introduced into the debate by Kalle Pihalinen, who argues that historiography is a genre of writing in which the author and the reader are bound by a pact of reference. Under this agreement both acknowledge the separation between fact and fiction, and the author is committed to writing about what is real and true. Unlike fictional writing, in which, according to Pihalainen, the author commits himself primarily to providing meaning for his readers, historical writing provides extra-textual evidence for presented accounts of past reality (e.g. in the form of quotations from sources and footnotes) (Pihlainen 2017: 115–131).

Pihalinen uses this definition of the reading contract in historical writing to make several claims, aiming to first point out the distinction between historical and fictional writing (Pihlainen 2017: 62–81) and second to grasp the features of historical narrative. These confines of genre impinge on historical narrative because, he argues, a focus on referring to extra-textual evidence clogs up the story and prevents it from closing. This in turn makes historiography less entertaining than fictional writing, because it will not provide easy interpretations of the past (Pihlainen 2017: 62–81). It also means that the reader reads in a certain way; as if the text is supposed to tell the truth and represent reality. Because of this, the reader is conditioned to constantly question what is presented to them (Pihlainen 2017: 83–98). Finally, the reader is thus active here: they examine the account presented by the author and give meaning to the facts provided. They are not a passive figure but merely a recipient of the message or lesson designed by the historian. This means that the reader, together with the author, becomes a co-creator of the text: they share with the author the task of presenting and giving meaning to the past. The reader is emancipated, endowed with agency and free to lead the reading process (Pihlainen 2017: 83–98).[1]

Egalitarian Historiography

The only comments from theorists of history participating in the debate on the relationship between historians and their readers is Simon's attempt to influence the addressee and mobilize them to act using the rhetoric of an unprecedented event. Meanwhile, discontinuous stories combining different time scales and experimenting with the inclusion of inhuman protagonists may, due to their complexity, multi-threading, and lack of closure, create an impression that historical writing is clumsy and inconclusive. This way of reconstructing the past makes it difficult to create strong, unambiguous stories that could serve as an instrument for political mobilization. This is because it will not produce powerful images of reality and visions of change that could integrate people and drive their political activity.

I claim, however, that generating complex, voluminous, inconclusive and ambiguous narratives is a suitable practice for two reasons. Firstly, "strong" narratives can be ineffective, because the reader makes use of stories about the past beyond the author's control. Knowledge about the reception process generated by literary

---

[1] The reader contract in historical writing has also been dealt with by Ahlbäck (2007), who analysed the possibilities of using literary theory resources to investigate the reader issue in historical writing (Ahlbäck 2007) and Marek Tamm, who proposed a pragmatic approach to the question of historical truth with the concept of "truth pact" (Tamm 2014).

studies tells us that we do not know what the reader will do with the message. Thus, the most cleverly composed, persuasive historical narrative, which was supposed to propel people to action, will encounter the problem of indetermination of the reading process. It is based on the conditioned personal and collective cultural experience, combines processes and generates meanings, senses, and images which have a vast number of variants.

Since "strong" stories produce a hierarchical relationship, they can also arouse resistance on the part of addressee, who does not want to be in the position of the one who does not know, passive and unaware, and to whom the author will explain everything. Such narratives may be rejected, because they do not offer recognition and cooperation to the reader but the role of the addressee who passively absorbs the knowledge tabled.

Secondly, the coherent, persuasive stories with a "strong" thesis enter into the didactic logic that produce an asymmetric relationship with the reader. Building an asymmetric relationship between academia and society is an unsuitable solution, simply because an unequal future is not a better future. Responses to the environmental crisis should be democratic and not reproduce inequality.

To describe the equal and unequal relationships among the historian, the reader and actors represented in historical writing, the ideas of Jacques Rancière are highly relevant. He has offered an extended discussion of the relationship between politics and aesthetics, in which he addresses the issues of political engagement of art, the relationship between a form of artistic presentation and the social world, and questions of visibility, identity, and agency. He is concerned with the philosophy of aesthetics but rarely discusses the objects of visual art, and the main focus of his analysis and examples is literature and, less frequently, historiography. In his opinion, aesthetics is a specific sphere of activity that produces forms of "distribution of the sensible" (Rancière 2004a: 12–13). The distribution of the sensible, in turn, involves selecting, combining, dividing, exposing and obscuring perceptual content and thereby granting or not granting visibility, position and agency to individuals and groups. Aesthetics is thus strongly related to politics, as Rancière puts it, as it "defines what is visible or not in a common space, endowed with a common language." Politics, on the other hand, "revolves around what is seen and what can be said about it, around who has the ability to see and the talent to speak, around the properties of spaces and possibilities of time" (Rancière 2004a: 12–13). Articulations of the sensual can be democratic and produce images of egalitarian forms of relationships among things, people and events. Rancière treats historical writing as a space of aesthetics, as indicated by his analysis of the writing of the Annales school in Names of History, in which he reads the texts of historians in the same way as in his other works he reads Flaubert's novels and from which he expects similar practices of sharing the sensual as from art forms (Rancière 1994).

In Rancière's view, the pedagogical process is hierarchical, because it is based on two non-egalitarian premises that are reproduced in the course of teaching. First, central to the teacher is the premise of the student's ignorance. This is the first knowledge they pass on to the student: the knowledge that the student themself does not know and is not capable of acquiring on their own, and so everything must be explained by the teacher. Teaching is thus a continuous reproduction of inequality. In other words, the teacher's role is to bridge the gap between their own knowledge and the student's ignorance. However, to reduce this gap, the teacher must constantly reproduce it (Rancière 2009: 8–9).

Second, it also assumes that the teacher knows the path the student must take to move from ignorance to knowledge. Thus, the teacher transfers their knowledge or knowledge from the textbook to the student in a process of uninterrupted transmission in the right order and in the right doses. The teacher explains to the student how to understand images, texts and actions. The knowledge that the pupil acquires should be identical to the teacher's knowledge, as the teaching process assumes the homogeneity of cause and effect. At the same time, the knowledge that the student acquires on his own is the knowledge of the ignorant. It is the duty of the teacher to interrupt his disordered acquisition of knowledge (Rancière 2009: 8–9).

Rancière points out that the positions of teacher and student can be changed without altering the asymmetrical nature of this relationship. Knowledge remains on one side, ignorance on the other. It is possible to claim that students should replace teachers and, for example, workers tell the intelligentsia how to change the world. But social change begins with the principle of equality, which means that teachers and students, workers and intelligentsia, are equal. The opposition of passivity and activity must be rejected—the student is active, just as the teacher is. They observe, select, and interpret data. By a similar principle, the reader is active, and the opposition of writing and reading should be rejected (Rancière 2009: 17–23).

In reality, the student knows a lot of things that they have learned on their own. The knowledge acquired from various sources is mediated by images, texts and actions, which the pupil reads and explains to themself in their own way without reproducing the content and meaning contained in them by the author. The student may be accompanied in this by an ignorant teacher (i.e. a teacher who ignores inequality). They need not be ignorant but need only separate their knowledge from their status as master. The teacher does not pass on their knowledge to their students but seeks it with them (Rancière 2009: 10–11, 13–14).

Following Rancière's way of thinking, one can argue that historiography has a strong political potential: it can reconfigure the visibility of things, people and events, divide and communalize places, resources and modes of expression, and make various forms of collective life conceivable. What is not at stake, however, is history's ability to change social reality through critical instruments that expose and explain its mechanisms. Rancière has challenged this approach in a number of texts pointing to the undemocratic implications of thinking in terms of surface and depth, and higher and false consciousness (e.g. Rancière 2004b).

Rancière postulates a historiography that brings no message or lesson to contemporary society. In Rancière's view, the formulation and communication of such meanings would be inegalitarian, as it would introduce preferences and hierarchies of interests of groups or candidates for power (Rancière 1999: 98–100). Rancière does not define the recipient of historiography, so it can be anyone, and he is not afraid to entrust them with the task of making meaning or bringing closure to the ambiguous and unfinished historical representation. The attempt to control the reception of the presentation, to impose a political thesis, enters into a didactic logic that produces an asymmetrical relationship with the reader.

The accurate course of action would be to seek historiographical forms that offer a reading contract which establishes equality between the historian and their audience. It is important to consider what textual means make such contracts and how equality contracts are practiced. This would likely require more analysis of how historians construct reading contracts. Preliminarily, it can be argued that egalitarian texts have open-ended conclusions and do not provide explicit explanations of reality and lessons for the future but instead leave it to readers to make meanings of the representation and draw lessons from the past.

Therefore, it seems that history should not follow the expectations of those who seek historiography with strong moral lessons for the present day. Complex and ambiguous historiography produces a more equal relationship between historians and their audience. Due to the lack of closure of the story, and thus without ascribing to it holistic meaning or moral lessons, in this type of writing the task of the reader is to make the narrative coherent and assign meaning to it. It does not specify who can be the addressee, so it can be anyone. It transfers the disposition to close the story and create its meaning to the reader.

It should be added that, according to Pihlainen, due to the character of their practices, historians' standard writing is unattractive, complicated and ambiguous. This is because historiography is constrained by a referential pact that imposes the obligation to provide a detailed reconstruction of the facts. Pihlainen claims that historians digging in source material struggle to consider the degree of its credibility, as well as discuss its various interpretations. In turn, this generates writing that is heterogeneous, incoherent, and excessively long. Of course, there are historians who move away from the commitments of their field in favour of producing compelling narratives with seemingly greater political and commercial potential. However, they might become commercially successful at the expense of discarding a detailed reconstruction of the historical past and the representation of their complexity. Pihalinen uses these observations to present an argument that historians should invest less energy in looking for new narrative forms for representations of the past and examine those we already have, because, surprisingly, they seem to meet some of the requirements needed to challenge the new circumstances we find ourselves facing (Pihlainen 2016).

Conclusion

An analysis of historical theory's response to the challenges of the Anthropocene era suggests several adjustments to the production of representations of the past. The reading contract in historical writing requires changes but not radical ones. The new version of the contract with the audience would confirm the principle that has been gaining importance for several decades, which postulates the engagement of historians in responding to the social, political and environmental problems of the contemporary world. It would involve the participation of human and inhuman protagonists in the historical narrative on equal terms and assembling threads of radically different temporality, from a short-term story to a history counted in the millions of years. The old rule that the historian seeks to diligently reconstruct historical facts would be preserved. There would also be an article—resulting from this principle—about the discontinuity and complexity of historical narratives and the reluctance of authors to make their stories more attractive by simplifying them and adding significant meanings. Although this would be a new clause in the contract, it would nevertheless sanction the existing state of affairs, as I have already mentioned, of historians who are driven by the obligation to reconstruct facts and produce complex, discontinuous narratives devoid of lessons or morals. The last element of the contract and, at the same time, its political framework would be an equality imperative which establishes the relationship between the historian and the reader as equal.

Wynn's speech from 2019 partially fulfils the above contract. Just as the contract does, Wynn postulates the involvement of historians in the environmental crisis and in the search for visions of a better future. He rightly criticizes the apocalyptic rhetoric of shock and despair, which is often employed by scholars wishing to raise audiences' awareness and guide their actions. Wynn's narratives of hope could be justified if they offered discontinuous, open-ended stories and invited readers to co-create them. However, he explains them to the audience, gives instructions on how to relate them to our current problems, and transforms them into a lesson for the reader. The two coherent, teleologically ordered life stories of Dansereau and Macpherson, each with a clear message embedded in them, recreate the pedagogical logic with its hierarchical ordering of the author and the reader. Meeting the proposed contract, environmental history cannot offer a lesson in the Anthropocene era which projects the passive audience absorbing what it teaches. In these new circumstances, history can remain a teacher of life only insofar as the addressee, discussing the stories delivered, gives them meaning and draws their own lessons from them.

Acknowledgment This research was funded by National Science Centre, Poland, grant no 2020/39/D/HS3/01262.

References

Ahlbäck PM (2007) 'The reader! The reader! The *Mimetic* challenge of addressivity and response in historical writing. Storia Della Storiografia 52:31–48

Chakrabarty D (2009) The climate of history: four theses. Crit Inq 35(Winter):197–222

Chakrabarty D (2018a) Planetary crises and the difficulty of being modern. Millennium J Int Stud 46(3): 259–282.

Chakrabarty D (2018b) Anthropocene time. Hist Theory 57(1):5–32

Chakrabarty D (2019) The planet: an emergent humanist category. Crit Inq 46(Autumn):1–31

González JA (1995) Autotopographies. In: Brahm G, Jr, Driscoll M (eds) Prosthetic territories: politics and hypertechnologies. Boulder, Westview, pp 133–150

Glotfelty C, Fromm H (eds) (1996) The ecocritism reader: landmarks in literary ecology. University of Georgia Press, Athens

Gumbrecht HU (2013) After 1945: latency as origin of the present. Stanford University Press, Stanford

Lambert D, Lester A (eds) (2006) Colonial lives across the British empire: imperial careering in the long nineteenth century. Cambridge University Press, Cambridge

Le Goff J (2009) Saint Louis. trans. G. Gollrad. University of Notre Dame Press, Notre-Dame

Pihlainen K (2016) The distinction of history: on valuing the insularity of the historical past. Rethink Hist 20(3):414–432

Pihlainen K (2017) The work of history. Routledge, New York

Rancière J (1994) Names of history. On the poetics of knowledge. trans. H. Melehy. University of Minnesota Press, Minneapolis

Rancière J (1999) Disagreement: politics and philosophy. trans. J. Rose. University of Minnesota Press, Minneapolis

Rancière J (2004a) The politics of aesthetics: the distribution of sensible. Trans. G. Rockhill. Continuum, London

Rancière J (2004b) Philosopher and his poor. Trans. J. Drury, C. Oster, A. Parker. Duke University Press, Durham

Rancière J (2009) Emancipated spectator. trans. G. Elliott. Verso, London

Simon ZB (2015) History manifested: making sense of unprecedented change. Eur Rev Hist 22(5):818–834

Simon ZB (2017) Why the anthropocene has no history: facing the unprecedented. Anthropocene Rev 4(3):239–245

Simon ZB (2019) Two cultures of the posthuman future. Hist Theory 58(2):171–184

Simon ZB (2020) The limits of anthropocene narratives. Eur J Soc Theory 23(2):184–199

Simon ZB, Tamm M (2020) More-than-human history: philosophy of history at the time of the anthropocene. In: Kuukkanen J-M (ed) Philosophy of history: twenty-first-century perspectives. Bloomsbury, London, pp 198–215

Tamm M (2104) Truth, objectivity and evidence in history writing. J Philos Hist 8: 265–290

Wynn G (2020) Framing an ecology of hope. Environ Hist 25(2020):2–34

History and Utopian Thinking in the Era of the Anthropocene

Dariusz Brzeziński

Abstract The text analyzes the different ways in which historical narratives can contribute to changing human sensitivity and social practices in relation to climate threats. The term "utopian thinking" is problematized here as a combination of relativization of the existing reality with the desire to develop alternatives to it. Three types of historical discourses devoted to the Anthropocene will be analyzed and juxtaposed in the paper: post-naturalism (represented by Dipesh Chakrabarty), eco-Marxism (analyzed on the example of Jason W. Moore's theory of the Capitalocene) and eco-catastrophism (illustrated in relation to the works of Naomi Oreskes and Eric Conway).

Keywords The Anthropocene · Historical narratives · Utopian thinking · Post-naturalism · Eco-Marxism · Eco-catastrophism

> I had always thought that the purpose of More's *Utopia*.
>> was not to provide a blueprint for some future.
>> but to hold for inspection the ridiculous waste and foolishness of his times,
>> to insist that things could and must be better.
>> David Harvey, *Spaces of Hope* (2000: 281).

Introduction

Twenty-two years have now passed since Paul J. Crutzen and Eugene F. Stoermer introduced the term "the Anthropocene" into scientific discourse. In a short text published in the *Global Change Newsletter* (Crutzen 2000: 17–18) they discussed the substantial and still growing impact of human activities on the natural processes taking place on Earth and in the atmosphere. As far as they were concerned, the second

D. Brzeziński (✉)
Institute of Philosophy and Sociology, Polish Academy of Sciences, Warsaw, Poland
e-mail: dbrzezinski@ifispan.edu.pl

School of Sociology and Social Policy, University of Leeds, Leeds, England

© The Author(s) 2022
A. Izdebski et al. (eds.), *Perspectives on Public Policy in Societal-Environmental Crises*,
Risk, Systems and Decisions, https://doi.org/10.1007/978-3-030-94137-6_4

half of the eighteenth century marked the end of the Holocene and the beginning of the Anthropocene, as humankind then acquired the role of a geological force. Crutzen and Stoermer did not stop at this proclamation but stated that the current situation required mankind to take appropriate steps. They wrote: "To develop a world-wide accepted strategy leading to sustainability of ecosystems against human induced stresses will be one of the great future tasks of mankind, requiring intensive research efforts and wise application of the knowledge thus acquired in the noösphere, better known as knowledge or information society. An exciting, but also difficult and daunting task lies ahead of the global research and engineering community to guide mankind towards global, sustainable, environmental management" (Ibidem, p. 18). This statement was reflected in the nature of the discourse on the Anthropocene that has been dynamically developing in the twenty-first century (Bińczyk 2019: 3–18). On the one hand, it analyses the scale of destructive human activity in relation to the natural environment; on the other, it is characterised by a reflection on how a harmonious coexistence between human and nature could be achieved.

Although this discourse was originally developed mainly by natural scientists, over the years more and more representatives of social sciences and humanities have joined it (Trischler 2016: 309–335; Zalasiewicz et al. 2010: 2228–2231). This also applies to historians (see e.g.: Coen 2018; Moore 2016; Levene et al. 2010). The article examines this tendency and aims to show the different ways in which historical narratives can contribute to changing human sensitivity and social practices in relation to the natural environment. Thus, the paper examines the dialectic interplay between history and utopian thinking, the latter concept being understood as a combination of criticism of the existing social reality with an incentive to look for alternatives. In this context, I follow the contemporary approaches to utopia developed, *inter alia,* by Levitas (2013), Sargent (2010) and Jameson (2005). The two above-mentioned components of utopian thinking correspond to the two parts of the paper. The first part presents chosen historical narratives devoted to the criticism of the relationship between humans and the environment in the modern era, whereas the second part explores the same texts but focuses on their utopian potential with regard to counteracting the Anthropocene. Selection of the particular theories for analysis in this paper was related partly to their significance among historians and other intellectuals. This choice also results from a comparative dimension of the article. The paper juxtaposes three narratives of the Anthropocene that indicate the need to apply significant changes to historical discourse: post-naturalism (represented here by Dipesh Chakrabarty), eco-Marxism (analysed in the example of Jason W. Moore's theory of the Capitalocene) and eco-catastrophism (illustrated in relation to the works of Naomi Oreskes and Eric Conway). In the concluding part, I summarize my analyses and refer to the title question of the book: What sort of past does the future need?

History and Criticism in the Era of the Anthropocene

Criticism of the *status quo* is considered a constitutive feature of utopian thinking (Jacobsen 2012; Sargent 2010; Levitas 1990), because a process of relativisation of the current state of affairs paves the way for the need to shape the alternative paths of development. Bauman (1976: 11) wrote in this context: "[…] the capacity to think in a utopian way does involve the ability to break habitual associations, to emancipate oneself from the apparently overwhelming mental and physical dominance of the routine, the ordinary, the 'normal'". Relating these observations to the issue discussed in this article, it should be noted that historians who conduct research on the Anthropocene strongly criticize the way nature has so far been conceptualized. They emphasize the need to reformulate the general assumptions as far as analyses of the human–environment relationship are concerned, so that they fully reveal the destructive scale of human activity. In this part of the paper I will highlight that their criticism is aimed at revolutionizing the nature of historical discourse.

The first of the three narratives analysed in this paper, "post-naturalism", is in fact a very diverse set of theories focused on the reconsideration of the relationship between the human world and the natural world. Christophe Bonneuil, to whose typology I refer in this article, modifying it to some extent,[1] wrote about this narrative as follows: "While modernity had promised to emancipate society from nature's determinism, the Anthropocene proclaims the inescapable immersion of human destiny in the great natural cycles of the Earth, and the meeting of the temporalities of short-term human history and long-term Earth history that had been viewed as separated for the last two centuries" (Bonneuil 2015: 24). This assumption is reflected in the works of Dipesh Chakrabarty, whose article "The Climate of History: Four Theses" (2009: 197–222) introduced reflection on the Anthropocene to historical research. This paper, as well as the later works of this historian (e.g. 2021; 2019: 1–31; 2017: 39–43; 2016: 103–113; 2015: 137–188), are, to this day, points of reference—often critical, as will be shown later in the text—in the development of this discourse in history and beyond.

"The Climate of History" starts with a reference to Alan Weisman's book *The World Without Us* (2007), which presents what would happen if humans suddenly ceased to exist. Chakrabarty points out that this thought experiment is useful not only in terms of awakening ecological awareness but also in the context of reformulation of some assumptions taken for granted in historical research. Imagining a future without human beings breaks the common-sense view of the relationships among past, present and future. It also contributes to asking questions about the place of humans in the history of the planet. As far as Chakrabarty is concerned, in the era of the Anthropocene, the centuries-old division into natural history and

---

[1] Bonneuil distinguished four main narratives of the Anthropocene: "naturalism", "post-naturalism", "eco-catastrophism" and "eco-Marxism". I decided to focus on the last three ones in this article because they are most concerned with changes in historical narratives. What is more, as the example of Dipesh Chakrabarty's theory shows, it is difficult to clearly separate the "naturalist" and "post-naturalist" narrative of the Anthropocene. His utopian vision of "planetary perspective" is, however, definitely closer to the "post-naturalistic" narrative.

human history has to be fully transcended. He considered previous attempts to extend this distinction—made, for example, by environmental historians—insufficient, as humans are perceived in their context as a "biological agent", not as a "geological agent", which does not allow an adequate illustration of humans' impact on the natural processes taking place on Earth and in the atmosphere. Chakrabarty claims (2009: 206) that "Humans are biological agents, both collectively and as individuals. They have always been so. [...] To call ourselves geological agents is to attribute to us a force on the same scale as that released at other times when there has been a mass extinction of species. We seem to be currently going through that kind of a period". As a consequence, Chakrabarty postulates that the advent of the Anthropocene requires history researchers to think of human beings from the perspective of "geological time" and "planetary space" (Chakrabarty 2019: 1–31; 2017: 39–43). It does not imply the need to abandon customary historical research but to supplement them with this new, widely extended approach. This perspective will make it possible to analyse the consequences of human activity in relation to other organisms, both living and non-living. It will also serve as a basis of historical research into how this transformed natural environment affects humans as a species. Both levels of analysis are also related to the utopian dimension of Chakrabarty's thought, which will be analysed in the next part of this paper.

Among historians who reflect on the destructive activity of humans toward nature, there is no agreement that the epoch following the Holocene should be called the "Anthropocene". The opponents of this term argue that it assigns the responsibility for degradation of the Earth to all humankind, which does not reflect reality. Not only were not all humans to blame for this crisis, but also most were its victims, they claim. This issue is discussed within the "eco-Marxism" narrative and characterized by Bonneuil as follows: (2015: 27–28): "While Marx theorised on the first contradiction of capitalism, its inability to reproduce the labour force, the eco-Marxist narrative sees the Anthropocene as a result of a second contradiction of capitalism: its inability to maintain nature". One of the most renowned representatives of this narrative is Jason W. Moore, who calls the present epoch the "Capitalocene". Moore published many papers on this issue (e. g. 2018: 237–279; 2017a: 594–630; 2017b: 175–202) and edited a book, *Anthropocene or Capitalocene? Nature, History and the Crisis of Capitalism* (2016).

Unlike Chakrabarty and the vast majority of the Anthropocene researchers, Jason W. Moore states that the beginning of human destructive activity in relation to nature took place long before the industrial revolution. In this context, he focuses on the "long" sixteenth century,[2] when the constitutive features of capitalism were developed. Taking into account, *inter alia*, such phenomena as agriculture expansion, rapid deforestation and a several-percent increase in coal production, Moore stated that it was that time when the natural environment began to be exploited on an unprecedented scale. In this context he introduced the concept of "Cheap Nature".

---

[2] This term – introduced by Ferdinand Braudel (1972) – refers to a period in history from the mid-fifteenth century to the mid-sixteenth century when the expansion of the European world-economy took place (Moore 2003: 431–458).

"For capitalism, Nature is 'cheap'", wrote Moore (2016: 2–3), "in a double sense: to make Nature's elements 'cheap' in price; and also *to cheapen*, to degrade or to render inferior in an ethico-political sense, the better to make Nature cheap in price. These two moments are entwined at every moment, and in every major capitalist transformation of the past five centuries". Moore claims that the emergence of "Cheap Nature" was accompanied by the appearance of a similar attitude toward humans. The most significant manifestation of this phenomena was the development of slavery, but there were also many other examples of this tendency. That is why Moore claims that history must be reconsidered to analyse how humanity and nature are bound together within the web of life. This conviction is reflected in his concept of the "Capitalocene", which he defines as "a multispecies assemblage", or "a world-ecology of capital, power, and nature" (Moore 2016: xi). Thus, Moore shares Chakrabarty's conviction in the fallacy of creating binary divisions between nature and society, and, therefore, also natural history and human history. He does so, however, from a different, Marxist perspective. This is also reflected in the distinct approaches to utopian thinking between the two.

The third narrative discussed in this paper is eco-catastrophism. According to Bonneuil (2015: 26), it: "[…] views the Anthropocene as an age in which modernity's project of indefinite growth and progress hits the wall of the planet's finitude" (see also: Rothe 2020: 148–151). A very good example of such an approach to the Anthropocene is a book, *The Collapse of Western Civilisation: A View from the Future*, which was authored in 2014 by Naomi Oreskes and Erik M. Conway. Similar to Alan Weisman's *The World Without Us*, their work presents the consequences of a future global ecological catastrophe. They do not, however, depict the world after the extinction of humans but the reality in which the Western model of civilization fails to cope with the climate catastrophe.

Oreskes and Conway juxtapose science fiction writing with historical analyses to indicate how the current challenges of the Anthropocene could be perceived from the point of view of the distant future. Their vision is presented as a narration of a historian living in the 'Second People's Republic of China' in 2393. He reflects on the twenty-first century processes that led to "the Great Collapse" of Western civilization. He analyses both the destructive impact of humans on the environment and the inability to make bold decisions within the West (in contrast to Chinese civilisation) that would enable adaptation to the changes taking place. What puzzles this future historian the most is the fact that these tragic events were fully predictable, and yet no measures were put in place to prevent them. He wrote: "Indeed, the most startling aspect of this story is just how much these people knew and how unable they were to act upon what they knew. Knowledge did not translate into power" (Oreskes and Conway 2014: 2). Oreskes and Conway point out in the commentary to this book that their aim was to highlight the fallacy of the aforementioned way of thinking. It is worth emphasising that they focus mainly on the normative and institutional spheres of social life and indicate that the lack of significant changes in these areas will inevitably lead to disaster.

The post-naturalist, as well as the eco-Marxist and the eco-catastrophic narrative, share the desire to relativize both the current ways of writing history and, at the same time, the perception of history. Regardless of all the differences between them, they focus on the destructive dimensions of the processes that have shaped modern civilization and thus open the way for the development of alternative thinking. In the book by Zygmunt Bauman cited at the beginning of this part of the article, he also wrote: "One cannot be critical about something that is believed to be an absolute. By exposing the partiality of current reality, by scanning the field of the possible in which the real occupies merely a tiny plot, utopias pave the way for a critical attitude and a critical activity which alone can transform the present predicament of man" (Bauman 1976: 13). In the next part of the article I will continue my analysis, presenting and comparing the visions of the post-Anthropocene future within these three narratives.

History and the Future in the Era of the Anthropocene

With reference to Jacoby's (2005) thought, the various visions of the future could be placed on the continuum between "blueprint" and "iconoclastic" utopianism.[3] The former is characterized by a very detailed and holistic presentation of the future. What is more, this approach assumes that there is one perfect, universal model of life, and that it is legitimate to strive by all means to realise it. At the other end of the continuum Jacoby placed "iconoclastic" utopianism, which, as its name suggests, does not involve mapping out the future. Instead, it is open to various alternative opportunities, none of which is considered final. Iconoclastic utopianism is characterised by the process of outlining directional guidelines with regard to the future without drawing the ultimate result of these efforts. The evolution of utopian thinking in recent decades consists of moving from blueprint visions towards iconoclastic ones that many intellectuals consider relevant to the reality of the twenty-first century (Levitas 2007: 289–306; Sargent 2005: 1–14). In this part of the paper I will emphasise that this statement also applies to utopian thinking in the era of the Anthropocene in all of the three narratives discussed here.

The turn from the blueprint to the iconoclacstic form of utopianism is related, *inter alia*, to the association of the former with the tradition of Enlightenment, which has been strongly criticised by many representatives of the social sciences and humanities since the second half of the previous century. Such prominent intellectuals as Arendt (1951), Popper (1945) and Hayek (1944) argued that there is a close relationship between the projects of a perfect society created on the basis of rationalism, and the origins of violence and totalitarianism. Therefore, the iconoclastic model of utopian thinking is shaped in opposition to the assumptions attributed to the tradition of Enlightenment (Jacoby 2005). In this context it is worth stressing that Chakrabarty's attitude towards this formation is ambivalent. On the one hand, he points out that the

---

[3] Similar distinctions between the two opposing types of utopianism were made, *inter alia*, by: Jameson (2005), Alexander (2001: 579–591), Bauman (2003: 11–25) and Bloch (1986).

current climate crisis emerged as a result of the processes initiated in the Enlight-enment. On the other hand, he states (2009: 210) that "[…] it is […] clear that for humans any thought of the way out of our current predicament cannot but refer to the idea of deploying reason in global, collective life". Chakrabarty advocates the devel-opment of technology, as well as organisational solutions, aimed at counteracting climate change, but he does not advocate the blueprint of utopia.

As far as he is concerned, the above mentioned efforts should occur in parallel with the emergence of a "planetary perspective". This concept reflects his view of the changes that, as he claims, needs to be made in historical narratives. Chakrabarty points to the necessity of transcending anthropocentrism and extending human responsibility to all organisms on the Earth. "Man will have to be placed in the larger context of the deeper history of life on this planet", he wrote (2017: 42). "[…] anthropocentrism may be necessary but will increasingly seem inadequate if one looks at the impact of the human ecological footprint on other forms of life and on the planet itself. So, our inevitable anthropocentrism will need to be supplemented (not replaced) by 'deep time' perspectives that necessarily escape the human point of view". It should be emphasised that Chakrabarty's utopian thinking is entirely iconoclastic in this matter. He neither presents a plan for how to achieve the revo-lutionary change in social imaginary nor describes the details of the reforms that should be implemented in relation to the anticipated "planetary" perspective. Instead, Chakrabarty indicates the direction mankind should pursue and also highlights the sense of historical necessity.

The term "planetary justice" is also a constitutive element of Moore's utopian thinking (Moore 2020: 161–182; 2019: 49–54), but he defined it in a different way to Chakrabarty. "In my view", he stated (2020: 176), "a radical strategy of planetary justice proceeds through that connective critique of capitalism, such that we can make clear—and organize around—the conditions of capitalogenic climate change. In world-ecological perspective […] the history of climate crisis, modern imperi-alism, the world colour line, and globalizing patriarchy open vistas through which to see today's crisis politics in ways that reveal the constitutive lines between global domination and empire, and the endless accumulation of capital". *Ergo*, Moore's utopian thinking resembles his historical analyses of the emergence and develop-ment of the Capitalocene. In analogy to his thesis that the emergence of Cheap Nature was accompanied by the subordination of reality to the principle of endless accumulation, he claims that only through rejection of the logic of capitalism can the climate crisis be solved. Moore's view on this matter reflects Marxist inclina-tions of his thought: he advocates the radical, holistic change, and considers all other solutions as definitely insufficient.

Moore's vision of the future remains closer to the iconoclastic than to the blueprint model of utopianism. He does not strive to outline the details of the post-Capitalocene world he postulates and points out only its fundamental principles. Among them are such imperatives as, *inter alia*, decarbonisation, decommodification and democrati-sation, which are in fact the negatives of capitalist logic. Moore argues that the shape of the future order should be discussed in a wide forum among scholars, artists and activists (Moore 2019: 54). The concept of the Capitalocene is in fact an invitation to take part in this debate on planetary justice.

In contrast to Jason W. Moore, Naomi Oreskes and Erik M. Conway do not postulate the need to establish an entirely new socio-economic order. Also, they do not share Dipesh Chakrabarty's belief that it is necessary to question anthropocentrism. Their utopian thinking is far less radical than that of those two intellectuals. They even claim that "Our story is a call to protect the American way of life before it's too late" (Oreskes and Conway 2014: 79). However, they also believe it is necessary to implement the normative and institutional changes in order to minimize the processes of climate destabilization. In this context, they indicate, *inter alia*, the need for more control mechanisms against political and economic neoliberalism. They also consider it indispensable to raise ecological awareness among the general public. A few years before *The Collapse of Western Civilisation* was published, Naomi Oreskes and Erik M. Conway published *Merchants of Doubt* (2011), which analyses the process of spreading doubt and confusion by some scientists wishing to discourage society from taking action aimed at changing given practices, attitudes or policies. One of the topics it discusses is the questioning of global warming. Oreskes and Conway returned to this issue in *The Collapse of Western Civilisation*, arguing that in order to counteract the Anthropocene it is necessary to restore the authority of science. Only then science will have a decisive influence on both political and individual choice.

Oreskes and Conway's utopian thinking has, in a similar way to that of Chakrabarty's and Moore's, an iconoclastic character. This is clearly demonstrated by a metaphor they use at the end of the book discussed here. They wrote that "[…] you can't predict what your readers will take away. Books are like a message in a bottle. You hope someone will open it, read it, and get the message. Whatever that is" (Oreskes and Conway 2014: 79). Thus, *The Collapse of Western Civilisation* should be interpreted as an inspiration to reflect upon both the challenges of the Anthropocene and upon appropriate ways to deal with them. In this context, it is also worth mentioning that this book has been written with a wide range of readers in mind. Oreskes and Conway not only refrained from using the hermetic scientific language but also chose a science fiction formula to attract readers. They left "a message in a bottle"[4] in the hope that it would be found and read, thus contributing to the spread of utopian thinking.

All three historical narratives discussed above, although differing in their details, are excellent examples of the transformations that have been taking place in the field of utopian thinking for the last few decades (Rüsen, Fehr, Rieger 2005; Jacoby 2005). Firstly, they do not visualise the post-Anthropocene or the post-Capitalocene world but present general assumptions upon which it should be founded. Secondly, they focus on the critical revision of the norms, practices and institutions that led to the current climate crisis, and indicate that it is necessary to transcend them. Thirdly,

---

[4] Zygmunt Bauman commented on this metaphor in his *Liquid Life* in the context of critical theory. He wrote: "The message in a bottle is a testimony to the *transience of frustration* and the *duration of hope*, to the *indestructability of possibilities* and the frailty of adversaries that bar them from implementation" (2005: 143).

they emphasise that mankind is at a critical point but at the same time inspire hope and desire for change. Last but not least, they are aimed at expanding the scope of human responsibility in terms of the "planetary" or "global" perspective.

Conclusions

The construction of this paper reflects the way the classic work of Thomas More is structured (1965 [1516]). This book is divided into two parts, the first of which presents the social, economic and political problems that took place in Tudor England, and the second of which is a description of an idealised island community. Both parts of this work form an inseparable whole: the vision of the Utopian Republic is a reference point for critical assessment of the situation in More's homeland, while the relativization of the *status quo* paves the way for the creation of alternative visions of the future (see e.g.: Cousin and Grace 1994). Following Moore, I analysed in this article how historical narratives entwine criticism of human's attitude towards nature with visions of a better future. On the one hand, I emphasised that writing on the history of modernity, not from the point of view of progress but from the perspective of the deepening climate catastrophe, may affect ecological awareness and stimulate related activities in this regard. On the other hand, I analysed how a presentation of a more or less detailed vision of a post-Anthropocene future may contribute to the relativization of those norms or practices that are considered to be responsible for the ecological crisis. These interpenetrating functions of historical discourse were illustrated in one example of three competing narratives: post-naturalism, eco-Marxism and eco-catastrophism. It has been shown that, despite significant differences among them, they all represent an iconoclastic model of utopian thinking.

The above reflections contribute to the answer to the more general question of what sort of past the future needs. I pointed out in this paper that, faced by the challenges of the contemporary world—the Anthropocene is, of course, only one of them—the historical narratives may play a very important role. They may, *inter alia* stimulate criticism toward the *status quo*, awaken individual and social awareness, extend the scope of human's responsibility, shape moral imperatives and inspire hope. The fulfilment of these functions will depend, however, on whether these narratives acquire a utopian dimension. My vision of the liaisons between history and utopian thinking is quite similar to the concept of the "rescue history" of the Polish historian Domańska (2014: 12–26). She wrote that "Rescue history as an existential history is a research perspective that via the inquiries of the past reflects on the meanders of the human condition. [...] It is a perspective that treats the problems undertaken by the researcher as a starting point for reflection (and self-reflection) on the human condition, the condition of the planet, the condition of the humanities, and history as a discipline" (Ibidem, p. 18). I agree with Domańska that future-oriented historical narratives should raise critical hope rather than indicate particular solutions. As far as I am concerned, the idea of the "iconoclastic utopia" may serve as an important source of inspiration in this regard. It encourages involvement in the historical process while emphasizing its complexity and never-ending changeability.

References

Alexander JC (2001) Robust utopias and civil repairs. Int Sociol 16(4):579–591
Arendt H (1951) The origins of totalitarianism. Schocken Books, New York
Bauman Z (1976) Socialism. The Active Utopia. George Allen & Unwin, London
Bauman Z (2003) Utopia with no Topos. Hist Hum Sci 16(1):11–25
Bauman Z (2005) Liquid life. Polity Press, Cambridge
Bińczyk E (2019) The most unique discussion of the 21st century? The debate on the Anthropocene pictured in seven points. Anthr Rev 6(1–2):3–18
Bloch E (1986) The principle of hope, vol I-III. Basil Blackwell, Oxford
Bonneuil C (2015) The geological turn. Narratives of the Anthropocene. In Hamilton C, Gemenne F, Bonneuil C (eds) The anthropocene and the global environmental crisis: rethinking modernity in a new Epoch. Routledge, London, pp 15–31
Braudel F (1972) The Mediterranean and the Mediterranean world in the age of Philip II, vol 1. Harper & Row, New York
Chakrabarty D (2009) The climate of history: four theses. Crit Inq 35(2):197–222
Chakrabarty D (2015) The human condition in the anthropocene. The Tanner Lectures in Human Values, Delivered at Yale University (February 18–19, 2015), pp 137–188. https://tannerlectures. utah.edu/Chakrabarty%20manuscript.pdf
Chakrabarty D (2016) Whose anthropocene? A response. In Emmett R, Lekan T (eds) Whose anthropocene? Revisiting Dipesh Chakrabarty's "Four Theses". RCC perspectives: transformations in environment and society, rachel carson center for environment and society, Munich, pp 103–113
Chakrabarty D (2017) The future of the human sciences in the age of humans: a note. Eur J Soc Theory 20(1):39–43
Chakrabarty D (2019) The planet: An emergent humanist category. Crit Inq 46(1):1–31
Chakrabarty D (2021) The climate of history in a planetary age. University of Chicago Press, Chicago–London
Coen DR (2018) Climate in motion. Science, empire and the problem of scale. University of Chicago Press, Chicago–London
Cousins AD, Grace D (eds) (1994) More's Utopia and The Utopian Inheritance. University Press of America, Lanham–New York–London
Crutzen PJ, Stoermer EG (2000) The "Anthropocene." Global Change Newsletter 41:17–18
Domańska E (2014) 'Historia Ratownicza' ['Rescue History']. Teksty Drugie 5:12–26
Harvey D (2000) Spaces of hope. University of California Press, Berkeley–Los Angeles
Hayek FA (1944) The road to serfdom. Routledge, London–New York
Jacobsen MH, Tester K (eds) (2012) Utopia: social theory and the future. Ashgate, Aldershot
Jacoby R (2005) Picture imperfect. Columbia University Press, New York, Utopian Thought for an Anti-Utopian Age
Jameson F (2005) Archaeologies of the future: the desire called Utopia and other science fictions. Verso, London
Levene M, Johnson R, Roberts P (eds) (2010) History at the end of the world? History, climate change and the possibility of closure. Humanities-Ebooks, Penrith
Levitas R (1990) The concept of Utopia. Philip Allan, London-New York
Levitas R (2007) Looking for the blue: the necessity of utopia. J Polit Ideol 12(3):289–306
Levitas R (2013) Utopia as method. The imaginary reconstitution of society. Palgrave MacMillan, Basingstoke

More T (1965) [1516]. Utopia. Penguin Book, London–New York

Moore JW (2003) Capitalism as a world-ecology. Braudel and marx on environmental history. Organ Environ 16(4):431–458

Moore JW (ed) (2016) Anthropocene or capitalocene? Nature, history and the crisis of capitalism. PM Press, Oakland CA

Moore JW (2017a) The capitalocene, Part I: on the nature and origins of our ecological crisis. J Peasant Stud 44(3):594–630

Moore JW (2017b) World accumulation and planetary life, or, why capitalism will not survive until the 'last tree is cut.' IPPR Progres Rev 24(3):175–202

Moore JW (2018) The Capitalocene Part II: accumulation by appropriation and the centrality of unpaid work/ energy. J Peasant Stud 45(2):237–279

Moore JW (2019) Capitalism and planetary justice. Maize 6:49–54

Moore JW (2020) 'Capitalism and planetary justice in the web of life' (interview with J. W. Moore conducted by: M. Gaffney, C. Ravenscroft and C. Williams). Polygraph 28:161–182

Oreskes N, Conway EM (2011) Merchants of doubt. How a handful of scientists obscured the truth on issues from tobacco smoke to climate change. Bloomsbury Press, London

Oreskes N, Conway EM (2014) The collapse of western civilisation: a view from the future. Columbia University Press, New York

Popper K (1945) The open society and its enemies, vol. I&II. Routledge, London

Rothe D (2020) Governing the end times? Planet politics and the secular eschatology of the anthropocene. Millennium J Int Stud 48(2):143–164

Rüsen J, Fehr M, Rieger TW (eds) (2005) Thinking Utopia. steps into other worlds, Berghahn Books, Oxford–Nowy York

Sargent LT (2005) A necessity of Utopian thinking: a cross-national perspective. In Rüsen J, Fehr M, Rieger TW (eds) Thinking Utopia. Steps into other worlds, Berghahn Books, Oxford–Nowy York, pp 1–14

Sargent LT (2010) Utopianism. Oxford University Press, Offord, A Very Short Introduction

Trischler H (2016) The anthropocene: a challenge for the history of science, technology, and the environment. NTM Int J Hist Ethics Nat Sci Technol Med 24(3):309–335

Weisman A (2007) The world without us. Virgin Book, London

Zalasiewicz J, Williams M, Steffen W, Crutzen P (2010) The new world of the Anthropocene. Environ Sci Technol 44(7):2228–2231

Potentials and Risks of Futurology: Lessons from Late Socialist Poland

Lukas Becht

Abstract In the decades after 1945, the future gained unprecedented prominence as an object of scientific anticipation and state planning in both capitalist and socialist countries of the Cold War world. In Poland, future studies or futurology emerged in the course of the 1960s in reaction to Western intellectual trends, the post-stalinist political Thaw, as well as the domestic socio-economic situation. The Polish futurology turned out to be one of the most productive, institutionally and personally stable research collectives when compared to other socialist countries. This research community generated various approaches to the problem of how to anticipate the unknown future. This chapter examines three of them: making the future an object of knowledge; subjecting it to conscious (political) control; imagining alternatives to the status quo. Re-examining these historical examples of anticipatory knowledge provides a mirror to discuss our current efforts at predicting and controlling the future.

Keywords Futurology · Future research · Poland · Late socialism · Historical epistemology

Planetary Crises, Historiography and the Futurology of the Past

How can you anticipate an unknown future? This age-old question gains renewed urgency in times of threats and challenges related to climate change, ecological crisis, the Anthropocene, and the Covid-19 pandemic. Under these circumstances, anticipatory ideas and practices such as statistical extrapolations, complex model simulations, and scientific and science-fiction scenarios play a crucial and omnipresent role in contemporary strategies of society for coping with perceived threats and instabilities (Robinson 2020; Geulen 2020). Although they are often grounded by faith in

L. Becht (✉)
RECET, University of Vienna, Vienna, Austria
e-mail: lukas.becht@univie.ac.at

Institute of Eastern and South Eastern European History, LMU Munich, Munich, Germany

© The Author(s) 2022
A. Izdebski et al. (eds.), *Perspectives on Public Policy in Societal-Environmental Crises*, Risk, Systems and Decisions, https://doi.org/10.1007/978-3-030-94137-6_5

45

science and the capability of technology to mitigate those challenges, notions of catastrophe dominate contemporary cultural imaginaries of the future (Horn 2014). To an observer, the multitude of claims about anticipatory knowledge, although undoubtedly necessary and often unquestioned tools for collective and individual decision making, appear to be taking part in struggles over (re)defining imaginations of the future as well as concepts of the future, present and past (Beckert 2018: 133; Andersson 2018: 17–18).

These re-negotiations have been argued to challenge fundamental notions of historical thinking and the relevance it assigns to knowledge of the past for present and future societies (Chakrabarty 2009; Nycz 2014). Historians' answers to these challenges will be neither definite nor useful in a utilitarian sense. But they can help to cultivate reflexivity and greater awareness of the contingencies inherent in contemporary ways of coping with the future. With the aim of better understanding the genealogies, dilemmas, and alternatives of present struggles, this paper suggests studying a historical example that highlights how anticipatory knowledge in a particular context did indeed renegotiate notions of the future and historical time.

How can and should we refer to the future? How do we anticipate it? How should anticipation be used to shape behaviour and political decisions in the present? These questions gained an unprecedented urgency around the middle of the twentieth century for various reasons, resulting in the emergence of future studies or "futurology", a transnational scholarly field revolving around the systematic study of the future in a long-term perspective (Andersson 2018; Seefried 2015a). This essay explores how, in the specific historical context of late socialist Poland of the 1960s, 1970s and 1980s, such scholarly activities shifted notions of the future by addressing it first as an object of knowledge, second of social engineering, and third of sociological imagination.[1] In its attempt to historicize past futurology, this essay draws on recent debates on how to write and conceptualize twentieth-century "histories of the future" (Rosenberg and Harding 2005) that go beyond Koselleck's (1979) classical concept of "vergangene Zukunft". The focus is less on past actors' visions of the future, but on concepts, epistemological categories and political circumstances that made such visions possible (Mallard and Lakoff 2011; Graf and Herzog 2016; Daston 1994; Rheinberger 2007; Roelcke 2010).

Future Research in Late Socialist Poland

According to Stanisław Lem, the most severe transformation in the decades after the Second World War was a "deep-going cultural re-orientation […]. Ever since, our culture stared only at the historical past and the present of humans. Now, however, it begins to turn a part of its diligent and solemn attention to the future of men." (Lem 1977: 11–12) Future-orientated preoccupations were a global phenomenon, particularly in the 1960s and 1970s. Popular culture, literature including Lem's science fiction itself, the mass media, and public, political, and scholarly discourse were fascinated by mid- and long-term horizons such as the approaching year 2000, a symbol of either hope for renewal and technological optimism or of *fin-de-siècle* fears of decay. Political and intellectual elites were convinced that what they perceived as accelerated deep-going social and technological change in emerging industrial mass societies, combined with new threats such as nuclear annihilation and ecological disaster, called for anticipation and political steering based on scientific knowledge. This sense of dramatic change and an urgent need for action may remind us of present challenges, while categories and practices related to the future have changed fundamentally.

Since the nineteenth century the power of Western nation states rested on "seeing" (Scott 1999: 4) and coping with social complexity through positivist claims of scientific prevision. The practice of state planning of national economies had spread across the globe inspired both by the record of Soviet industrialization and the capitalist New Deal in the United States. Yet, after the two World Wars the future among intellectuals was a contested and problematic idea stimulating various reconfigurations and new approaches from positivist, to critical, explicitly normative and utopian (Andersson 2018: 2; Seefried 2015a; 2015b). Absorbing new theories and scientific paradigms such as game theory, cybernetics, and systems theory, approaches to scientific anticipation based on non-linear and probabilistic thinking spread in the US and Western Europe and formed what came to be known as the eclectic field of "future(s) research" or "futurology". Here, the idea of multiple possible futures, whose anticipation was meant to contribute to more "rational" and more "conscious" decisions in a given present, was directed against Soviet-style central planning and the Marxist–Leninist doctrine of historical evolution. Socialist countries had their own traditions of future-related thinking and anticipation drawing on national traditions and Soviet controversies over different approaches to long-term planning in the 1920s.

Yet, it was only after 1956 that renewed contacts and transfers of ideas between the US, Western Europe, the Soviet Union and Central European socialist countries allowed people, activities, and ideas around "future research, management and evaluation, prognostication, planning and philosophy of the future" (Flechtheim 1968: 8) to connect on both sides of the Iron Curtain (Seefried 2017; Andersson 2018: 123–181). Before, highly centralized planning and the doctrine of historical materialism strongly tied practices of future-related thinking to five-year plans and the Marxist-Leninist *telos* of socialist transformation laid out by the party leadership's directives. Economists in the late 1950s, however, embraced ideas of decentralization, mathematical tools of optimal planning and the earlier forbidden language of cybernetics (Kochanowicz 2010). At the same time, according to the sociologist Stanisław Ossowski (1967: 180), revisionist social thought with its calls for democratization and a renewed search for socialism's future path of development, did not alter the substantial tenets of the "monocentric" concept of the future and social order dictated hierarchically by the centre of power (the Communist Party leadership). Hence, most paradoxically, in Poland after 1956 an interest in new approaches to scientific anticipation and the problem of the future emerged among social scientists and economists as a combination of contacts and scholarly exchange with colleagues, books and theories from the West and reform-minded Marxist and non-Marxist scholars' aim to re-formulate the nexus between socialist modernization, future-thinking and planning.

More precisely, systematic research on the future was debated by Polish social scientists in the early 1960s in their attempts to grapple with what they regarded as new phenomena under socialism; namely, the lifestyles and cultural practices created by mass-media and consumption. This led them to announce the establishment of "Polish 'futurological' studies" (Siciński 1967: 243) at a large conference in May 1967, where more than seventy participants from the social, cultural, economic, natural, medical and technical sciences gathered to discuss methods of "predicting the future and a model of culture"[2] for socialist Poland (Kurczewski and Lutyk 1967). Studies on the future of socialist culture were meant to guide socialist planning beyond its fixation on economic figures and parameters of investment, productivity and work effort, yet the full realization of this initiative was interrupted by the events of March 1968.

Nevertheless, the idea was continued by the Committee for Research and Prognostics "Poland 2000"[3], which was created at the Polish Academy of Sciences in 1969 and still functions today. The research this committee has organized since then established a tradition of interdisciplinary debate among scholars from various fields, who pool their expertise and research in order to draft holistic outlooks on Poland's future in relation to global trends, analyses of specific social problems and areas of policy intervention. It was less a workshop for new methods than a stable collective of scholars who devised expert opinions for the government, mainly on request, but with considerable room for injecting reform proposals. Furthermore, many translations of foreign works related to future studies, such as the "limits to growth" report to the Club of Rome, go back to the initiative of this collective, who also held regular methodological seminars. These activities brought together more than 300 scholars and intellectuals with different worldviews, including opposition-minded intellectuals, technocrats and supporters of the communist regime. During the reform processes after 1989, the most learned future researchers in Poland played a rather marginal role, as the new elites were often recruited from a younger generation. In fact, political and public attention for futurology had been diminishing since the mid-1970s. Nevertheless, with methods such as scenario planning, model simulations, and Delphi panels diffusing into business consultancy, market and consumer research, NGO's and the state administration, contemporary activities of "foresight" in Poland as elsewhere have their roots in the futurology of the 1960s and 1970s.

Studying the future under state socialism was an activity constrained by (self-) censorship and the "monocentric" party control over political decisions (Sułek 2009). However, it should be seen as an ambivalent resource in the context of a late socialist society in which science, technology and expertise, in general, were typically referred to by an official rhetoric that had given up on big, revolutionary ideological projects and was instead searching to consolidate the regimes' power through piece-meal reforms (Klumbytė and Sharafutdinova 2013; Plaggenborg 2010). This, in turn, allowed scholars to gain considerable room for manoeuvre for ambivalent or even critical stances, reform proposals and the transnational exchange of ideas, either by paying lip-service to, or truly believing in, the "modernity" of socialist society. Their evolutionary optimism prevailed also in the face of catastrophic global predictions in the 1970s, fading only with the severe crisis of state socialism in the 1980s. The following three examples are aimed at depicting how future research produced and re-negotiated the future as an object of epistemology, governance and imagination.

Epistemology: Meta-Prognostic Modelling of the Future

In the twenty-first century, it is an unquestioned practice that complex scientific models simulate future natural and social courses of events. That these models trans-form the unpalpable future into an "object" of knowledge is seldom problematized explicitly. Yet epistemological dilemmas arise when the future is objectified. They become visible when looking at a peculiar historical example, in which modelling the future served to determine the general limits of the "sphere of reasonable planning" (Rolbiecki 1972: 52) in the late socialist Polish context.

This problem was elaborated in an article published in 1972 in the Polish journal *Zagadnienia Naukoznawstwa* read by historically and philosophically interested students of science. Its author, the rather unknown philosopher Waldemar Rolbiecki (1927–2002), worked at the research planning office at the Polish Academy of Sciences. He attended workshops led by Tadeusz Kotarbiński, a philosopher of science and the main representative of the Polish school of praxeology, the philosophy of effective human actions (Skarbek 2003). Fascinated by the newly emerging future research, Rolbiecki wrote several articles and a book, in which he developed what he called "prognoseology" (Rolbiecki 1970: 239–261), a meta-prognostic inquiry into the foundations of "the prognostic activity of man" (Rolbiecki 1967). At the core of this reflection, as with every anticipatory activity, was an intellectual opera-tion, which is exemplified by the model displayed below?: the transformation of the unknown future into an object of (scientific) knowledge.

Rolbiecki defined the future as the compound of all events that are possibly going to happen and have not yet occurred; thus, not as a time span or a phenomeno-logical horizon. The resulting topological model (Fig. 1) included a core (α) which

Fig. 1 An "ergonal" model of the future and the limits of planning [Greek letters added by the author, L.B.] (Rolbiecki 1972: 52)

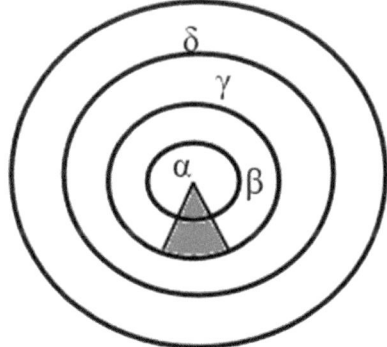

represents all intentional human actions that will be undertaken in the future. Moving in an outward direction, the next circle (β) is defined by consequences of these actions, which themselves are not intentional deeds; followed by all (γ) events resulting from past—and thus irreversible—human actions; and finally, the largest circular sphere (δ) of events contains events which neither result from nor are themselves human actions.

Rolbiecki's (1972: 49) argument, which set out to delineate limits to predictability and to planning the future in this very general manner, makes the strong assumption that "in principle the whole future (i.e., every future event) can be discerned." Yet his conclusion was visiualized by a triangular cone depicting how narrow "the scope of reasonable planning of the future" (Rolbiecki 1972: 52) was in comparison to the full range of future events. This conclusion should be read against the background of socialist politics in Poland in the early 1970s, which still placed high hopes on central planning as the major instrument of creating a technologically advanced and affluent society. Equally remarkable, however, is that Rolbiecki explicitly suggested to move future thinking away from questions of instrumentality, teleology and effective action towards an inquiry into human (and non-human) agency.[4] He (Rolbiecki 1972: 46–47) noticed that it was becoming ever more difficult to neatly separate human from non-human future events, as their causal entanglements were growing with interventions into the ecosphere and the use of technology.

Social Technology: Future Research as an Instrument of State Planning

Notwithstanding such sobering claims, the aim of efficiently shaping social relations and behaviour was and still is the most important rationality behind the will to anticipate. As part of a complex transnational history of debates and practices of social engineering and state planning in the nineteenth and twentieth centuries (Matejka et al. 2018; Etzemüller 2009; van Laak 2008), future researchers shared the aim of a "rational" design of social relations and of using scientific evidence to more efficiently achieve desired goals of social, economic or technological "progress". In fact, the emergence of future studies in the People's Republic of Poland was intertwined with debates about socialist planning and its methods in the 1960s and 1970s. New tools and paradigms were introduced into the planning process and time horizons extended through cybernetics and the mathematical techniques of optimization. Yet the future research conducted in Poland was less a mathematized practice than an interdisciplinary framework for thinking about desired social and cultural goals. Among the planning officials, at least among those curious in the voices of social and humanist scholars from outside the planning office's apparatus, this did not necessarily lead to a change in the concrete planning techniques or their impact on real economic consequences. However, their interest in future studies translated into shifting notions of the future and how to address it via planning.

This can be exemplified by a presentation given by Józef Pajestka (1924–1994) at a seminar in Tokyo in the autumn of 1967, which was concerned with "The World in 2000". He travelled there as one of the leading planning officials from the People's Republic, formally as director of the Institute for Planning, a research institute within the structure of the state socialist Planning Commission.[5] In his talk on "The State and Approaches to Future Studies in the Socialist Countries" (Pajestka 1967), Pajestka reflected on the instrumentality of future research for planning and how it had contributed to a shift of approaches compared to highly centralized hierarchical planning that had been practiced under Stalinism (Ellman 2014: 22–47). It should be said that this was neither indicative of a substantial change in the system's overall functioning nor its efficiency to produce economically desired outcomes. Yet the metaphors by which planning was described by elites such as Pajestka had shifted considerably.

As part of an evolutionary typology of approaches to planning, Pajestka's paper argued that the traditional model, exercised through a strictly hierarchical system of decision-making and designed to maximize industrial production, had been based on "a set of clearly defined goals (formulated in the long-range plan) which are to be attained whatever the circumstances. (This is similar to guiding a missile which is to attain the Moon)." (Pajestka 1967: 160). According to this metaphor, plans functioned as instructions devising long-term goals that were to be achieved as a result of consecutive short-term plans. Pointing out deficiencies when it came to consumption and efficient use of resources, the debates after 1956 experimented with cybernetics, complex simulations, mathematical programming, and futurological projections as a way to include the expertise of sociologists and culturalists. Although historians have convincingly shown that this was hardly able to revolutionize the modus operandi of the socialist economy (Ellman 2014: 366), the categories, based on which planners were addressing the future were now aiming at "optimal" decisions among many possible alternatives, and targets moved between conflicting social, economic, or cultural optima. Pajestka and his colleagues described the role of future research for planning as a cone, the cone of a flashlight (Secomski 1971: 6). The new "type of steering has no direct, built-in quantitatively determined goals which have to be absolutely attained. It follows, of course, certain rules and tends towards certain directions, it implies also a look ahead (anticipation) […]. This is somewhat similar to motoring for pleasure, where the headlights serve to choose the direction one wishes to follow and the best and safest road." (Pajestka 1967: 160) The idea of futurology providing headlights—at least in theory—opened the model of planning via centralized and unquestionable instructions to limited deliberation about desirable goals *and* means, including those of non-socialist origin. This also meant that planning was no longer "programmatic steering" but instead "adjustment" through "rolling plans" (Pajestka 1967: 158–161).

Despite the change of terminology future research remained an instrument of social technology, a means of achieving goals defined outside of itself. Furthermore, the practice remained rather hierarchical, and the use of prognostic research in the 1970s did not in fact live up to its promises of more rational, efficient, and socially balanced plans. Similarly, technocratic hopes of converging with Western capitalist economies in terms of growth, innovation and supply of consumer goods while diverging in social, cultural and political orientations, did not survive the severe crises of socialist economies in the 1980s.

Sociological Imagination: Future Research as Contemporary Utopia

It was against the background of the political and economic crisis in Poland of the late 1970s and early 1980s that a fundamentally different understanding of future research regained relevance among scholars of Polish society. Although this opposition is easily overestimated in retrospect, it is worth noting that the "past futurologies" depicted here were not only means of efficient adjustment and prolongation of the status quo but could also provide a framework for thinking about desired goals and imagining a radically different future.

A somewhat surprisingly pessimist paper presented by the sociologist Andrzej Siciński (1924–2006) to a workshop in Mexico City in 1978, which was later re-worked several times for presentations in Poland, illustrates this alternative. The workshop was a joint meeting of an international research group gathered under the auspices of the World Futures Studies Federation (WFSF) and the United Nations University, aimed at the formulation of visions of world order and social organization that, in a world after de-colonization, the oil crisis, and perceived limits to economic and demographic growth, could be "desirable" to a vast transcultural majority of people and nations (Masini and Steenbergen 1983). Siciński travelled to Mexico as a professor of the Polish Academy of Sciences specializing in empirical life-style research. His scholarly interest in future studies had emerged in the early 1960s, when he was one of the most active in promoting its methods and transnational debates in Poland (Filipkowski 2017; Gliński and Kościański 2009). Having expressed hopes that scholarly interest in the future would expand social imaginaries beyond the mere sociotechnical goals of planners and politicians (Siciński 1972), Siciński also sat on an expert panel for the government of Edward Gierek (Mazurek 2015) in the late 1970s. His presentation at the Mexican WFSF conference, however, documents an intellectual and political frustration with future research's ability to contribute to desired social improvements, while at the same time the sociologist began to support the emerging opposition movement back in Poland.

Siciński (1983: 101–2) opened his paper by stating the radical impossibility to design a vision of a desirable society that would meet with acceptance from different cultural standpoints and, at the same time, have a realistic chance of realization. However, he then set out to sketch principles which he deemed most fundamental to such a vision. Siciński's desirable society had to balance freedom and equality, something which could nowhere be better realized than in a cooperative. One would have to "introduce cooperative principles into the administration of society in towns and regions, and, finally, on a global scale" (Siciński 1983: 104). Experts and planners, including sociologists like Siciński, would perform the role of advisors to the public, not to central authorities. Decisions were to be taken based on the participation of those affected by them. However, global convergences of excessive materialistic consumerism, processes of cultural standardization and the centralization of decision-making procedures like the "monocentric" party rule in state-socialist Poland, made Siciński doubt the realization of his vision.

In fact, the introduction of martial law by the Communist Party in December 1981 to crush the Solidarity movement led Siciński to radicalize his conclusion. After these events, the situation in Poland was, in a critical way, closed to the realization of a collectively desirable social order. As Siciński put it, future research should, first and foremost, practice utopian thinking (Siciński 1985). While faith in future thinking as an instrument of social engineering had dwindled away, Siciński insisted on its power to imagine alternatives to the status quo.

Conclusion

How to know the future in advance, and how to control and imagine it, are acute questions in the face of the Anthropocene and climate crisis. Historicizing futurologies of the past means asking how historical actors and collectives dealt with those questions under particular historical circumstances. Arguing that this may help to better understand the genealogies and contingencies inherent in contemporary anticipatory ideas and practices, the short episodes from the history of future studies in late socialist Poland in the 1960s, 1970s and 1980s, which this essay presented above, may sharpen today's awareness of at least three major dilemmas related to those questions.

First, anticipatory knowledge always objectifies the future. One has to carefully observe, then, how this is done, and whether it creates greater awareness of one's own and others' agency or whether it is primarily geared towards enhancing efficient action. As the second and third examples have shown, anticipation is an element of political strategies that govern social relations. It serves socio-technical purposes. It is thus very important to examine closely how predictions construct their relationship to historical time and agency: Do they adjust to an externally defined future telos, or do they serve to formulate visions of radical change? Which actors articulate them following which objectives? One should thus evaluate contemporary practices of anticipation as to whether they are able to expand and broaden our imagination towards alternative futures instead of narrowing them down to inevitable processes. Also, one should be critically aware of the social positions and political agendas behind anticipations that claim to be scientific and objective. Obviously, these are no binary dilemmas, neither in practice nor in theory.

Being aware of those questions and their historical manifestations may help to historicize and critically evaluate the contemporary sense of crisis and the role of scientific anticipations within them. Shades of optimism and pessimism have changed considerably since the 1970s. Also, in terms of complexity, systematicity and refinement, contemporary anticipatory ideas and practices largely excel in the historical future research of the 1960s and 1970s. A striking difference to today's projections can also be seen in the size and complexity of their data sets and algorithms. However, these futurologies of the past established precedents for the orientation of anticipatory ideas and practices towards historically changing scientific standards. Their analysis, as this essay hopes to have shown, is a worthy effort for historians, because it contributes to a deeper and critical understanding of what a futurology of the Anthropocene will (and should) look like. However, it is not historians who will decide whether such ascience of the Anthropocene's future is ever going to emerge.

Notes

1. This is not a chronological order but an analytical distinction. I will use the terms "future research" and "future studies" interchangeably. For historical works on the little-known case of future research in Poland, see Kiecko (2018), Sułek (2009), and Becht (2017). In general, future research under state socialism was rather referred to as "prognostyka" to distinguish it from the "futurology" in "the West". Sommer (2016; 2017), Catanus (2015), Rindzeviciute (2016a, b), Rocca (1981), Andersson (2018: 122–140).
2. All translations in the text, including potential mistakes, are from the author, L.B.
3. Its full original name is *Komitet Badań i Prognoz "Polska 2000" przy Prezydium Polskiej Akademii Nauk*, later renamed *Komitet Badań i Prognoz "Polska w XXI wieku"* ("Poland in the 21st Century"), and after 2000, *Komitet Prognoz "Polska 2000 Plus"* (Komitet Prognoz "Polska w XXI Wieku" przy Prezydium PAN 1999).
4. In fact, Rolbiecki (1972: 43) himself made a sharp distinction between his own approach concerned solely with the agency (*sprawczość*) of human actions, and an efficiency-related perspective (dealing with *sprawność*) which in Polish sociology at the time was commonly associated with praxeology and the concept of social technology or social engineering (*socjotechnika*) by Adam Podgórecki (1966).
5. In Polish, *Instytut Planowania*. When his presentation was published, Pajestka had already become deputy head of the Planning Commission (*Komisja Planowania przy Radzie Ministrów*); hence, formally ranking as a deputy minister and member of government.

References

Andersson J (2018) The future of the world. Futurology, futurists, and the struggle for the post cold war imagination. Oxford University Press, Oxford
Becht L (2017) From Euphoria to Frustration: institutionalizing prognostic research in the polish people's republic, 1969–76. Acta Poloniae Historica 116:277–299

Beckert J (2018) Imaginierte Zukünfte. Fiktionale Erwartungen und die Dynamik des Kapitalismus. Suhrkamp, Frankfurt

Catanus AM (2015) Official and unofficial futures of the communist system. Romanian future studies between control and dissidence. In Andersson J, Rindzeviciute E (eds) The struggle for the long-term in transnational science and politics. Forging the future. Routlegde, New York/Abdington, pp 170–194

Chakrabarty D (2009) The climate of history: four theses. Crit Inq 2(35):197–222

Daston L (1994) Historical epistemology. In Chandler J, Davidson AI, Harootunian H (eds) Questions of evidence. Proof, practice, and persuasion across the disciplines. The University of Chicago Press, Chicago/London, pp 282–289

Ellman M (2014) Socialist planning. Cambridge University Press, Cambridge

Etzemüller T (2009) Social engineering als Verhaltenslehre des kühlen Kopfes. Eine einleitende Skizze. In Etzemüller T (ed) Die Ordnung der Moderne. Social Engineering im 20. Jahrhundert. transcript, Bielefeld, pp 11–39

Filipkowski P (2017) Back to future's past. Andrzej Siciński's scientific futurology. Acta Poloniae Historica 115:267–271

Flechtheim OK (1968) Futurologie–Möglichkeiten und Grenzen. Edition Voltaire, Frankfurt/Berlin

Geulen C (2020) For future. Zum Problem des vorauseilenden Denkens Geschichte der Gegenwart 25.11.2021. https://geschichtedergegenwart.ch/for-future-zum-problem-des-vorauseilend endenkens/. Accessed 5 Feb 2021

Gliński P, Kościański A eds (2009) Socjologia i Siciński. Style życia, społeczeństwo obywatelskie, studia nad przyszłością. IfiS PAN, Warszawa

Graf R, Herzog B (2016) Von der Geschichte der Zukunftsvorstellungen zur Geschichte ihrer Generierung. Probleme und Herausforderungen des Zukunftsbezugs im 20. Jahrhundert. Gesch Ges 3(42):497–515

Horn E (2014) Zukunft als Katastrophe. S. Fischer, Frankfurt

Kiecko E (2018) Przyszłość do zbudowania. Futurologia i architektura w PRL. Fundacja Nowej Kultury Bęc Zmiana, Warszawa

Klumbytė N, Sharafutdinova G (2013) Introduction. What was late socialism? In: Klumbytė N, Sharafutdinova G (eds) Soviet society in the Era of Late Socialism, 1964–1985. Lexington, Lanham, pp 1–14

Kochanowicz J (2010) 'Początki planowania w Polsce po drugiej wojnie światowej w perspektywie porównawczej. In Mączyńska E, Wilkin J (eds) Ekonomia i ekonomiści w czasach przełomu. PTE, Warszawa, pp 203–222

Komitet Prognoz "Polska w XXI Wieku" przy Prezydium PAN ed (1999) Studia nad przyszłością w Polsce w pracach Komitetu Prognoz "Polska w XXI wieku". Komitet i jego twórcy (w 30-lecie powstania Komitetu) 1969–1999. Elipsa, Warszawa

Koselleck R (1979) Vergangene Zukunft. Zur Semantik geschichtlicher Zeiten. Suhrkamp, Frankfurt

Kurczewski J, Lutyk A (1967) Futurological conference. Polish Sociol Bullet 2(8):127–130

Lem S (1977) Phantastik und Futurologie. 1. Teil. Insel Verlag, Frankfurt

Mallard G, Lakoff A (2011) How claims to know the future are used to understand the present: techniques of prospection in the field of national security. In: Camic C, Gross N, Lamont M (eds) Social Knowledge in the Making. Chicago University Press, Chicago, pp 339–378

Matejka O, Kott S, Christian M (2018) Planning in cold war Europe: introduction. In: Matejka O, Kott S, Christian M (eds) Planning in cold war Europe: competition, cooperation, circulations (1950s–1970s). De Gruyter, Boston/Berlin, pp 1–17

Masini E, van Steenbergen B (1983) Introduction. In: Masini E (ed) Visions of desirable societies. Pergamon Press, Oxford et al., pp 3–8

Mazurek M (2015) Rewizyta etnograficzna. Jak się wytwarza wiedzę socjologiczną. Kultura i Społeczenstwo 3(59):31–62

Nycz R (2014) Wstęp: Humanistyka przyszłości. Teksty Drugie 5(14):7–11

Ossowski S (1967). Koncepcje ładu społecznego i typy przewidywań. In Ossowski S, Dzieła, vol 4, O nauce. PWN, Warszawa, pp 173–193

Pajestka J (1967) The State and Approaches to Future Studies in the Socialist Countries. In: The Japan Economic Research Center (ed), The World in 2000. JERC, Tokyo, pp 152–176

Plaggenborg S (2010) Verstetigte Gegenwart. Über das Zeitverständnis im real existierenden Sozialismus. In Schulze Wessel M, Brenner C (eds) Zukunftsvorstellungen und staatliche Planung im Sozialismus. Die Tschechoslowakei im ostmitteleuropäischen Kontext 1945–1989. R. Oldenbourg, München, pp 19–32

Podgórecki A (1966) Zasady socjotechniki. Wiedza Powszechna, Warszawa

Rheinberger H-J (2007) Historische Epistemologie zur Einführung. Junius, Hamburg

Rindzeviciute E (2016a) The power of systems. Cornell University Press, Ithaca, N.Y, How Policy Sciences Opened Up the Cold War World

Rindzeviciute E (2016b) A struggle for the soviet future. The birth of scientific forecasting in the Soviet Union. Slav Rev 1(75):52–76

Robinson KS (2020) The coronavirus is rewriting our imaginations. The New Yorker, 1.5.2020. https://www.newyorker.com/culture/annals-of-inquiry/the-coronavirus-and-our-future. Accessed 29 Dec 2020

Rocca GL (1981) A second party in our midst: the history of the soviet scientific forecasting association. Soc Stud Sci 2(11):199–247

Roelcke V (2010) Auf der Suche nach der Politik in der Wissensproduktion: Plädoyer für eine historisch-politische Epistemologie. Ber Wiss 2(33):176–192

Rolbiecki W (1967) Prognostication and prognoseology. On the need of systematic inquiries about the prognostic activity of man. In: Jungk R, Galtung J (eds), Mankind 2000. Universitetsforlaget, Oslo, pp 278–285

Rolbiecki W (1970) Przewidywanie przyszłości. Wiedza Powszechna, Warszawa

Rolbiecki W (1972) Ergoniczny aspekt przyszłości. Cztery rodzaje zdarzeń przyszłych. Zagadnienia Naukoznawstwa 1(8):42–53

Rosenberg D, Harding S (eds) (2005) Histories of the future. Duke University Press, Durham and London

Scott JC (1999) Seeing like a state. Yale University Press, New Haven and London, How certain schemes to improve the human condition have failed

Secomski K (1971) Prognostyka. Wiedza Powszechna, Warszawa

Seefried E (2017) Der kurze Traum von der steuerbaren Zukunft: Zukunftsforschung in West und Ost in den 'langen' 1960er Jahren'. In: Hölscher L (ed) Die Zukunft des 20. Jahrhunderts. Dimensionen einer historischen Zukunftsforschung. Campus, Frankfurt and New York, pp 179–220

Seefried E (2015a) Zukünfte. Aufstieg und Krise der Zukunftsforschung 1945–1980. De Gruyter, Berlin and Boston

Seefried E (2015b) Reconfiguring the future? Politics and time from the 1960s to the 1980s–introduction. J Modern Eur Hist 3(13):306–316

Siciński A (1972) Future research. Polish Perspect 1(15):10–16

Siciński A (1967) Polskie studia "futurologiczne." Kultura i Społeczenstwo 2(11):243–244

Siciński A (1983) How is a vision of a desirable world possible today? In: Masini E (ed) Visions of desirable societies. Pergamon Press, Oxford et al., pp 101–108

Siciński A (1985) O wizji pożądanego społeczeństwa. Przyczynek do konstruowania współczesnej utopii. In Mokrzycki E, Ofierska M, Szacki J (eds) O społeczeństwie i teorii społecznej. Księga poświęcona pamięci Stanisława Ossowskiego. PWN, Warszawa, pp 299–312

Skarbek J (2003) Waldemar Rolbiecki (1927–2002). Kwart Hist Nauki Tech 3–4(48):191–202

Sommer V (2016) Forecasting the post-socialist future. Prognostika in Late Socialist Czechoslovakia, 1970–1989. In: Andersson J, Rindzeviciute E (eds) Forging the future. Transnational perspectives on the history of prediction, Routledge, London et al, pp 144–168

Sommer V (2017) From socialist post-industrialism to market economy: Futurology in Czechoslovakia (1960s–1980s). Bohemia 1(57):55–81

Sułek A (2009) On the unpredictability of revolutions. Why did polish sociology fail to forecast solidarity? Polish Sociol Rev 4(168):523–537

Van Laak D (2008) Planung. Geschichte und Gegenwart des Vorgriffs auf die Zukunft. Gesch Ges 3(34):305–326

Globalization as Adaptive Complexity: Learning from Failure

Miguel Centeno, Peter Callahan, Paul Larcey, and Thayer Patterson

Abstract Our modern global civilization has been facilitated by increasingly technologically-advanced, interconnected, and interdependent systems. These systems have been constructed at an ever-increasing scale and level of complexity without an awareness of the risky mechanisms inherent in their design. At first glance, one may find few similarities between our modern globalized present and ancient civilizations. When we see past civilizations as complex adaptive systems, however, we can begin to recognize patterns, structures, and dynamics that have remained consistent through the centuries. Mechanisms like tipping points, feedback loops, contagions, cascades, synchronous failures, and cycles that can be responsible for systemic collapse are fundamental characteristics of any complex adaptive system, and can therefore serve as an effective common denominator from which to examine collapses through the ages. We argue for an analytical framework that incorporates these systemic characteristics for the study of historical collapse with the belief that these common mechanisms will help illuminate and expose relevant vulnerabilities in historical systems. In the end, we hope to learn from past societies and civilizations and allow our modern systems to benefit from lessons of systemic failures that historians may share with us. We believe these insights could inform how we see our systemic vulnerabilities and help to build a more resilient future.

Keywords Systemic risk · Historical collapse · Systems theory · Globalization · Complexity · Fragility

M. Centeno
Department of Sociology, School of Public and International Affairs, PIIRS Global Systemic Risk, Princeton University, Princeton, USA
e-mail: cenmiga@princeton.edu

P. Callahan (✉) · P. Larcey · T. Patterson
PIIRS Global Systemic Risk, Princeton University, Princeton, USA
e-mail: pwcallah@princeton.edu

P. Larcey
e-mail: plarcey@princeton.edu

T. Patterson
e-mail: patterson@princeton.edu

© The Author(s) 2022   59
A. Izdebski et al. (eds.), *Perspectives on Public Policy in Societal-Environmental Crises*,
Risk, Systems and Decisions, https://doi.org/10.1007/978-3-030-94137-6_6

Introduction

There is a specter haunting globalization and modern life: the potential for widespread civilizational collapse. Stories of dystopian fiction and apocalyptic futures have never been more popular, with audiences flocking to big budget disaster movies (Roberts 2020). Our world is existentially anxious because we sense that our trajectory is not sustainable (Ord 2020). Even the most optimistic possibilities of scientific and technological progress cannot guarantee our collective ongoing stability and prosperity. Global systemic shocks like 9/11, the Global Financial Crisis of 2008, and COVID-19 have heightened the awareness of the fragility of our increasingly globalized and interdependent way of living.

With the goal of understanding this precariousness of our modern world, we investigate "failures" in history and examine whether there are insights from systemic risk that can illuminate.

patterns of historical collapse. A certain teleological triumphalism has dominated modern social science where victors wrote their histories and survivorship biases led us to focus on the civilizations that remain standing. We see value in reversing this view by attempting to learn from failed civilization. We may have access to something that past doomed societies lacked—self-awareness of our own trajectory towards destruction, access to historical hindsight, and an understanding of themes and patterns that led to systemic failures in societies. We seek to identify systemic causes and mechanisms for breakdown that can provide historians with a systemic perspective for analyzing the past and can allow these past collapses to serve as cautionary tales for our present.

The paper begins with a discussion of hubris as a theme in social development. We then present a summary of the structure of globalization as a complex adaptive system. We follow by defining collapse, then moving on to its most significant causes. The penultimate section discusses some of the mechanisms through which isolated failures could lead to systemic collapse. We end with consideration of the governance required, if not to avoid the risk, then to mitigate the consequences with the goal of creating a more stable present and future.

> The Enemy Is Us.
> "We have met the enemy and he is us"
> (Kelly 1971)

Hubris has been a cautionary theme in mythology, literature, and religion throughout history. Humans have a habit of taking a few successes as a sign of continued and future prosperity, often extrapolating it into a perception of infallibility. This has led to building ever taller edifices on fragile foundations. Dubai's Burj Khalifa may serve as a contemporary analog to the Tower of Babel. What would possess us to build something so incongruous with its natural environment, and surrounded by an unsustainable set of city states? Believing that we have it all under control, and that tomorrow will be just like today, humans create systems on which we depend and then neglect the need to build enough robustness and resilience into the system to assure it can survive crises (Pastor-Satorras et al. 2015; Taleb 2007).

We argue that this characteristic overconfidence also characterizes globalization (Brauer 2018). The global system is a set of tightly coupled interactions that together allow for the continued flow of information, money, goods, services, and people. We have clear evidence that—environmental hazards aside for the moment—globalization has actually been a very good thing for humans collectively. Life expectancy has increased globally by more than two decades since 1960 (Roser et al. 2013) and there is continuing evidence that the science of longevity will sharply accelerate (Schwab 2019). We now produce enough food to feed the planet, and enjoy an unprecedented economic and technological standard of living. Like a Roman during the reign of the Antonines, we can look around us and marvel at what we have created (Birley 2000).

Much of this advance has come through our ability to create systems that are technologically advanced, complex, interdependent, and constructed at massive scale. Expansive networks of telecommunication, transportation, energy, agriculture, trade, among others, have facilitated this progress, but have given rise to new and unprecedented risks (Manheim 2020; Oughton et al. 2018). We see these networks as complex adaptive systems (CAS), where the interactions of components create new dynamics that cannot be explained by the characteristics of the constituent parts. Because of this complexity, the risks associated with maintaining a CAS are non-linear and impossible to predict (Helbing 2009). Emergent risks in such a system are the threats that originate not in any single component, but from the collective structure and dynamics of the system in its entirety. In the case of CAS, risk of systemic failure when looking at the whole may be far greater than when the system is viewed simply as the sum of its parts (Crucitti et al. 2004). This is particularly true of "systems-of-systems" that rely on the coordination of various domains. The agricultural system, for example, relies on networks of finance, trade, water, labor, energy, electricity, transportation, communications and others to efficiently plant, grow, harvest, transport, and sell foodstuffs in a globalized society (Centeno et al. 2015a, b). A miscoordination in any of these underlying and interdependent systems could be catastrophic.

Globalization requires the continued flow of people, money, commodities, goods, services, and the co-operation of vast numbers of individuals (Danku et al. 2019; Foreman-Peck 2007). COVID-19 has shown us that none of us is isolated from the rest of the globe. A novel virus can rapidly emerge to bring down economies, change elections, and humble even the most powerful. Even with warnings, foresight, and suggested mitigation strategies, overconfidence and failure of imagination enabled such a deadly scenario (Cambridge Global Risk Index 2019: Executive Summary 2018; Epstein 2009; Nuzzo et al. 2019).

We now live in a global system-of-systems where a failure in one part could lead to disaster across the whole structure. The sheer quantity and breadth of possible interactions requires a shift in our analysis of interdependence. Moreover, to this complex system we have added a pursuit of optimization and efficiency that leads to short term gains, but lays the foundation for longer term catastrophe (Centeno et al. 2015a, b). Global systems, much like the Burj Khalifa, are wonders to behold, but the increase in complexity and tight coupling make a "normal accident" ever more likely and more dangerous (Perrow 1984; Ledwoch et al. 2018). That is, we have created systems which we can never truly comprehend, whose risk profiles we cannot understand (Wildavsky and Dake 1990), over which no one has responsibility, and on which we have staked our continued survival.

Our hubris lies not only in our overconfidence in our increasingly fragile systems, but more so in our belief that our 21st Century civilization is immune to the tragic fates of fallen societies in history. While our modern societies and the systems upon which they rely are at a scale, scope, and degree of complexity far greater than their historical counterparts, the mechanisms of systemic failure and collapse remain the same. In this way, lessons from the past that relate to the fundamental systemic characteristics still remain relevant today. Because of the unimaginable magnitude of potential contemporary collapse, the study of past systems for insights is more urgent and compelling than ever.

Looking to History

Perhaps the one constant in recorded history is that even the most apparently powerful and successful systems inevitably break down. History demonstrates over and over again that the second law of thermodynamics applies to human created systems: we cannot escape entropy—the inexorable trend toward greater chaos in nature (Meyer and Ponthiere 2020). Consequently, no form of social order is eternal.

Several years ago, we began a scholarly project on Global Systemic Risk (risk.princeton.edu). We sought to bring insights from complexity (Holovatch et al. 2017) and network theory (Barabasi and Frangos 2002) to identify: (1) how the system of globalization works and (2) what are the risks associated with this global complexity. Soon enough, we realized that it was important not just to analyze the system and identify key vectors, but also to imagine how it might all come apart (Vespignani 2010). We began to see that the risks to our increasingly interconnected and globalized systems are substantial and that its widespread failure could be catastrophic. To anticipate the future, we looked to lessons from the past to discover shared characteristics of doomed civilizations, which might offer warning signs for modern collapse.

We held two workshops with scholars whose knowledge and perspective could contribute to a better understanding of what we meant by collapse (risk.princeton.edu/collapse). Participants included biologists, historians, physicists, modelers, mathematicians, and even authors of dystopian essays and fiction. By assembling this group, we sought a more clear image of the modern world and its possible endings.

Analyzing a data set of historical collapses (see Peter Turchin's work for an excellent example: seshatdatabank.info), we sought to distinguish causality from correlation, and exogenous causes from endogenous ones. What do previous "falls of empire" have to teach us about how we might best prepare for an uncertain and potentially perilous future (Taleb 2012)? We look to our experience from our study of global systemic risk and systemic collapse to provide insights and perspectives that historians may apply to their study of historical collapse. While we do not go in depth into different case studies in this chapter, the systemic mechanisms of collapse we choose to illuminate are heavily influenced by these historical perspectives.

Defining Collapse

We begin with a standard definition of a system: "a regularly interacting or interdependent group of units forming an unified whole" (Merriam-Webster). We are interested in how the structure and dynamics of a system decline over time—how this "unified whole" decreases in scale or scope, and how the central axis of action of the system moves from the system itself to its constituent parts.

If there is one central theme in the collapse literature it is that there is notable disagreement about the meaning of the term "collapse" (Middleton 2013; Yoffee 2005; Yoffee and Cowgill 1988). One area of debate is what exactly constitutes a collapse. Another criticism is that our historical view of collapse tends to have multiple cognitive biases—cultural bias, availability bias, confirmation bias, etc. Another critique is that speaking only of collapse ignores comparable scenarios where civilizations survived hardships and shocks, and thereby conceals or ignores significant elements of robustness and resilience that are important to identify (Nicoll and Zerboni 2019). We have to also address *what* exactly collapses. One standard measure is the level of social complexity of a society (e.g., level of interdependence, control, coordination Renfrew 1973; Tainter 1988)). Another is the level of political control or simple performance measures, such as nutrition or life expectancy (van Zanden et al. 2014).

We recognize that the term "collapse" is somewhat ambiguous with different definitions across academic disciplines. In this analysis we propose borrowing the usage of the term "collapse" from the literature on networks, systems, and complexity. In these fields, collapse refers to the disaggregation and breaking apart of a connected network. Collapsing complex systems break down or fragment into smaller units requiring less order, complexity, coordination, and organization to function. The systemic dynamics are thus reduced on a macro scale. The best question for such network analyses may be: has the system lost a significant part of its aggregate functionality (Hernández-Lemus and Siqueiros-García 2013)?

In this type of analysis, what collapses is not necessarily an entire society or civilization, but instead the larger organizational framework (Yoffee 2005; Kauffman 1993). So, the Maya late-classic city states decentralized and became smaller agglomerations of farmers, but the larger Maya "civilization" remained. Similarly, the radius of collapse may also differ such that the failure of some systemic elements does not imply the collapse of the entire *status quo ante*. The Western Roman Empire became a collection of much smaller political units—with significantly fewer interactions between them—yet the Eastern portion maintained its structure and societal identity in a semblance of its former self for a further thousand years (Mango 2002). We are therefore interested in the dynamics of aggregation and fragmentation over short to long-term periods.

For this fragmentation to be relevant to our discussion of collapse, it must have significant negative long-term consequences or costs associated with it. That is, social collapse must involve the loss of basic structure or function, or at the very least a decline in critical measures such as nutrition, life expectancy, or peace. Ibn Khaldûn's term *asabiyyah*, meaning "group feeling" or "social cohesion," might represent the antithesis of collapse. Interestingly, in 1377, Ibn Khaldûn wrote in his *Muqaddimah* that all social systems have collapse written into their structures and that the cycles of rise and decline may be inevitable (Khaldûn 2015).

We need to differentiate between the gradual cycles of rise and decline and the more rapid collapses. The two may have very different causes. Most discussions of collapse focus on a dramatic event or moment when key indicators begin to mark social breakdown. For our purposes, we identify collapse as a clear inflection point followed by a perceived fall in living conditions for those within the system (see Haldon et al. 2020 pp. 3–7).

We also need to distinguish a simple crisis from a collapse. While the Global Financial Crisis of 2008 was a major systemic shock with significant costs, it did not precipitate a collapse of the entire financial system (Coggan 2020, 338–47). The "Second 30 Years War" (1914–1945), however, did produce a collapse in the global, social, and political order (Ferguson 2006). This last example should also remind us that one person's collapse may be another's opportunity. Similarly, the collapse of 19th Century colonial empires might be lamented in some parts of the world, but celebrated in others. The most dramatic collapse of all, caused by an asteroid 65 million years ago, was certainly a disaster for the dinosaurs, but it provided an ecological opening for mammals.[1]

It is therefore important to remove, in a sense, the normative aspects of collapses and analyze these systemic transformations descriptively as ecological phenomena with niches disappearing and appearing. In ecological systems, for example, "collapse" or "release" is a critical phase of cyclical regeneration, which allows for new reorganization as different feedbacks and competition within the system allow for new systemic characteristics to emerge with regrowth (Gunderson and Holling 2002).

Identifying the Causes of Collapse

Given the intractable complexity in many historical civilizational systems, it is often difficult to reach consensus on the causes of various collapses. Historian Alexander Demandt (1984) famously counted 210 different explanations given for the collapse of the Roman Empire (Demandt 1984). Others argue that the Roman Empire never

---

[1] Note that there remains some discomfort with the Asteroid theory with others focusing on the volcanic activity in the Deccan Plateau.

truly collapsed but instead fragmented and slowly faded away (Brown 1976).[2] More recently, environmental change or failure has become a frequent explanation for the collapse of certain complex societies (Middleton 2017). As with any tragic denouement, it may be impossible to string together the various episodes, mistakes, and challenges that might have led to the loss of *asabiyyah*. Each observer might choose a different critical moment or decision that precipitated the downfall.[3]

Generally, we can distinguish between two broad categories of explanations for collapse: exogenous, and endogenous. Narratives of exogenous causality—where a shock from outside the system is responsible for its downfall—are the most common. Volcanic eruptions, earthquakes, and sudden climate shifts are common in human history and are often associated with systemic collapse (Bostrom and Cirkovic 2008; Ord 2020). Similarly, much of history is the story of human invasion, conquest, and brutality, viewed as an exogenous cause by all of the civilizations conquered. From the 15th Century onwards, the "rise of the West" (Hoffman 2015) meant catastrophe for almost all of the world outside Europe, which led Stephen Hawking to his famous caution against searching for life in the universe whence "such advanced aliens would perhaps become nomads, looking to conquer and colonize whatever planets they can reach" (Stephen Hawking's Universe 2010).

Exogenous explanations, however, while salient and dramatic, can be misleading as they often neglect the importance of internal systemic characteristics. For example, two distinct systems could experience a comparable exogenous shock, leading one system to succumb while the other survives and prospers. Thus, of greater relevance than the exogenous shock itself are the qualitative differences between the two systems that explain these disparate results. In this way, viewing collapse as simply the consequence of unlucky exogenous shocks makes for an unsatisfying account (Bailey 2011).

By contrast, narratives of endogenous causality—where a system's internal characteristics are responsible for causing or perpetuating a downfall—perhaps offer greater explanatory power for past societal collapses. The most relevant causes of collapse may not be the specific factors that initiate the process, but the structure that allows perturbation and contagion to amplify through the system as in a chemical reaction. Thus, collapse may not be precipitated by the failure of any single component, but instead by the unexpected dynamic interactions of these nodes in a complex network. Instead of individual causes, we might better focus on the systemic mechanisms that escalate local challenges into existential crises.

Unlike the randomness and *sui generis* qualities of many exogenous shocks, endogenous collapses seem to share a common theme: managerial failure. These failures include loss of legitimacy, unsustainable inequality, hyperbolic discounting, overuse of resources, misplaced faith in the reliability of advanced technologies, and an overemphasis on efficiency. Managerial failures of human agency and fiduciary leadership reduce the system's internal capacity to withstand shocks (both exogenous

---

[2] A similar argument has flared concerning the Meghalayan Age beginning around 2200 BC with disputes about the extent of global civilizational collapse.

[3] For an excellent overview of the literature see (Haldon et al. 2020).

and endogenous), making them more vulnerable to the mechanisms of collapse we identify below.

Systemic Mechanisms of Collapse

All ecological and human-made systems share components (Siskin 2016)and behaviors that allow for growth or development (Kauffman 2013). However, like healthy cells that become cancerous, these once beneficial features may evolve into threats that can ultimately destroy the system as a whole. Below, we discuss common mechanisms of collapse.

Tipping Points

Every complex social system contains thresholds beyond which social cohesion falls apart. These tipping points are the levels of tolerance within a system that, when exceeded, mark the rapid transition to a new state or equilibrium. For societies, this could be the moment when a behavior in which the person or society has engaged in for many years suddenly has more drastic consequences than expected. This is a consequence of inertia, force, stress, or momentum building up that leads to a phase change, causing the system to transition into a different equilibrium state of structure or dynamics. Tipping points can serve as gateways of opportunity or pathways to failure (Milkoreit et al. 2018). Examples of tipping points with negative consequences include a final straw breaking a camel's back or a rubber band stretched beyond its breaking point: after a minor additional stress, it loses its functionality. An example of a tipping point that leads to greater systemic resilience is that of herd immunity—after a certain level of inoculation is reached, the population is protected from further infection. The key in all instances is the persistence or irreversibility of the transformation (Dakos et al. 2019; Bentley et al. 2014).

Different individuals or groups will have different attitudes towards such points of inflection (Bossomaier et al. 2013). Some might be totally unaware of the negative consequences—or again, the positive contributions—that could be precipitated by this small additional stress. In these instances, the inflection point comes as a total surprise, which adds to the precariousness of the transition to the new state. Other individuals or societies might be well aware of the potential change and have established a monitoring system, a sort of "Geiger counter" offering reassurance. Yet other societies may actually be aware of the danger, yet appear to do little to avert it—global pandemics and climate change are two such examples.

It may well be impossible to predict the tipping point or even identify it post hoc, because tipping points may be contextual—only becoming critical under certain circumstances. The most frequent examples of tipping points may be found in the start of wars where antagonism, fears, and perceived injustices lead to the creation of a spiral into violence. In Thucydides' account of the negotiations between Sparta and Athens, the debate in Corinth may be seen as the tipping point leading to the resolution of the Thucydides trap (Robinson 2017). Or consider a doomsday weapon a la *Dr. Strangelove*, established precisely as a public "red line," or "line in the sand," the crossing of which begins a chain-reaction that cannot be stopped. Caesar's crossing of the Rubicon can be regarded as one such tipping point. Again, the central lesson from tipping points is that an apparently small perturbation can set off a series of events that leads to irreversible change, or in the worst scenario, collapse.

Feedback Loops

Stable social systems are fundamentally cooperative and reciprocal, with systemic dynamics that reinforce (or undermine) this social cohesion through feedback loops. These are structures that use the measure of output from a process to determine the subsequent input back into the beginning of the cycle (Martin 1997). The critical element is that the start of some action or process is at least partly determined by its previous ending. Perhaps the simplest feedback loop in our everyday lives is a faucet, where the last felt or monitored temperature of the water is used to recalibrate the subsequent flows. Many incentive systems are forms of feedback loops: as a result of a certain level of performance, rewards or punishments are determined for the next round. The human nervous system (Lessard 2009) is an example of a feedback loop where we are encouraged or discouraged from certain forms of behavior by signals of pain or pleasure.

Within social systems, social norms and even institutionalized rules are forms of feedback loops, as they determine responses to individual actions. It is also possible for feedback loops to operate within "black boxes"—hidden from the observer or obscure enough so as to not be part of a conscious strategy, but the result of a pre-programmed set of responses. Feedback loops are critical to establishing equilibria. The relationship between supply, demand, and price may be seen as a constantly iterative loop of inputs and outputs, feeding back on each other.

Positive feedback loops can move a system away from its equilibrium, while negative feedback loops can diminish the effects of perturbations, returning the system to its steady state. In this way, negative feedback loops can lead to important improvements in behavior or functionality by reinforcing the current steady state. Unhealthy positive feedback loops exist when a society may reward a kind of behavior until such a point at which that behavior is no longer appropriate (Kolmes 2008). This is often the case when feedback loops for some are misaligned with a collective good. When studying instances of historical collapse it may be important to identify feedback loops, with either positive feedback loops causing a stable society to spiral into disorder, or negative feedback loops enabling a social system to absorb otherwise catastrophic shocks.

Contagions

COVID-19 has made the phenomenon of contagion far too familiar and highlights the inexorable systemic risk inherent in globalization (Smil 2019). From a network science perspective, contagion involves the passing of objects, effects, or characteristics from one node to another transmitted through contact or a systemic connection. A person may infect a group by coughing, someone shouting "Fire!" may lead to the spread of alarm or panic, or the failure of one part of a system may lead to malfunction elsewhere or even of the whole. Similar to tipping points and feedback loops, contagion might be considered beneficial or positive if the resulting behavior is one that is considered valuable. Inventions and their subsequent "viral" diffusion are an example of a beneficial contagion.

Cascades

A cascade—or uncontrollable domino effect—might be best thought of as a combination of contagion and tipping points. A normally regulated flow may transition into a cascade if it results in an increase in output that overwhelms its neighboring nodes that are receiving these flows. When a gradually spreading contagion reaches a tipping point and triggers failure in some components or regions of a system, this can precipitate a cascading sequence of failures—the magnitude of each failure is significant enough to touch off an even greater neighboring failure. In this way, complex systems can contain within them leverage that increases the magnitude of the failure at each step in the cascade.

Perhaps the best known cascading failures in our modern systems are within highly coupled energy infrastructures (Korkali et al. 2017). In political systems, the assassination of Franz Ferdinand in 1914 is perhaps the most infamous instance in modern history of a cascading failure. Interdependent nations tightly coupled through alliances designed to increase geopolitical systemic resilience created pathways and dynamics for a cataclysmic cascading failure. One domino fell and ultimately led to a world-transforming global conflict.

Synchronous Failures

While complex systems may be designed to survive individual localized failures, a certain number of such simultaneous failures will overwhelm any system. Such a "perfect storm" of events is considered a synchronous failure. Probability theory dictates that random events will eventually occur simultaneously, or at least in close proximity of time or location. Such a clustering of these failures, or the simultaneous and synergistic interaction of several failures, may result in a challenge unimagined by designers, and for which the system is not prepared (Homer-Dixon et al. 2015).

Charles Perrow's concept of a "normal accident" illustrates how such an apparently innocuous confluence of events can lead to catastrophe (Perrow 1984). In tightly coupled and complex systems, two apparently unrelated events can lead to a disastrous outcome. Natural disasters are particularly dangerous because they often involve the failure of various social systems simultaneously. The response to the failure of one part of a system might then lead to a strain in another part that leads to systemic breakdown.

Synchronous failures are particularly threatening because no individual or society can prepare for the infinite number of disastrous combinations and consequences (West 2017). We might be able to create mechanisms to deal with individual problems, but in the face of multiple failures, resources may be taxed beyond their limits. In the case of complex systems, the interaction of failures may lead to consequences not expected from the isolated failure of each. Invaded societies weakened by novel pathogens found themselves fighting two battles instead of one. While either invasion or pandemic may have been manageable shocks on their own, the confluence of both served as a coup de grâce.

Cycles

The notion of civilizational or biological cycles is central to natural and behavioral sciences. The organic cycle of death and life is one that dominates our planet (Walker et al. 2017). Without death and decomposition, new biological life may be impossible. For over a half-century, economic policy has been guided by attempts to regulate the cyclical nature of inflation and unemployment, booms and busts. The central notion of Keynesianism is to avoid the deep troughs of the cycle through monetary and fiscal intervention. Ecological systems experience oscillating cycles of population growth and decline based on factors such as predator–prey dynamics. Similarly, climate systems experience natural cycles through the activity of sunspots, and astronomical interactions, resulting in temperature fluctuations and drought. These cycles in the environmental systems can be catastrophic for civilizations that are unexpectedly deprived of food or water (Parker 2013).

Many societies, such as the Mexica, organized their lives in accordance with a calendar of rise and decline. Cultures and religions have embraced the notion of reincarnation reflecting their social faith in the inevitability of the cycle of death and rebirth. Since at least the Enlightenment, or even the Renaissance, European and associated societies have sought to escape the inevitability of cyclicality and have constructed the expectation of linear progress. While this aspirational desire to transcend cyclicality may account for economic and social dynamism (Sweezy 1943), it also makes the cyclical decline a threatening prospect. Much like Shelley's *Ozymandias*, even the mightiest of civilizations that expected to prosper eternally have ultimately declined, transformed, or collapsed (Shelley 1818).

Resilience and Mitigation

At its core, resilience refers to the capacity of any system—a human body, a building, a city, a tropical forest—to survive shocks and disruptions (Walker and Cooper 2011). This concept has migrated from engineering and ecology into all disciplines in which systems are studied (Evans and Reid 2013; Levin and Lubchenco 2008). While a distinction is sometimes made between robustness and resilience, for our purposes in this discussion, we use the term resilience to cover attributes of both. In this framing, resilience is a combination of two general qualities: resistance to shocks with the ability to remain unchanged (what is often referred to as robustness); and flexibility, recoverability, or the ability to change enough to survive after a disruption (most commonly referred to as resilience). An example of withstanding shocks would be building a dam to prevent flooding; that of the latter, the provision of boats in case of flooding.

The central difference between these two aspects of resilience might be best understood as a system's ability to prevent a crisis versus its capability to bounce back from one. In regulatory terms, we can either try to prevent failures in our physical, infrastructural, economic, or epidemiological systems, or we can design triage protocols, contingencies, and recovery plans to mitigate the damage.

Why not focus on both prevention and mitigation? The two qualities of resistance and flexibility are complementary, but also represent a series of tradeoffs; it helps to be strong and supple, but building resilience requires resources and you cannot maximize both. The ideal system design or evolution will weigh, balance, and combine these two qualities along some "golden mean" depending on preferences and contexts. It is in these tradeoffs and balances that we find the most challenging policy dilemmas.

Systemic resilience is a "public good" that is eroded by managerial failures that make collapse through endogenous mechanisms more likely. One modern managerial failure that increases systemic fragility is the focus on ever-increasing efficiency, where cost savings and just-in-time management have replaced redundancy, slack, and reserves. This efficiency has created greater systemic interdependence through increased reliance on one's suppliers and neighboring nodes, making modern systems more susceptible to the mechanisms of endogenous collapse.

As another example of managerial failure threatening resilience, decision-makers within systems often focus on their own interests and their relationships with those to whom they are immediately connected, neglecting to consider the inherent endogenous systemic risks beyond the control of any one participant. This creates negative externalities such as the "tragedy of the commons," where short-run self-interest—rather than coordination and cooperation—can lead to collapse (Hardin 1968).

Governance strategies must be devised with an awareness of these mechanisms of collapse, and the managerial failures that can allow these mechanisms to threaten the viability of our modern systems. Resilience can be prioritized through regulations and standards for safeguards, monitoring, and risk management.

Conclusion

Globalization at an ever-increasing scale and level of complexity is a modern tale of hubris. Building increasingly technologically-advanced, interconnected, and inter-dependent systems without an awareness of the risky mechanisms inherent in their design will inexorably lead to endogenous failures and potential collapse. These risks of globalization have brought us to our study of systemic risk, and to our interest in learning insights about systemic collapse from history.

At first glance, one may find few similarities between ancient civilizations and our modern globalized present. When we see these civilizations as complex adaptive systems, however, we can begin to recognize patterns, structures, and dynamics that have remained consistent through the centuries. Mechanisms like tipping points, feedback loops, contagions, cascades, synchronous failures, and cycles that can be responsible for systemic collapse are fundamental characteristics of any complex adaptive system, and can therefore serve as a useful common denominator from which to examine collapses through the ages. We offer this systemic framework for the study of historical collapse with the belief that these common mechanisms will help illuminate and expose relevant vulnerabilities in historical systems. In the end, our hope is that we may learn from past societies and civilizations and allow our modern systems to benefit from lessons of systemic failure that historians may share with us. We believe these insights could inform the way we see our own systemic vulnerabilities and help to build a more resilient future.

References

Bailey M (2011) Risk and natural catastrophes: the long view. In: Skinns L, Scott M, Cox M (eds) Risk. Cambridge University Press, Cambridge

Barabasi A-L, Frangos J (2002) Linked: the new science of networks science of networks, 1st edn. Perseus Books Group, Cambridge, Mass

Bentley et al., 2014 Bentley RA, Maddison EJ, Ranner PH, Bissell J, Caiado CCS, Bhatanacharoen P, Clark T et al (2014) Social tipping points and earth systems dynamics. Front Environ Sci 2. https://doi.org/10.3389/fenvs.2014.00035

Birley AR (2000) Hadrian to the Antonines. In: Bowman AK, Rathbone D, Garnsey P (eds) The Cambridge Ancient History: Volume 11: The High Empire, AD 70–192, edited by, 2nd ed, 11:132–94. The Cambridge Ancient History. Cambridge University Press, Cambridge. https://doi.org/10.1017/CHOL9780521263351.004

Bossomaier T, Barnett L, Harré M (2013) Information and phase transitions in socio-economic systems. Complex Adapt Syst Model 1. https://doi.org/10.1186/2194-3206-1-9

Bostrom N, Cirkovic MM (2008) Global catastrophic risks, Illustrated. Oxford University Press, Oxford

Brauer D (2018) Theory and practice of historical writing in times of globalization. In: Brauer D, Roldán C, Rohbeck J (eds) Philosophy of globalization, pp 397–408. De Gruyter. https://doi.org/10.1515/9783110492415-029/html

Brown P (1976) The making of late antiquity. 5th Printing edition. Harvard University Press, Cambridge, Mass

Cambridge Global Risk Index 2019: Executive Summary (2018) University of Cambridge: Cambridge Centre for Risk Studies. https://www.jbs.cam.ac.uk/faculty-research/centres/risk/publications/managing-multi-threat/cambridge-global-risk-index/cambridge-global-risk-index-2019-executive-summary/

Centeno MA, Nag M, Patterson TS, Shaver A, Jason Windawi A (2015a) The emergence of global systemic risk. Ann Rev Sociol 41(1):65–85. https://doi.org/10.1146/annurev-soc-073014-112317

Centeno M, Callahan P, Patterson T (2015b) Systemic risk in global agriculture: conference report. PIIRS Global Systemic Risk & Agriculture and Food Security Center, Princeton, New Jersey. https://risk.princeton.edu/img/Princeton-Columbia_Agriculture_Conf_Report_2014-10-24_(v2016-09-27).pdf

Coggan P (2020) More: the 10,000-year rise of the world economy. Main edition. Economist Books

Crucitti P, Latora V, Marchiori M (2004) Model for cascading failures in complex networks. Phys Rev E Stat Nonlinear Soft Matter Phys 69:045104. https://doi.org/10.1103/PhysRevE.69.045104

Dakos V, Matthews B, Hendry AP, Levine J, Loeuille N, Norberg J, Nosil P, Scheffer M, De Meester L (2019) Ecosystem tipping points in an evolving world. Nat Ecol Evolut 3(3):355–362. https://doi.org/10.1038/s41559-019-0797-2

Danku Z, Perc M, Szolnoki A (2019) Knowing the past improves cooperation in the future. Sci Rep 9(1):262. https://doi.org/10.1038/s41598-018-36486-x

Demandt A (1984) Der Fall Roms: Die Auflösung des römischen Reiches im Urteil der Nachwelt. Beck, München

Epstein JM (2009) Modelling to contain pandemics. Nature 460(7256):687–687. https://doi.org/10.1038/460687a

Evans B, Reid J (2013) Dangerously exposed: the life and death of the resilient subject. Resilience 1(2):83–98. https://doi.org/10.1080/21693293.2013.770703

Ferguson N (2006) The war of the world: history's age of hatred. Allen Lane, London

Foreman-Peck J (2007) European historical economics and globalisation. J Philos Econ 1(1):23–53

Gunderson LH, Holling CS (2002) Panarchy: understanding transformations in human and natural systems. Island Press

Haldon J, Chase A, Eastwood W, Medina-Elizalde M, Izdebski A, Ludlow F, Middleton G, Mordechai L, Nesbitt J, Turner BL (2020) Demystifying collapse: climate, environment, and social agency in pre-modern societies. Millennium 17(November):1–33. https://doi.org/10.1515/mill-2020-0002

Hardin G (1968) The tragedy of the commons. Science 162(3859):1243–1248. https://doi.org/10.1126/science.162.3859.1243

Helbing D (2009) Managing complexity in socio-economic systems*. Europ Rev 17(2):423–438. https://doi.org/10.1017/S1062798709000775

Hernández-Lemus E, Siqueiros-García J (2013) Information theoretical methods for complex network structure reconstruction. Complex Adapt Syst Model 1(April):8. https://doi.org/10.1186/2194-3206-1-8

Hoffman PT (2015) Why did europe conquer the world?, 1st edn. Princeton University Press, Princeton

Holovatch Y, Kenna R, Thurner S (2017) complex systems: physics beyond physics. Eur J Phys 38(2):023002.https://doi.org/10.1088/1361-6404/aa5a87

Homer-Dixon T, Walker B, Biggs R, AS Crépin, Folke C, Lambin E, Peterson G et al (2015) Synchronous failure: the emerging causal architecture of global crisis. Ecol Soc 20(3). https://doi.org/10.5751/ES-07681-200306

Kauffman S (2013) Evolution beyond Newton, Darwin, and Entailing Law: the origin of complexity in the evolving biosphere. In: Lineweaver CH, Davies PCW, Ruse M (eds) Complexity and the arrow of time. Cambridge University Press, Cambridge. https://doi.org/10.1017/CBO9781139225700

Kauffman SA (1993) The origins of order: self-organization and selection in evolution. Oxford University Press, New York

Kelly W (1971) Pogo cartoon. Ink and blue pencil on paper

Khaldûn I (2015) The Muqaddimah. Dawood NJ (ed) Translated by Franz Rosenthal. Princeton University Press, Princeton N.J

Kolmes SA (2008) The social feedback loop. Environ Sci Policy Sustain Dev 50(2):57–58. https://doi.org/10.3200/ENVT.50.2.57-58

Korkali M, Veneman JG, Tivnan BF, Bagrow JP, Hines PDH (2017) Reducing cascading failure risk by increasing infrastructure network interdependence. Sci Rep 7(1):44499. https://doi.org/10.1038/srep44499

Ledwoch A, Brintrup A, Mehnen J, Tiwari A (2018) Systemic risk assessment in complex supply networks. IEEE Syst J 12(2):1826–1837. https://doi.org/10.1109/JSYST.2016.2596999

Lessard C (2009) Basic feedback controls in biomedicine: synthesis lectures on biomedical engineering, 1st edn. Morgan and Claypool Publishers, San Rafael, Calif.

Levin SA, Lubchenco J (2008) Resilience, robustness, and marine ecosystem-based management. Bioscience 58(1):27–32. https://doi.org/10.1641/B580107

Mango C (ed) (2002) The oxford history of byzantium. Oxford University Press, Oxford

Manheim D (2020) The fragile world hypothesis: complexity, fragility, and systemic existential risk. Futures 122:102570.https://doi.org/10.1016/j.futures.2020.102570

Martin LA (1997) An introduction to feedback. Massachusetts Institute of Technology

Merriam-Webster. n.d. System. In Mirriam-Webster. https://www.merriam-webster.com/dictionary/system

Meyer P, Ponthiere G (2020) Human lifetime entropy in a historical perspective (1750–2014). Cliometrica J Hist Econ Econ Hist 1:129–167

Middleton G (2013) That old devil called collapse. E-International Relations (blog). https://www.e-ir.info/2013/02/06/that-old-devil-called-collapse/

Middleton GD (2017) Understanding collapse: ancient history and modern myths. Cambridge University Press, New York, NY

Milkoreit M, Hodbod J, Baggio J, Benessaiah K, Calderón-Contreras R, Donges JF, Mathias JD, Rocha JC, Schoon M, Werners SE (2018) Defining tipping points for social-ecological systems scholarship: an interdisciplinary literature review. Environ Res Lett 13(3):033005.https://doi.org/10.1088/1748-9326/aaaa75

Nicoll K, Zerboni A (2019) Is the past key to the present? Observations of cultural continuity and resilience reconstructed from geoarchaeological records. Quaternary International. https://agris.fao.org/agris-search/search.do?recordID=US201900215845

Nuzzo JB, Mullen L, Snyder M, Cicero A, Inglesby TV (2019) Preparedness for a high-impact respiratory pathogen pandemic. The Johns Hopkins Center for Health Security. https://www.centerforhealthsecurity.org/our-work/publications/preparedness-for-a-high-impact-respiratory-pathogen-pandemic

Ord T (2020) The precipice: existential risk and the future of humanity. Hachette Books

Oughton EJ, Usher W, Tyler P, Hall JW (2018) Infrastructure as a complex adaptive system. Research article. Complexity. Hindawi. https://doi.org/10.1155/2018/3427826

Parker G (2013) global crisis: war, climate change and catastrophe in the seventeenth century. Yale University Press, New Haven. http://www.yalebooks.com/book.asp?isbn=9780300153231

Pastor-Satorras R, Castellano C, Van Mieghem P, Vespignani A (2015) Epidemic processes in complex networks. Rev Mod Phys 87(3):925–979. https://doi.org/10.1103/RevModPhys.87.925

Perrow C (1984) Normal accidents: living with high-risk technologies. Basic Books, New York

Renfrew C (1973) Explanation of culture change: models in prehistory. University of Pittsburgh Press, Pittsburgh

Roberts A (2020) It's the end of the world: but what are we really afraid of? Elliott & Thompson

Robinson E (2017) Thucydides on the causes and outbreak of the peloponnesian war. In: Forsdyke S, Foster E, Balot R (eds) The oxford handbook of thucydides. Oxford University Press. https://doi.org/10.1093/oxfordhb/9780199340385.001.0001

Roser M, Ortiz-Ospina E, Ritchie H (2013) Life expectancy. Our world in data. https://ourworldindata.org/life-expectancy

Schwab K (2019) Globalization 4.0: a new architecture for the fourth industrial revolution. Foreign Affairs. https://www.foreignaffairs.com/articles/world/2019-01-16/globalization-40

Shelley PB (1818) Ozymandias. Poem

Siskin C (2016) System: the shaping of modern knowledge, 1st edn. The MIT Press, Cambridge, Massachusetts

Smil V (2019) Growth: from microorganisms to megacities, Illustrated. The MIT Press, Cambridge, Massachusetts

Stephen Hawking's Universe (2010)

Sweezy PM (1943) Professor Schumpeter's theory of innovation. Rev Econ Stat 25(1):93–96. https://doi.org/10.2307/1924551

Tainter J (1988) The collapse of complex societies. Cambridge University Press

Taleb N (2007) The black swan: the impact of the highly improbable, 1st edn. Random House, New York

Taleb NN (2012) Antifragile: things that gain from disorder. Random House

Vespignani A (2010) The fragility of interdependency. Nature 464(7291):984–985. https://doi.org/10.1038/464984a

Walker J, Cooper M (2011) Genealogies of resilience: from systems ecology to the political economy of crisis adaptation. Secur Dialogue 42(2):143–160. https://doi.org/10.1177/0967010611399616

Walker SI, Packard N, Cody GD (2017) Re-Conceptualizing the origins of life. Philos Trans Royal Soc Math Phys Eng Sci 375(2109):20160337. https://doi.org/10.1098/rsta.2016.0337

West G (2017) Scale: the universal laws of life, growth, and death in organisms, cities, and companies. Reprint edition. Penguin Books

Wildavsky A, Dake K (1990) Theories of risk perception: who fears what and why? Daedalus 119(4):41–60

Yoffee N (2005) Myths of the archaic state: evolution of the earliest cities, states, and civilizations. Cambridge University Press, New York

Yoffee N, Cowgill GL (1988) The collapse of ancient states and civilizations. University of Arizona Press

Zanden JL, Baten J, Mira d'Ercole M, Rijpma A, Smith C, Timmer M eds (2014) How was life?: Global well-being since 1820. OECD Publishing, Paris. https://www.oecd.org/statistics/how-was-life-9789264214262-en.htm

Disjunctures of Practice and the Problems of Collapse

Rowan Jackson, Steven Hartman, Benjamin Trump, Carole Crumley,
Thomas McGovern, Igor Linkov, and AEJ Ogilvie

Abstract This chapter asks what insights long-term historical information from before the Great Acceleration and Anthropocene might offer to policy and practice in the twenty-first century. Conventional sustainability research usually focuses on shallower time horizons that could miss insightful environmental and social processes evolving over centuries to millennia. Although we push for increased engagement with historical researchers, parallels between pre-modern and contemporary environmental and societal challenges need to be treated with caution. So-called cases of societal collapse—often associated with environmental calamities—provide limited or at best flawed parallels with challenges faced today. The pitfalls of reductionism and determinism that often attend collapse discourse account for social agency and complexity in incomplete and unconvincing ways. Instead, we argue that historical evidence should serve as context to environmental problems faced today, as antecedents of the accelerated environmental change of later modernity rather than as direct analogies. Historical antecedents can be understood, to an extent, as previous

R. Jackson
Edinburgh University, Edinburgh, UK

S. Hartman (✉)
University of Iceland, Reykjavik, Iceland
e-mail: hartman@hi.is

Arizona State University, Tempe, AZ, USA

B. Trump
University of Michigan, Ann Arbor, MI, USA

B. Trump · I. Linkov
USACE-ERDC Risk and Decision Science, Vicksburg, MI, USA

C. Crumley
University of North Carolina, Chapel Hill, NC, USA

T. McGovern
Hunter College CUNY, New York, NY, USA

AEJ Ogilvie
University of Colorado, Boulder, CO, USA

Stefansson Arctic Institute, Akureyri, Iceland

75

A. Izdebski et al. (eds.), *Perspectives on Public Policy in Societal-Environmental Crises*,
Risk, Systems and Decisions, https://doi.org/10.1007/978-3-030-94137-6_7

experiments against which to test and improve theory or to structure possibilistic scenarios that help anticipate unexpected social and environmental challenges. In concluding, we suggest that researcher in historical sciences and the humanities require resources, space and incentives to explore sticky questions of uncertainty, risk, and vulnerability to environmental change together with global change researchers, policymakers, and environmental practitioners.

Keywords Anthropocene · Collapse · Environmental Humanities · Global Change · Environmental History · Archaeology

Introduction

Archaeology and history have long research traditions focusing on human–environment interaction (Trigger 2006). Environmental history and environmental archaeology research traditions, in particular, cover a substantial body of material evidence, reconstructing how humans perceived and changed their environments, how different cultures utilized natural resources, and how societies responded to short and long-term environmental change (Haldon et al. 2018; Riede 2017; Kintigh et al. 2014a, b). This latter focus has received sustained attention since the second half of the twentieth century and has been dominated by cases of so-called collapse (Tainter 1988). However, as Butzer (2012: 3632) explains, "the concept has intuitive appeal but ambiguous meaning" that draws attention to historical disciplines, but without sufficient clarification of relevance (Richer et al. 2019) or validity under closer scrutiny. This chapter addresses the pervasiveness of collapse and resilience concepts with the aim of reconsidering how historical data, case studies and 'lessons from the past' may be applicable in environmental science, as well as in various policy and governance (e.g. planning and emergency response) contexts. If we are indeed now living in a no-analog age, then some of the most significant challenges for historical disciplines today include identifying, unpacking and demonstrating the relevance of pre-modern social and environmental cases not only to present-day vulnerabilities, but also to scenario-building efforts intended to better prepare our societies for potential social-ecological crises and risks in the future.

Among other questions, this chapter asks what insights and information from before the Great Acceleration and the Anthropocene (as most widely defined) may hold for the wicked and messy environmental challenges facing twenty-first century globalized society. In turn, we consider the insights and richness that deep-time perspectives can offer to global change research, particularly now in the first years of the United Nation's Decade of Action to deliver the Sustainable Development Goals as the international science community and intergovernmental bodies attempt to address, with increasing urgency, climate change, biodiversity loss, wealth disparity, unsustainable consumption and resource usage as well as a great many other global challenges (UN SDG 2021). These high-level actors, internationally and nationally, have begun to be more active and vocal in efforts to engage scientific domains such as qualitative social sciences and the humanities, as well as indigenous knowledge communities, in the very processes of knowledge production, scientific assessment and policy advisement that have effectively excluded them for the past half century (Castree et al. 2014). The Club of Rome's report *The Limits to Growth* (Meadows et al. 1972) helped to consolidate the environmental turn that had already begun as a grass-roots cultural and political preoccupation in the 1960s into an ever more coherent policy agenda of national and international prominence from the 1970s and 1980s onward (Blewitt 2018). However, the role of humanities and qualitative social sciences in large-scale efforts to assess relevant scientific knowledge for the purposes of policy planning on questions of environment, climate, conservation and sustainability has until the past few years tended to regard these knowledge domains (historical studies and critically examined data sources of the past in particular), as *othered* bedfellows so strange they have essentially had no place in the bed, reminiscent of the famous opening line of L.P Hartley's novel *The Go-Between*: "The past is a foreign country; they do things differently there" (1953).

There are some very notable signs recently that the situation has changed and that active collaboration is genuinely being encouraged between mainstays of the global change agenda, such as geosphere and biosphere research communities, and humanities disciplines, including historical studies, anthropological/archaeological disciplines, cultural heritage, arts and philosophy. These include: (1) UNESCO's formalization in spring 2021 of the humanities-led BRIDGES Coalition as the Sustainability Science arm of its international science programme Management of Social Transformations; (2) the first ever UNESCO-ICOMOS-IPCC International Co-Sponsored Meeting on Culture, Heritage, and Climate Change and in December 2021 and publication of three white papers scoping the crucial intersection of heritage and climate in early 2022 (UNESCO 2020a, b; ICOMOS 2019).

While these developments are certainly welcome and long overdue, recent high-level international initiatives promoting the value of integrated humanities, social sciences, and natural sciences research for sustainability, including co-production of knowledge and solutions-orientated action on global challenges with diverse (non-academic) societal partners, carries its own risks tied to no small degree with the very ambitions on which many of these scientific, epistemic and social innovations rest (Castree 2014, 2016; Jackson et al. 2018). Enhanced expectations of the strategic benefits that historical case studies and data can provide to policymakers, in tandem with previously siloed wisdom from non-academic communities (e.g. knowledgeable local citizen scientists partnering with scholars and scientists, inhabitants of threatened social and environmental systems, including indigenous communities), can run the risk of producing shallow or scientifically questionable results if scientific integrity and quality take a back seat to shorter term political or social agendas, however meaningful and justifiable these priorities may be on their own or in a wider societal context. This is the note of caution on which we conclude, together with a series of questions proposed for further investigation.

Historical Experiments: Primacy, Principle and Practice

What role can past cases of social-ecological system disturbance play in helping scientists, policymakers and environmental managers address present and future vulnerabilities facing societies in the twenty-first century? Why and how societal crises typically develop, as well as what shapes their generalized outcomes, are questions of central interest not only to the risk and crisis research community but to planning, governance and response agencies. These are questions where historical research disciplines, it would stand to reason, should be in a strong position to contribute knowledge and useful case studies capable of having real world impact. How well does such an expectation hold up? Can case studies of particular past societies that underwent extreme exogenous or endogenous stress demonstrate whether certain communities or socioeconomic structures are more amenable to change without losing vital system capacities or integrity? Efforts to address such questions may understandably tempt us to think in rather reductive values of success or failure. Such logic, along with the master narratives that undergird it, have exerted a powerful hold on the popular imagination in recent decades (Diamond 2005; see Middleton 2017) and have even influenced policy and governance agendas to some degree (see IPCC AR5, Chap. 16, 2014).

Why are some past societies considered as more successful, having adapted to major human or natural disturbances, while others have become the historical poster children of societal collapse and failure? In the clarity of hindsight, the causes of what has been described as societal collapse in many of these narratives can appear predestined. In some cases, unsustainable resource use can expose societies to long-term deprivation in what has been termed world systems theory—the transition of societies from core, hegemonic to more marginal, peripheral political economies (Wallerstein 2004)—or exacerbated existing inequalities potentially leading to what often gets called 'collapse' (Tainter 1988; Diamond 2005; Kohler and Smith 2018). For other societies, exposure to acute shocks, such as extreme weather events (e.g., flooding, drought, extreme heat or frost), natural hazards (e.g., volcanic eruptions, earthquakes, tsunamis) or warfare, can possibly trigger violent shifts in operational and governance capacities at their prior levels of complexity. Such shifts in complexity—or threshold crossing events—have been conceptualized using a range of theoretical frameworks and have occupied significant debate in archaeology and history, particularly in the last 15 years (Tainter 2003; McAnany and Yoffee 2010). How to usefully characterize sudden disruptions or more chronic stress and their respective effects on past societies have been central to the discourses of *collapse*, and more recently *resilience*, including discussions of what these concepts specifically mean and entail, and if they can indeed help us to understand the dynamics of long-term social and environmental change (Aimers 2007; Middleton 2012, 2017; Jackson et al. 2017). Just what such system disruptions may be able to tell us about challenges associated with future environmental change, even the prospect of social-ecological system collapse, are questions of another kind and order altogether. However, in numerous sectors of political and social debate, in technocratic interventions focused on innovation in the science-policy interface and indeed even within some nascent fields such as sustainability science, earnest efforts are underway to plan for mitigation of the risks our societies could face if pathway dependencies or system shocks contribute to future structural or functional changes in societies and environments from which these systems are unable to recover (Dow et al. 2013; Wise et al. 2014). Discussions and efforts unfolding in each of these sectors give meaningful space to rational consideration of how incentive structures underpinning human behavior, among other things, can be altered to avoid disastrous worst-case outcomes within such systems (O'Brien 2018).

Anticipating future hazards is a significant challenge that is both compounded and undermined by environmental complexity, technological innovation and the interaction of social systems with and within these systems (Bostrom and Cirkovic 2008). Popper (1956) termed this difficulty *The Poverty of Historicism*: the complexity of interacting natural and social systems making historical prediction intractable. From the perspective of risk science, this is due to two critical deficiencies. First, we cannot know the full range of future hazards that societies may face, nor characterize the extent to which they may yield harm (Hochrainer-Stigler et al. 2020). Poorly identified or poorly characterized hazards can leave societies open to surprise, making it difficult for stakeholders to adequately respond to threats as they arise, and for societies more broadly to understand how best to act and overcome associated disruptions (IRGC 2018; Hynes et al. 2022). One example of hazard uncertainty is captured in November's (2008) characterisation of the spatiality of risk, where new social configurations create ill-defined 'grey areas' for risk management. Attention has been drawn to the Geneva fire department's broadening of the traditional fire risks from poorly maintained apartments to incorporate risks on industrial estates that have been repurposed for residential and nightlife but may also include hazardous chemicals and machinery (November 2008). Second, even if we may have some knowledge of the hazards we could face in a certain area, there is a near infinite number of societal vulnerabilities that can influence or impact these hazards. Societal systems, from infrastructure to the environment, from commerce to governance and culture, involve a wide range of nested dependencies that, if disrupted, can generate sudden, cascading disruptions, breakdown or even failure (Hoffman and Oliver-Smith 2001). Even if we were to harden one potential vulnerability (e.g., safeguarding public health through agricultural sustainability), other unexpected exposure points remain (e.g., the sudden arrival of a deadly human pathogen). Moreover, conversation in the 'risk and vulnerability' research community emphasizes the cultural construction of disasters, proposing that societies generally tend not to be truly prepared or spared from the extreme effects of system disruptions, but merely manage shocks and longer-term stress on the system by distributing the risks internally to the most socially vulnerable groups and the most redundant structures and functions (see, e.g. Oliver-Smith and Hoffman eds. 2020). This is a view consistent with environmental humanities scholar Rob Nixon's concept of *slow violence* (Nixon 2011) and calls to mind Joseph Tainter's oft-noted observation: "Some people and some ecosystems benefit from sustainability efforts, while others don't. When confronted with the term "sustainability," therefore, one should always ask: Sustain what, for whom, for how long, and at what cost?" (Tainter 2003, 214–15).

From an historical perspective, path dependency and lock-in can have significant ramifications for social, political and technological transformations towards sustainable and resilient societies (Wise et al. 2014; Adamson et al. 2018; Jackson et al. 2018; O'Brien 2018). System path dependence, especially in environmental policy, can reinforce a tendency to favor status quo operations, making it difficult for societies to anticipate risk and build capacity (Samuelson and Zeckhauser 1988; Yudkowsky 2008; Riede and Jackson 2020). An example of this is the identification of socio-technical lock-in, such as car culture, that undermines the wider transition to public transport and cycling in urban environments (Urry 2007; Geels 2012). Lacking a catalyzing incentive to change (or a disincentive to move away from existing behaviors, e.g., unsustainable resource consumption), societies can become locked into path dependencies that leave them increasingly vulnerable to crisis. A prominent example of such pathways can be seen in the Representative Concentration Pathway projections of IPCC AR5 (2014), which plot a range of possible emissions trajectories that map onto climate risk projections in what have come to be known as the 'burning embers' diagrams.[1] Path dependent behaviors have become ossified in common assumptions about economic logical and rational choice. Such assumptions of *Homo economicus* fail to take account of irrational practices that damage public health and fail to take anticipatory action to address environmental damage (Dietz et al. 2003; Thaler and Sunstein 2008; Palma-Oliveira et al. 2018). This may be associated with insufficient information to inform choices or social norms that create path dependent behaviors—with notable examples including dietary choices due to lack of choice or information, personal hygiene such as washing hands, and driving rather than taking public transport—all of which are informed by choice of architecture (i.e., infrastructure, social context) (Thaler and Sunstein 2008). Adapting from one paradigm to another often requires a forceful driver that, while reorienting societal and environmental system properties, can also lead to highly unfavorable outcomes (Allen et al. 2019). Even when extreme systemic change generates growth and normatively preferable long-term outcomes, the lived experience of those in the time of crisis may be violent, unjust, chaotic and prone to deprivation.

Policymakers and scientists need insight not only to identify the characteristics that promote societal resilience to disruption, but also realistic approaches to how such resilience might be generated without the risk of sweeping societal harm. Historical cases, because they may be viewable from a distance as "completed experiments of the past" (Speilmann et al. 2016; Nelson et al. 2017), provide narratives and insight into both needs, yet lessons from historical scholarship only reach decision makers with considerable difficulty and historical expertise may be ignored or mis-applied. In the post WW2 period, the "lessons of history" for many policy makers were largely restricted to the perceived lessons of the Munich appeasement, leading directly to

---

[1] As Mahony and Hulme (2012) explain, the "burning embers" diagrams that have featured in the IPCC Assessment Reports have become prominent visualisations of abstract conceptualizations of future risk.

prolonged and deadly intervention in conflicts in Vietnam, Laos, and Cambodia that produced long-term human and ecological damage that continue to the present (Hess 1994; Hendrickson and McMaster 1997; Khong and Yuen 1990). Pathway dependency and failure to apply appropriate locally scaled historical perspectives have had similarly disastrous outcomes for US, Soviet, and other outside military and nation-building efforts in Afghanistan where Afghan and Central Asian long-term history and the political dynamics of kin-based society regulated by feud were ignored and inappropriate social, economic, and military models were initially employed and never seriously reconsidered over decades of conflict (Loyn 2009; Waldman 2013). Neither Western nor Soviet leadership applied the locally scaled "lessons of history" of prior social and economic structures in both South-East and Central Asia that promoted long-term resilience in the face of both local factional and ethnic conflict and resistance to external hegemonic threats (Mongol, Chinese, French, British, Russian). The input of historians and social scientists aware of the actual local historical backgrounds, the likely future range of social strategies of resilience and repeating local patterns of social durability or fragility were ignored or undervalued by a succession of policymakers applying inappropriate historical lessons.

Theory as Tool: Complex Adaptive Systems

Better connecting appropriate historical cases with other data and the analytical tools of complex systems theory or other scenario exercises (Rounsevell and Metzger 2010) could help to bridge such gaps. To understand change in the past, the traditional approach is to establish a temporally ordered chain of events, termed causation, which allows generalization about a behavior. Causation is a dependent relationship among events/properties/variables. If we are to learn from the past, the establishment of causation is both fundamental—in that a chronology of events is necessary—and problematic, because the past, like the present, is the sum of many events, properties, and variables that operate at various temporal and spatial scales and have relevant properties (e.g., slow/fast, change in rate of change). The adaptive cycle common to resilience serves more often as a model or metaphor for change over time rather than a road map for complex interrelations.

Complex adaptive systems (CAS) can carry research into deep time and offer a more nuanced and practical interpretation of the past. CAS offer several ways to analyze and incorporate time into historical analysis: initial conditions, path dependency, a tendency to undergo irreversible processes (thus creating system history). Complex systems (such as the human–environment relation) are comprised of both linear (predictable) and non-linear (emergent) properties. This fundamental dynamism makes CAS the very essence of change over time (Sinclair et al. 2018). The establishment of this more robust form of causation requires a meta-theoretical approach that considers the properties of dynamic systems (Allana and Clark 2018). Research in this arena focuses on how the entire system operates and how its design and operation affect risk (Webster 2005). A tight chronology of events before the fact is a first step in the study of dramatic change and remains indispensable. However, many ways of knowing about the past (e.g., documents, environmental data, archaeological materials, individual and collective experience) add to the diversity of information, offer cross-checks to its interpretation, and contribute to a more holistic reading of key factors that shaped decisions.Interactions can be influenced in several ways, including path dependency, feedback, and memory. Path dependency means that past events amplify through positive feedback to strongly affect interactions today. For instance, the loss of Finland to Russia in the Swedish-Russian war of 1808–9 led to border closures, which continue to influence the development of reindeer husbandry in Sweden (Moen and Keskitalo 2010). The tendency of decision makers to follow and amend earlier decisions is powerful, minimizing costs and disruption and adhering to tradition. Feedbacks are chains of events that influence themselves, either positively or negatively. A classic example of a positive feedback is that the higher temperatures of climate change melt ice and snow at high latitudes, changing the albedo (the reflectivity of the surface) and trapping more heat, which in turn increases temperatures, and melts more snow. Memory and short system history may also affect interactions. One example is the common use of baselines in fisheries management to assess the wellbeing of preferred species, using recent data and individual experience, thus missing the tell-tale signs of earlier mismanagement (Engelhard et al. 2015). All these modifications have strong effects on dynamic systems.

Diversity, Flexibility and Durability: An Alternative Nomenclature with Alternative Implications

The concept of *durability* has been introduced (Murphy and Crumley 2021) in contrast to the more familiar terms of resilience and sustainability. The term *resilience* is often used simply as the ability to withstand a shock without a fundamental change of functions, whereas forms that permit reorganization at diverse scales and contexts of time and space, while others remain unchanged, can be examined more usefully, and perhaps less opaquely, using the concept of durability. Most importantly, resilience does not offer a robust means for dealing with change over time.

The term *sustainability* also tends to fail the needs of different communities of practice because it often carries the connotation of 'able to be continued indefinitely' (e.g., an activity that does not acknowledge the inevitable reality of a resource's deterioration or depletion) and does not fit well with complex dynamic systems that endure diverse challenges. To apply lessons from the past to today's issues, we must keep in mind that most social systems—ancient and contemporary—have suffered the impacts of climate change, population fluctuations, resource depletion, pestilence, and greed.

Durability, in contrast, is the positive outcome of practices and strategies that were undertaken and refined over time as the context changed (Murphy and Crumley

2021). A durable system is the result of a long-term process that is characterized by continuous development, accumulation of knowledge, and incremental experimentation and observation. It includes how societies regenerated after episodes of 'collapse' or managed the 'art of not collapsing'. By studying the trajectories of durable systems that we can investigate in the archaeological and historical record and in communities of practice, we can avoid some of this trial and error (Crumley 1994, 2007).

Working through case studies that span geographies, cultures, social systems, and time periods, we identify common threads that represent various levels of success in past efforts to cope with changing environmental conditions. These approaches often combine mitigation and adaptation activities undertaken in close collaboration with local and indigenous communities.

Durability does not, however, imply lasting forever. Durability introduces the idea that things will not last forever and must be maintained, while simultaneously advocating for investing in the right moves that will help these things last longer. The characteristics of management strategies that lead to durability are worth examining and establishing more fully. It is necessary to study further the effects of short-term decision making on long-term durability, including how decisions made by distant policymakers may ignore thoughtful strategies that provide the 'non-declining utility' (a common definition of sustainability) of key resources (e.g., soil, organic matter, fresh water) over relatively long timescales. Recent work has begun to detail which factors contribute to the building of durable systems and which introduce vulnerabilities over great time scales, from several centuries to millennia (Brewington et al. 2015, Hicks et al. 2016).

Two such factors are diversity and flexibility. For long-term societal survival, archaeological data and historical records show us that *diversity* is key. A recurring strategy found in durable systems includes the *flexibility* that diversity provides. These reinforce one another: while diversity (of resources, strategies, and perspectives) is the basis for wider choices, flexibility is the ability to alter management and governance to better fit the situation. Biocultural diversity is the basis for flexible social, political, economic, and other strategies. Diverse and flexible strategies within a social-ecological system that has some slack—where each key variable need not be 'just right' for the system to function—offer risk management options that provide vital latitude in the face of external or internal changes (Dugmore et al. 2013).

Many of the long-lasting examples in the archaeological and historical record involve political systems that were dramatically transformed over time, but which overlay a social system that lasted millennia (Meyer and Crumley 2011). These underlying systems were diverse and flexible; they were also labor-intensive, and often imbued with a worldview that honored and protected key resources. They contrast dramatically with today's widespread clear-cutting, fossil-fuel dependency, high energy requirement, use of chemical fertilizers and pesticides, monoculture, and so on, to produce food (Hilding-Rydevik et al. 2018; Iuga et al. 2018). Past practices and management strategies are reservoirs of knowledge that provide viable options for today and can point to alternative strategies for the future.

Learning from Systems Under Stress: Antecedents and Anticipation

Modern societies are afflicted with a variety of stressors that may yield substantial and lasting harms if not ameliorated. The SARS-CoV-2 pandemic (hereafter COVID-19) contains an abundance of potential cases where we can see clearly divergent outcomes: nations/sectors of short-term or blunted impact, nations/sectors with extensive disruption yet rapid recovery, and nations/sectors with sustained and grievous losses in complexity. In some societies the economic, social, and environmental sectors, and their assets, have recovered quickly from disruption, while others lag well behind their pre-pandemic levels of functionality. The core variable informing system performance in this regard is often described as the systems' *resilience*, or system capacity to recover from and adapt to disruption (Hynes et al. 2020). Systemic resilience can be influenced by history, nature, and current culture, as evidenced by the very different outcomes of COVID-19 response in island states like Iceland and New Zealand with high levels of contemporary political cooperation (social capital) and investment in science education and deep historical memories of disastrous epidemics as part of the national historical narrative in comparison with nations experiencing depleted social capital and easy international and inter-regional travel when confronted by pandemic. In Iceland and New Zealand, the impacts of smallpox in the eighteenth and nineteenth centuries form part of school curricula while the story of the 1917–21 Spanish Flu was a specialist subject in the US until 2020 (e.g., Moxnes and Christopherson 2008; Flecknoe et al. 2018). It may be that resilience can be driven by accumulated social capital and the presence of an ocean "moat" in the recent pandemic examples of Iceland and New Zealand, but effective policy still needs the support of a fully and appropriately mobilized historical consciousness for policymakers.

The study of complex systems is a growing field of scholarship driven by a desire to understand how various societal, infrastructural, and economic interconnections collectively influence actions and outcomes (Urry 2003). Such systems thinking is usually driven by a desire to understand susceptibility to extreme events or shocks in modern complex societies increasingly reliant on interconnected digital systems to facilitate functionality (Walker and Salt 2012). Though this sort of scholarship may certainly be helpful in enabling societal resilience in the face of a variety of uncertain and complex disruptive events, its emphasis on exploring modern systems to the exclusion of the past ignores one promising avenue of inquiry—the ability to understand and explain why certain societies or institutions in the past survived and recovered from significant disruptions, and why others appear to have collapsed or declined into reduced forms for extended periods of time.

Systems theory and resilience are philosophies and analytical strategies that can help explain how internal system structure and characteristics (or "endogeneities") influence the capacity of systems to prevent, mitigate, and recover from external shocks and stresses of an extrasystemic nature (or "exogeneities"). Systemic threats are those understood to have consequences or outcomes that can reverberate throughout various elements of society, such as an epidemic which has the capacity to disrupt local economies or governance procedures. States with rigid, inadaptive or brittle institutional, political, and economic systems may be more prone to lasting disruption or even collapse in the most extreme cases. Similarly, those with the capacity for recovery and adaptation in the face of a systemic exogenous shock may be far more likely to survive and even thrive in the aftermath of such disruptions.

The potential importance of historical cases and popularly perceived "lessons of history" in driving current understanding of societal challenges and emergencies cannot be understated. As the COVID-19 pandemic began to spread, everyone from senior policymakers to individual households searched for metaphors and lessons to mitigate possible harms and anticipate what might happen next. Lacking comparable cases from the recent past, experiences of the 1917–20 Spanish Flu pandemic a century before were initially ignored, and some lessons (extensive masking, moving classes outdoors, reducing crowding) were applied late and unevenly despite the clear historical record of the impact on local mortality in a much deadlier pandemic event. Failure to understand and act upon the practical historical lessons of the 1917–20 pandemic certainly exacerbated systemic losses from panic, fear, misunderstanding and ultimately self-destructive actions at national and smaller-scale community levels, as well as at the level of individual response.

Though it is impossible to predict all future hazards, an improved synergy from historical cases to present and future challenges can help to: (a) equip many with more nuanced understandings of how complex societal systems function under duress, recognizing the tradeoffs between optimizing short term efficiency and long-term durability and resilience (Hegmon 2017); and (b) provide some useful lessons and strategies—a so-called usable past—that can contribute to more normatively favorable outcomes (Cooper and Sheets 2012).

What Relevance Can History Have if We Are Living in a No-Analog Age?

The Anthropocene has been subject to intense discussion and debate in the environmental and social sciences and humanities for its implications for the nature of human social life and ecosystem processes (Chakrabarty 2009). It is most commonly defined as commencing in 1950, marked by a global radionuclide marker and the expansion of extractive industries, mass consumerism and population growth (Steffen et al. 2011, 2015) as well as rapidly escalating emissions of carbon dioxide into the earth's atmosphere owing to numerous interrelated developments, such as increased urbanization, industrial expansion and globalization of commerce. Other suggestions place the beginning of the Anthropocene in the eighteenth century, at the start of the industrial era (Steffen 2003) or, more controversially, 5–8 millennia BP (Ruddiman 2003, 2007). Ruddiman's (2013) early Anthropocene claim rests on the hypothesis that land-use change associated with the transition to agriculture and animal husbandry was responsible for the stabilization of climate. But, as Erlandson and Braje (2013) write, the arbitrary date for the start of this epoch is less relevant than the antecedent factors that shaped it. Both archaeology and history provide evidence of the shaping of the human niche in the earth system and how humans have adapted to long-term climate variability (Smith and Zeder 2013). This contextual knowledge has a significant role to play in unravelling the complexity of human impacts on and entanglement within natural ecosystems (Boivin et al. 2016). This contextual knowledge is also important in understanding the efficacy of long-term adaptation to environmental change and how different scenarios may have played out when societies were faced with acute stress. However, context also helps to highlight the dilemma of commensurability of scales, tools, methods and data (both their availability and resolution) when we seek to compare cases and engage multiple disciplines across the spectrum of scientific domains.

The subject of 'relevance' in historical and archaeological research—including the associated discipline of palaeoecology—has received significant attention in recent years (Hudson et al. 2012; Richer et al. 2019). This is a corollary of a broader shift towards practical, impact-driven knowledge production in what has come to be known as 'Mode 2' science (Barry and Born 2013; Nowotny et al. 2013). The shift from 'Mode 1' to 'Mode 2' knowledge production has redirected the emphasis from knowledge driven by theory and academic audiences (Mode 1 knowledge) to a wider audience in civil society, policy and practice—so-called transdisciplinary and co-produced knowledges (Mode 2 knowledge) (Collins and Evans 2002). For archaeologists, this development has increasingly favored efforts to draw civil society into knowledge production through community archaeology (Dawson et al. 2017; Stump and Richer 2017), citizen science (Smith 2014; Dawson 2015) and museum exhibitions (Jackson et al. 2017; Riede 2017), and has prompted debate and discussion on the role that archaeology can and should play in addressing environmental and social problems (Dawdy 2009; Riede et al. 2016; Rockman and Hritz 2020). The same can be said of other historical studies disciplines generally, even if the methodological realities of data retrieval and culture of scholarship in long established or traditional approaches to historical research have tended on the whole to make archival documentary research more of an individual and intradisciplinary pursuit than has typically been the case in field-based archaeological and paleoenvironmental research, which have long favored interdisciplinary methodologies and transdisciplinary forms of collaboration and community cooperation. There have been calls for some time to consider the advantages of team-based interdisciplinary and transdisciplinary approaches in particular as a viable model for conducting integrated historical research, and these calls have increased in recent years especially within projects and approaches such as historical ecology, integrated environmental humanities and cross-field intrahistory study (historical-archaeological and paleoenvironmental consilience research) (Crumley 2007; Haldon et al. 2018; Hartman et al. 2017; Hartman 2015, 2016, 2020; Izdebski et al. 2016).

In some respects, the question of what relevance archaeology and history have is fairly straightforward if not muddied by the popularity of collapse discourse in academia and civil society (see Jackson et al., *this issue*). The most obvious relevance of these disciplines, as well as those closely adjacent (e.g. historical geography, historical anthropology, historical climatology, even paleoecology and paleoclimatology), is their combined ability to help reveal the story of human history and the co-evolution of society and environment (Haider et al. 2021). Popular non-fiction literature, such as Jared Diamond's *Guns, Germs and Steel* (1997) and *Collapse* (2005) and Yuval Noah Harari's *Sapiens* (2011) and *Homo Deus* (2015), provide sweeping narratives of the rise of modern societies and the resulting impacts on environments past, present and future. Without intense archaeological and historical research, these highly influential texts would not be possible, but it should also be recognized that these texts are also widely contested (see for example McAnany and Yoffee 2010) and apply particular frameworks (e.g., determinism, reductionism) that articulate history in particular configurations of cause and effect.

But if we are to learn from history, how one reads the past should be considered. *Which* lessons of history become embedded in popular culture and become privileged in scenario building for current and future policy making have critical importance. If the only memory of conflict avoidance is Munich and Appeasement, then the pathway to frequent military interventions and massive investment in standing military deterrence on hair trigger readiness becomes immediately deepened and widened. A wider reading of historical examples might flag up the recurring problems of past states over-investment in militaries increasingly unsuited to their actual missions leading ultimately to state instability and painful hegemonic transition.

Archaeological and historical records can be broadly divided into two categories of relevance: discrete and continuous. Discrete information includes examples of so-called completed experiments of the past, where societies have a discontinuity with the present. Abandoned settlement or societies that came to an unrecorded end fit within this category and have often been examined for retrospective evidence of social vulnerability, social-ecological resilience and adaptive capacity and flexibility (Diamond and Robinson 2010; Butzer and Enfield 2012; Speilmann et al. 2016). Continuous information, by contrast, has a continuity with contemporary environmental and social problems (Sachs 2020). Mass extinction, land-use change, species domestication and the evolution of societies and technological innovations can all be understood through a retroactive analysis of human–environment interaction (Boivin et al. 2016) that remains ongoing. Example of these continuous data are the vast archives of animal bones and historical records of commercial fishing in the North Atlantic since ~AD 1000 (Kwok 2017; Barrett 2018). Such records provide evidence of the scale of fishing, the average catch size and the spatial distribution of fish over the last millennium, improving our understanding of human impacts on marine ecosystems at the local and regional scale and offering a revised baseline for pristine marine ecosystems (Hambrecht et al. 2018; Hilding-Rydevik et al. 2018).

Though underutilized, archaeological data that have a continuity with contemporary environmental impacts has a clear relevance for policy and practice (Cooper and Sheets 2012). Legislation on marine protected areas and broader conservation and resource-use practices can benefit significantly from a deeper understanding of the long-term impacts of human activities on ecosystem structure (Dunne et al. 2016).

But if discrete archaeological and historical records are to have a relevance in policy and practice, the lessons need to be interpreted carefully and translated with a clear acknowledgement of the opportunities and limitations of such datasets (Table 1). A consistent criticism of Diamond's historical narrative has been his lack of training in archaeology and history, but for a clear recognition of relevance to be grasped, historians and archaeologists must provide a clear exposition of opportunities and limitations of using such records (d'Alpoim Guedes et al. 2016). This can permit the location of usable archaeological information that is relevant to policy priorities (see Stump 2013), *usable knowledge.*

However, realizing this ambition way well depend on stronger concerted efforts among relevant scholarly communities to align historical studies, in the broadest and most usefully inclusive sense, with open science ambitions. This could be done, for example, by investing in expanded and robust cyber infrastructure enabling management and sharing of data more effectively across disciplinary communities without counter-productive contextual knowledge losses (i.e., with glosses and critical apparatuses that could help to translate data with greater nuance for use and distillation across scientific and scholarly communities). Such efforts would depend on building and enhancing research infrastructures, and targeting funding priorities that could enable them, that almost by definition would need to address wider scientific domains extending far beyond the needs of individual subjects. Such a vision of research design might well involve working backward from identification of present or future societal challenges to be addressed through integrated/consilient research on comparable challenges faced by societies in the past, and then identifying the knowledge needs to be facilitated in order to address those challenges through a more successful integration of data and methods than is generally observed today across distinct scientific/scholarly disciplinary communities. Such open science investments can also have further benefits by improving the science/policy interface.

What Deep Time Perspectives Can Offer to Contemporary Debates

Natural experiments are used increasingly in historical disciplines to formalize comparison because it is often not possible, for practical or ethical reasons, to conduct controlled laboratory-based experiments (Dunning 2008, 2012). In archaeology, this method has been used to examine, among other things, variation in artefacts styles to assess the linkage between technological change and environmental variables (Riede 2006, 2014), to examine the role of humans in the deforestation of Pacific islands (Rolett and Diamond 2004), and to examine patterns of subsistence, settlement and exchange using computation models (Kohler et al. 2012). Advances in computational modelling and data resolution in particular have improved our ability to link changes in human societies with environmental change (d'Alpoim Guedes et al. 2016). For example, such models have been used to examine the effects of climate change on

Table 1 Summary of the pros and cons of historical and archaeological studies with lessons of climate change impacts and adaptation

Table 1	Climate change adaptation studies of the past	
	Pros	Cons
Record length and resolution	Deep-time perspectives provide an extensive longitudinal analysis of human populations prior to, throughout and in the aftermath of climate perturbation, natural hazard events and other broader socio-political and economic stressors (Kintigh et al. 2014a, b; Riede 2014b; Butzer 2012)	Data resolution is limited to material remains, written records, and environmental proxies as evidence of human activities. Evidence may be incomplete, lack precision or be subject to bias given the methods, interpretation, and varied preservation of the record (Ogilvie 2010)
Multiple sites, cultures and rates of change	Deep-time records provide evidence of human responses (positive and negative) to the differential impacts of climate change. Multiple sites across space and between different cultures provide evidence of different strategies adopted to adapt to changing environmental, economic and political conditions (Nelson et al. 2006; Spielmann et al. 2016)	Selective human sampling of the environment plus differences in rates of preservation across archaeological sites can lead to an inconsistent and partial record across space and through time (Dawson 2015)
Distributed settlement networks and regional-scale environmental change	Distributed observational networks of the past offer new ways of combining local records of human activities in the past. Combining accurate datasets across entire regions provides a multidimensional long-term record of human adaptation to climate change (Nelson et al. 2016; Dugmore et al. 2013)	Societies of the past have inherent differences to societies in the twenty-first century. Revolutions in science, automobility, economic development and health distinguish modern societies and cultures from the past (Butzer 2012)
Analog and Analogy	The archaeological record offers a 'completed experiment' of the impacts on and responses of human populations to climate change (Dugmore et al. 2013)	The absolute size of population, settlement, economic networks, migration and infrastructure in the twenty-first century is without historical parallel (Butzer 2012)
	Processes and responses hold similar characteristics in contemporary and past societies. Equality, sharing and traditional forms of knowledge are similar processes in modern and pre-modern societies (Kintigh et al. 2014b)	Anthropogenic climate change is having effects on the rate and magnitude of global change on a scale never witnessed by human societies, though processes operating at the local scale are, in some cases, similar (Boivin et al. 2016)

maize cultivation and turkey raising in the prehispanic Pueblo societies of the US Southwest (Kohler et al. 2012; Bocinsky and Kohler 2014). Biocultural changes and human-environmental impacts have also undergone extensive analysis, comparing environmental variables with resource accessibility as a way of assessing societal resilience (Hegmon 2017; Spielmann et al. 2016; Rolett and Diamond 2004).

Studying social and environmental change from a deep time perspective offers useful context scales (from multigenerational to millennial-scale) to understand the origins of modern societal problems (Redman 2005; Marks 2015; Burke 2015). Kohler (2012) stresses this point to emphasize the importance of historicity to social evolution and response to social and environmental hazards. This requires consideration of the impacts that social and environmental phenomena have on the resilience of societies to hazards in the long term (Kohler 2012). For example, the investment in canal irrigation systems by the Hohokam in the US Southwest was successful over a multi-century timescale of sustaining farming, increasing settlement concentration, supporting investment in public architecture, and evolving hierarchical modes of governance (Nelson et al. 2012, 2016). Irrigation systems allowed settlements to manage short-term variations in precipitation, but this also enhanced reliance on the socio-economic systems supported by irrigation. As environmental conditions deteriorated from the late-14 and fifteenth centuries, the irrigation system was damaged by unanticipated flooding and was subsequently unable to manage extensive drought. The investment in communal infrastructures enhanced path dependency, making it difficult to invest in alternative resources (Nelson et al. 2012). By contrast, the Mimbres settlement, also of the US Southwest, shows no evidence of pronounced hierarchy and little investment in public infrastructure, and it developed in very different ways to the Hohokam. Recent evidence of flexible settlement and pottery production suggests a greater capacity of the Mimbres to respond to drought and economic change (Nelson et al. 2006), albeit at lower population densities (Nelson et al. 2016). Such long-term datasets can provide the depth required for researchers to examine discrete cases in a broader diachronic context and to explain the various limitations on a given society's social and cultural capacity to adapt (Butzer 2012; Spielmann et al. 2016).

The analysis of past disasters also has the potential to demonstrate a causal relationship between the pre-impact vulnerability of populations to social and environmental change and evidence of post-impact cultural change (Cooper and Sheets 2012; Riede 2014). Box 1 provides examples of disaster risk reduction (DRR) in contemporary and palaeosocietal contexts. Understanding how societies responded differently to extreme events, such as volcanic eruptions, provides a basis not only to assess socio-cultural limitations, but also suitability of anticipatory or post-impact adjustments to risk. With sufficient data resolution, archaeology and other historical sciences can provide empirical evidence of challenges associated with long-term adaptation to disasters and environmental change, whereas contemporary research relies on informed speculation on societal capacity to adapt to future, and hence unknown, environmental change. In turn, such experiments of history, with their known outcomes, can feed into the evidence-based construction of realistic disaster scenarios (Mazzorana et al. 2009; Riede 2017).

Box 1: The temporality of risk: Disaster Risk Reduction (DRR)

Disaster risk reduction (DRR) is a field concerned with risk reduction in human settlements in proximity to hydrometerological hazards, such as hurricane hotspots and mountainous areas prone to avalanches (Keiler et al. 2006; Fuchs et al. 2007), geophysical hazards, such as volcanic eruptions and tectonic zones (Smith 2013), and anthropogenic hazards, such as industrial and infrastructure hazards (November 2008). These fields receive substantial government funding and attention from the natural sciences, civil engineering and the social sciences for their roles in reducing societal risk via monitoring and prediction (Earle et al. 2012), hazard-response protocols (Alexander 2010), infrastructure design, and post-disaster relief (Alexander 1995, 1997). There is also a significant archaeological literature on past disasters (Riede 2014a, 2016, 2017; Riede 2017), and the long-term cultural responses to environmental change (Dugmore et al. 2013; Dow et al. 2013). Riede's (2014a) science of past disasters emphasises the potential of long-term cases, or completed experiments, for assessing 'possibilistic' (Clarke 2007, 2008) outcomes of volcanic eruptions on human populations, evaluating the spatio-temporal impacts of eruptions, and the long-term interplay of social contextual factors, impact event, and capacity to respond (Riede 2017)

Deep-time perspectives can illustrate the vulnerabilities of populations before and after eruption events to explain both the 'spatiality' (November 2008) and 'temporality' of risk (Riede 2017). Iceland, for example, is alert to the immediate effects of eruption events on nearby settlements, farming, and aviation (Donovan and Oppenheimer 2011). In contrast, transatlantic aviation between Europe and North America was unprepared for impacts associated with the 2010 eruption of Eyjafjallajökull (Lund and Benediktsson 2011). This was largely due to the unanticipated synergistic effects of a volcanic plume, meteorological conditions and a lack of appropriate crisis management (Alexander 2013; Inkpen 2016; Donovan and Oppenheimer 2012). From a DRR perspective, effective crisis response requires the identification of the far-reaching effects of such events on vulnerable social groups (Linkov et al. 2022a), 'critical infrastructures' (Linkov et al. 2022b), essential resources, and environmental stability (Kuklicke and Demeritt 2016). Vitally, it is often the frequency of hazards that conditions populations to risks (Lawson et al. 2012); the more frequent events occur, the more a population is required to adapt its behaviour (Rockman 2003, 2012). Therefore, the vulnerability of a population to low-frequency yet high-magnitude events may not be apparent until such an event takes place (Dugmore and Vésteinsson 2012). Long-term perspective that can more effectively include the complex social responses and spatial and temporal impacts of rare events (such as volcanic hazards) on past societies adds important context to the societal perception of risk.

Possibilistic Reasoning, Counterfactuals and Scenario Modelling

In modern *adaptive management* strategies many agencies and organizations have been strongly influenced by the Resilience Alliance's widely disseminated *Assessing Resilience in Social-Ecological Systems: Workbook for Practitioners* (2010) which emphasizes practices aimed at breaking out of path dependent management structures and accepting levels of uncertainty and risk as part of an ongoing process of assessment, application, and re-assessment. An example below taken from Washington State's criteria for evaluating and prioritizing adaptation options provides a useful example of the kinds of questions practitioners and policy makers are now asking. Note that most of these relevance questions have a historical component and will in practice be difficult to address with only a few decades of records to consult. All would benefit from contributions by scholars of the past (Table 2).

Adaptive management strategies and the recommendations of the *RA Workbook* (2010) are regularly connected to the process of scenario building as a means of coping with uncertain futures, often deliberately contemplating a range of "what if" counter-factual pasts to get a better understanding of an actual range of potential future outcomes (Levy 2008; Ison et al. 2014). Climate change response is increasingly seen as one of the ultimate "Wicked Problems" which defy effective problem formulation and resist simple engineering solutions that ignore the social and practical limits imposed by prior interventions and longer historical trends. Current

Table 2 Criteria for evaluating adaptation options. Adapted from Table 2.11 in Washington State 2010 Climate Response Strategy "Potential Criteria for evaluating and prioritizing adaptation options" (http://www.ecy.wa.gov.climatechange/2010TAG.htm)

Criteria	Relevance
Importance	What is at stake if we do nothing? Are changes likely to affect unique or valuable species, ecological functions, keystone environments, watersheds, treaty rights?
Urgency	What ae the costs of delaying action? Is it likely to cost more to implement later than now? Are the consequences of not acting now irreversible?
Co-Benefits	Are there benefits beyond the immediate adaptation goal? Will total benefits long term exceed the costs of implementation? Are costs and benefits equitably distributed across communities?
Feasibility	How feasible is the proposed action given existing laws, policies, and the political climate? How technically feasible is the action? Is there an opportunity to repurpose existing actions and strategies or will a completely new approach be required?
Robustness	What is the likelihood that the proposed action will be effective across the range of possible future scenarios? Does it allow for effective adaptive management?
Cost	How costly will the proposed action be in terms of time, money, staff, or other resources? Will this investment preclude other adaptation options or aid them?
Other	What impacts on greenhouse gas emissions? Equity of impacts and benefits across communities? Consistency with national laws and policies?

best practice in scenario building for forward planning emphasizes the need for wide stakeholder and knowledge holder participation in framing the initial questions and range of possible outcomes to avoid automatically privileging "common sense" perspectives that in practice often fatally constrain effective scenario design. There is a need to push the historical perspective in scenario building back beyond the decadal scale and to incorporate more diverse and detailed "lessons of history" delivered by knowledge holding communities experienced in handling the multiple practical and theoretical issues of understanding the past (Rounsevell and Metzger 2010). The engagement of history, environmental humanities, archaeology, and paleoecology in agency and organizational scenario building exercises and their participation in regular follow up assessment meetings may be a critical contribution point for these disciplines. While the "no analog" future may in fact present challenges never before faced by society, some social responses can be shown to be recurring across time and cultures and others can be identified rarely. Many past societies have reacted to local resource shortage by migration, regularly involving violence, and thus planners should build mobility (controlled and uncontrolled) as a likely feature of future social response into most scenarios. A common feature of large-scale, long-lasting, hegemonic, imperial, multi-cultural societies has been the creation of an inclusive elite sub-culture with shared language, taste in art, and use of material symbols to interconnect what may be very different local cultures in a common system that can balance diversity and shared values. Might promoting similar mixes of diversity with competence in participation in a shared elite culture gain support from past cases as a positive objective for "favorable outcome" scenario end-points?

Why?

Why not?

The past thus represents both a storehouse of completed experiments in human ecodynamics and an active resource for "what if" scenario construction and testing.

Policy Implications

One of the significant challenges for adaptation in policy and planning is the interplay between key concepts and terminology associated with the human dimensions of climate change and measuring progress in adaptation. The Global Center on Adaptation's (2020) *State and Trends in Adaptation report 2020* highlight among its top policy recommendations the need to improve measures and metrics for progress and improving the knowledge base required to inform adaptive planning. A similar challenge is highlighted in Ford et al.'s (2018: 193) review of climate change vulnerability research, critiquing the 'conceptual vagueness', siloed nature of research, and the 'static' nature of monitoring. This report and study identified the lack of conceptual clarity and ability to test and measure adaptation and vulnerability as a significant barrier to effective adaptive planning. However, as we will argue, historical experiments offer analogues with which to examine, test and measure concepts over multiple temporal and spatial scales.

As explored in the previous section, natural experiments of history or, as some archaeologists say, completed experiments of the past, offer retrospective evidence of human–environment interaction over extended timescales (Redman 2005; Hartman et al. 2017). The benefit of this retrospective information is not only observable evidence of vulnerability and adaptation to climate variability and other exogenous changes (Nelson et al. 2016), but also the counterfactual questions that can be asked of different societies in the context of these changes (Carr 1964). Counterfactuals allow plausible questions to be asked regarding processes that may underly patterns of social and ecological change in a given context. These 'what if' questions also have an application in strategic planning, to determine plausible outcomes of different decisions and to plot a range of plausible scenarios (Jin et al. 2021).

One of the significant errors that has plagued historical and archaeological research in the early twenty-first century has been the assumption that deterministic and reductionist understandings of human–environment interaction will yield lessons about how to respond to environmental change in the future (*see for example* Diamond 1997, 2005). Such approaches have been critiqued for their limited emphasis on human agency, ingenuity and capacity in response to environmental challenges (McAnany and Yoffee 2010). But more importantly, the lack of dialogue with researchers of contemporary human vulnerability and adaptation has undermined the application of historical evidence. Jackson et al. (2018) emphasize the need for archaeologists to work in more active dialogues with researchers of contemporary global change, publish in interdisciplinary journals read by global change researchers and policymakers, and use their connections with museums and local communities to communicate archaeological relevance (Cooper and Sheets 2012; Hartman et al. 2017; Sigurðardóttir et al. 2019). First and foremost, archaeological and historical researchers need to carve out space for an effective dialogue with adaptation researchers and planners where relevant and insightful evidence about potential limits and barriers to adaptation can be identified. In this sense, knowledge of how to respond to uncertain futures can be co-produced.

Exercises in deep-time thinking have the potential to combine expertise from existing adaptation planning strategies and frameworks with historical evidence of adaptive continuity and change to co-produce qualitative scenarios (see Riede and Jackson 2020). Figure 1 visualizes the opportunity for collaborations that synthesize lessons from historical sciences with contemporary place-based research into vulnerability and adaptation. This figure illustrates the distributed evidence of long-term adaptation in the archaeological and historical record—evidence that has limited direct application to the present. By combining evidence of adaptative, capacities, limits and barriers, and path dependencies with existing planning, a range of potential scenarios could be tested. What if scenarios based on historical evidence could be tested using such scenarios in order to anticipate the unanticipated consequences of different adaptive strategies and physical and socio-economic constraints on the adaptive pathway (see Fig. 1). As noted earlier, the contingency of human responses

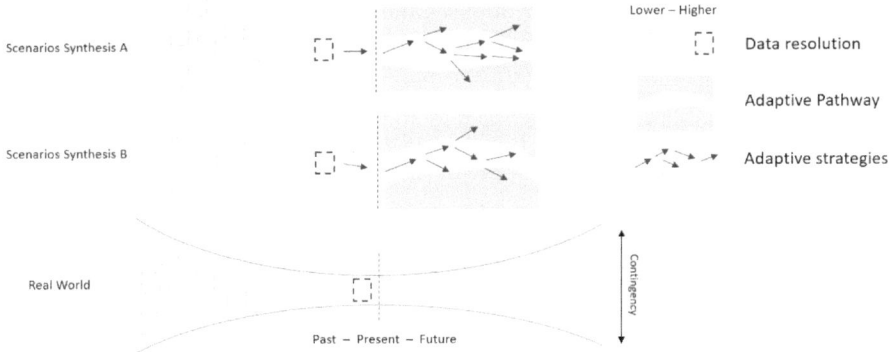

Fig. 1 Framework for scenario synthesis that combined historical scenarios and contemporary adaptive planning to consider a range of qualitative scenarios

to environmental change and the challenge of measuring vulnerability and adaptation are significant barriers to effective adaptive planning (Global Center on Adaptation 2020; Ford et al. 2018). Completed experiments of history can engage a largely unexploited network of disciplines capable of influencing co-production and planning of futures scenarios and anticipating barriers and limits to adaptation (see Hulme 2011).

Finally, it should be noted that adaptive planning, and environmental sustainability and sustainable development more broadly, are intergenerational challenges that, by definition, require an understanding of the interplay between continuity, including path dependent behaviors, and social and environmental change. Progressive approaches to policy and governance, including the Welsh government's Minister for Further Generations, already incorporate measures that think beyond current generations. Historical disciplines potentially have a significant and growing role to play in support of these foresight and planning efforts.

The Problems of Sustainability, Resilience, Transformation

Investigation of the human dimensions of climate change is a vast and contested field of study (Wigley et al. 1981; Huntington et al. 2007; Castree et al. 2014). The contestation of the field has been associated, in part, with the limited inclusion of the social sciences and humanities in global environmental assessments, such as the Intergovernmental Panel on Climate Change (IPCC) and the Intergovernmental Science-Policy Platform on Biodiversity and Ecosystem Services (IPBES) (Hartman 2015; 2020), which has spurred debate about the limited consensus on global change terminology and associated concepts, especially among marginalized disciplines and knowledge domains (Hulme 2011; Castree 2017; Castree et al. 2021). The result has been a succession of pervasive but often nebulous concepts that are indeed contested in specific disciplinary contexts, but insufficiently theorized and debated across disciplinary boundaries and sectors. *Sustainability, vulnerability, adaptation, resilience,* and *transformation,* as indeed the trope of *collapse,* are dominant concepts that receive significant credence and use, but minimal clarification across subjects focusing on impacts and adaptation to climate change (Blewitt 2018). Here we provide a brief micro-lineage of these contested terms and the challenges associated with their use.

Though the core ideas encapsulated in the concept of *sustainable development* were already current in the 1970s, arguably earlier in specific contexts (e.g. nature conservation and preservation), the term was first coined in 1980 in the *World Conservation Strategy: Living Resource Conservation for Sustainable Development,* published by the International Union for Conservation of Nature and Natural Resources, in cooperation with the World Wildlife Fund, the United Nations Environmental Programme, the Food and Agriculture Organization of the United Nations and UNESCO. The concept of sustainable development was distilled and popularized in more memorable form, however, in the 1987 report of the Brundtland Commission, *Our Common Future,* which characterized it as development that "meets the needs of the present without compromising the ability of future generations to meet their own needs" (WCED). Since then the term has been the subject of significant debate. One refinement of the concept has been a growing preference in many fields of study and social endeavor for the term *sustainability* over the original term owing to the colonialist, exploitative, extractivist First-World/Third-World or Developed/Undeveloped World dichotomies on which the concept of *sustainable development* is so evidently predicated. After all, who or what is developing in the implicit subjectivity of the phrase?—or into whom or what else is it being developed in its implicit non-agentive passivity? The answer is hiding in plain sight, as the so-called undeveloped or underdeveloped world (more fully represented in the Global South) contrasts with a developed world (the more highly industrialized Global North) whose unsustainable growth, rapacious appetite for overconsumption and poorly restrained fossil-fuel energy burning during much of the 20th and now the early twenty-first centuries have led to a projected monthly average of well over 420 ppm of carbon dioxide in the Earth's atmosphere in 2021 (NOAA 2022).

The inherent challenge for sustainability and sustainable development has been the contradiction between economic growth and ecological health and flourishing, as well as the balance of meeting human needs within ecological constraints (Mahoney et al. 2022). The challenge of balancing needs with sustainable limits is common throughout definitions of sustainable development and reflected in Stefanovic's (2000) 'mediative thinking,' which balances antecedent factors shaping ecological constraints, economic development, social organization, and environmental values. Blewitt (2018) argues that the ambiguity of sustainability and sustainable development is also a strength, as a heuristic for balancing society, economy, and the environment.

Since the turn of the millennium, the literature on adaptation and vulnerability to climate change has boomed (Bassett and Fogelman 2013), but the use of this terminology has remained pervasive. Although the literature has grown significantly in recent years, as Bassett and Fogelman (2013) explain, the characterizations of adaptation as 'adjustment' by the hazards school and 'reformist' by the political economy school have been dominant counterbalancing discourses since the 1970 and 1980s. These approaches to adaptation have been closely associated with vulnerability, with 'adjustment' adaptations tending to view vulnerability as the outcome of climate impacts, while 'reformist' adaptations are seen as the social and political contexts in which vulnerability to climate impacts are created (O'Brien et al. 2007). A more recent trend has been the view of 'transformative' adaptation as a corollary of an inadequate focus on the social context of vulnerability in political and economic systems (O'Brien 2018).

Transformation, as defined in IPCC *Assessment Report 5* (2014), is regarded as a "change in the fundamental attributes of natural and human systems… [reflecting] strengthened, altered, or aligned paradigms, goals, or values towards promoting adaptation for sustainable development". Feola (2015) draws attention to the conceptual vagueness and the remaining pervasiveness of the social transformation concept, noting that 'transformation' is largely used as a metaphor for fundamental change within a smaller problem-based literature, with the plural form reflecting a plural understanding of the concept. Transformation has, in many ways, inherited the same idiomatic pervasiveness that Anderson (2015) identifies in the concept of resilience, but it also has the potential to clarify concepts such as adaptation and address the weaknesses of resilience theory. The plurality and nuance that can be brought to these concepts by different disciplinary and theoretical perspectives has the potential to be a great strength, in both theoretical and applied contexts. However, these very qualities can also become derailing weaknesses when defining criteria are not carried over from one discipline or community of practice to another. Instead of stringent conceptual distinctions (reflecting how such terms may have been introduced in their original scientific contexts) that are rigorously upheld through close dialogue and mutual literacy-building activities among and across distinct disciplinary and research domains and other user communities, terms and concepts such as *sustainability, transformation, resilience* and *collapse* can be compromised by broad, loose or metaphorically fuzzy application in different parts of the science-policy-governance interface. The kinds of conceptual, semantic and even methodological slippages that occur as the terms cycle through a process from Mode 1 to Mode 2 research, feeding back into new Mode 1 contexts, effectively lead to a situation whereby these terms can mean a little of everything and, in other contexts, a whole lot of nothing the more they are used, generalized and popularized.

These terms have varied meanings for different users and communities of practice, which effectively leaves them devoid of universally comprehensible meanings over time the more they are coined in ever widening contexts and transactions. Series of semantic slippages occur between user communities, in the move from Mode 1 research to the science policy-interface, then onward to Mode 2 research. In new interfaces from science to policy to governance, planning and social-environmental management they become, progressively, ever more poorly signifying signifiers through these transactions of use and abuse. In the end, terms such as sustainability, resilience, transformation and adaptation move so far from their definitions as set out and refined when first they were coined in their original scientific contexts that, through various feedback loops (usage by policymakers, different societal stakeholders with different and sometimes even opposing interests and priorities), their meanings as established originally in rigorous scientific discourse/s is effectively washed away. Hence the phenomena of greenwashing (the 'Good Anthropocene', 'sustainable work,' etc.) or empty policy jargon easily co-opted for the transactional use *de jour*.

We highlight this pervasiveness and terminological confusion in order to signal an opportunity offered by the increased engagement of history, archaeology, and the environmental humanities in global change research (see Hartman 2015, 2020; Hartman et al. 2017; Holm et al. 2015; Jackson et al. 2018). The role of historical disciplines, including archaeology, has been highlighted in identifying cross-scale interactions (between society and the environment) and longitudinal (process-based) understandings of vulnerability that address some of the existing conceptual inadequacies (Ford et al. 2018). More active multilateral conversations and research collaborations among established global change research disciplines and historically under-engaged disciplines in qualitative humanities and social sciences, especially transdisciplinary site-based research approaches such as those found in historical ecology (Crumley 1994, 2007, 2018a, b) richly blending (qualitative and quantitative) research methodologies, can provide valuable footholds and models of collaboration to help overcome, and perhaps even begin to remediate, problems of knowledge exchange deficit through transactional semantic slippage of terminology and conceptual abuse.

Conclusion

The ambition to use past cases of environmental crisis and social-ecological change in efforts to address future risks has perhaps never been greater than in the present era of rapid environmental and social change. However, this ambition requires that researchers and scholars in the historical sciences examine, and face head on, the implications of any number of sticky questions. In concluding we suggest some open-ended questions as the basis for closer scrutiny, further analysis and continued creative engagement:

1. What role can past cases of social-ecological system disturbance play in helping scientists, policymakers and environmental managers address present and future vulnerabilities?

    The relevance of archaeological and historical information, as we have discussed in this chapter, has all too often been lost in the hyperbole of collapse discourse, but to learn from the past a clear dialogue between historical researchers and global change science, policy and practice is needed.

2. Why is it important to bring historical cases of social and environmental change into our efforts to address present and future challenges posed by processes of abrupt change?

    Historical studies scholars might regard the answer to this question as self-evident; however, in light of the insufficient involvement of historians, archaeologists and other deeper-time historical studies scholars in scientific assessment, the policy-science interface and governance efforts more generally focused on the challenges of global change, it seems anything but clear where and how historical casework may fit into broader efforts to apply knowledge of past changes to 21st century risk and vulnerability mitigation work as it is actually carried out today.

3. What clear limits might there be to the use of historical cases as guides to help define, address and plan for vulnerability scenarios?

    The most obvious limitations for historical disciplines have been the pitfalls of historicism and environmental determinism (see Popper 1956; Hulme, 2011), but they are all too easy to fall into. It is far easier to draw direct analogies that ignore context, nuance and socio-environmental complexity. However, there is far more value to be gained from a wide range of scenarios and social situations than (over) simplistic lessons that neglect human agency and the co-evolution of culture, society and the environment.

4. How can we as historical research scholars take account of those limitations to avoid producing noise rather than genuinely useful knowledge (a usable past) in efforts to better prepare for foreseeable risks, even regime-changing system shocks?

    Active engagement with researchers of contemporary global environmental change to co-produce useable knowledge is a pro-active opportunity for historical researchers to identify gaps and solutions using historically informed but contextualized knowledge (see Jackson et al. 2018).

5. What can we do to more readily recognize and avoid disjunctures in scientific/theoretical versus practical/applied use of central concepts from sustainability studies and resilience science, especially if these effectively lead to the prevalence of such noise among diverse key communities of practice?;

    Some obvious cases in point are the ways in which the concepts of "resilience" or "collapse" are employed in the respective fields of sustainability science or historical sciences, on the one hand, versus how they tend to be understood and acted upon by policymakers, planners or emergency response agencies on the other hand. As suggested in Ford et al. (2018), we should seize the opportunity to clarify and improve our understanding of concepts in sustainability studies and resilience science through engagement with historical disciplines. These completed natural experiments can test the applications of concepts to different geographical and social situations and extend the temporal horizon of concepts such as vulnerability, adaptation and resilience.

6. What steps can help refine the ways in which historical cases may be used to better understand and plan for vulnerability scenarios?

    Where, in other words, can we efficiently direct our efforts to operationalize historical cases so that they can help scientists, scholars from non-historical disciplines, policymakers, societal planners and environmental managers understand the human-dimensions of environmental change?
7. Can we build viable wider-purpose toolkits for emergency response scenarios or social-ecological system planning from specific lessons and case studies of past environmental change at various scales?

    Or do we need to be thinking in other ways that take better account of the multi-scalar complexity and irreproduceability of natural, social and technological causes and effects in specific (always contingent) contexts of social-ecological system change.
8. Can we provide different kinds of knowledge takeaways from historical cases to policymaking and governance actors?

    Is there a way to offer greater qualitative layering or contextual nuance to risk and vulnerability scenarios derived from models of system change according to the dominant models of resilience theory and sustainability science in use today?
9. What might these look like ideally and how might they function?

    What future does the past have in our efforts to better prepare for the major societal challenges that await us?

References

Adamson GC, Hannaford MJ, Rohland EJ (2018) Re-thinking the present: the role of a historical focus in climate change adaptation research. Glob Environ Chang 48:195–205
Aimers JJ (2007) What maya collapse? Terminal classic variation in the Maya lowlands. J Archaeol Res 15(4):329–377
Alexander D (1997) The study of natural disasters, 1977–97: Some reflections on a changing field of knowledge. Disasters 21(4):284–304
Alexander DE (1995). A survey of the field of natural hazards and disaster studies. In: Geographical information systems in assessing natural hazards. Springer, Dordrecht, pp 1–19
Alexander DE (2010) The L'Aquila earthquake of 6 April 2009 and Italian Government policy on disaster response. J Nat Resour Policy Res 2(4):325–342
Alexander DE (2013) Resilience and disaster risk reduction: an etymological journey. Nat Hazards Earth Syst Sci 13(11):2707–2716
Allana S, Clark A (2018) Applying meta-theory to qualitative and mixed-methods research: a discussion of critical realism and heart failure disease management interventions research. Int J Qual Methods. https://doi.org/10.1177/1609406918790042
Allen CR, Angeler DG, Chaffin BC, Twidwell D, Garmestani A (2019) Resilience reconciled. Nat Sustain 2(10):898–900
Anderson B (2015) What kind of thing is resilience?. Politics 35(1):60–66

Barrett JH (2018) Medieval fishing and fish trade. In: Gerrard C, Gutiérrez A (eds) The Oxford handbook of later medieval archaeology in Britain. Oxford, Oxford University Press

Barry A, Born G (2013) Interdisciplinarity: reconfigurations of the social and natural sciences. In: Barry A, Born G (eds) Interdisciplinarity: reconfigurations of the social and natural sciences. London, Routledge

Bassett TJ, Fogelman C (2013) Déjà vu or something new? The adaptation concept in the climate change literature. Geoforum 48:42–53

Blewitt J (2018) Understanding sustainable development. Routledge, Oxon

Bocinsky RK, Kohler T (2014) A 2000-year reconstruction of the rain-fed maize agricultural niche in the US Southwest. Nat Commun 5:5618. https://doi.org/10.1038/ncomms6618

Boivin NL, Zeder MA, Fuller DQ, Crowther A, Larson G, Erlandson JM, Denham T, Petraglia MD (2016) Ecological consequences of human niche construction: examining long-term anthropogenic shaping of global species distributions. Proc Natl Acad Sci 113(23):6388–6396

Bostrom N, Milan C (eds) (2008) Global catastrophic risks. Oxford UP

Brewington S, Hicks M, Edwald et al (2015) Islands of change vs. islands of disaster: managing pigs and birds in the Anthropocene of the North Atlantic. The Holocene 25(10):1676–1684

Burke P (2015) The French historical revolution: the annales school 1929–214, 2nd edn. Polity, Cambridge

Butzer KW (2012) Collapse, environment, and society. Proc Natl Acad Sci 109(10):3632–3639

Butzer KW, Endfield GH (2012) Critical perspectives on historical collapse. Proc Natl Acad Sci 109(10):3628–3631

Carr EH (1964) What is history? Penguin, Hammondsworth

Castree N (2014) The anthropocene and the environmental humanities: extending the conversation. Environ Humlties 5(1):233–260

Castree N (2016) Geography and the new social contract for global change research. Trans Inst Br Geogr 41(3):328–347

Castree N (2017) Global change research and the "People Disciplines": towards a new dispensation. South Atlantic Quarterly 116(1):55–67

Castree N, Adams WM, Barry J, Brockington D, Büscher B, Corbera E, Demeritt D, Duffy R, Felt U, Neves K, Newell P (2014) Changing the intellectual climate. Nat Clim Chang 4(9):763–768

Castree N, Bellamy R, Osaka S (2021) The future of global environmental assessments: making a case for fundamental change. Anthropocene Rev 8(1):56–82

Chakrabarty D (2009) The climate of history: four theses. Crit Inq 35(2):197–222

Clarke L (2007) Thinking possibilistically in a probabilistic world. Significance 4(4):190–192

Clarke L (2008) Possibilistic thinking: a new conceptual tool for thinking about extreme events. Soc Res 669–690

Collins HM, Evans R (2002) The third wave of science studies: studies of expertise and experience. Soc Stud Sci 32(2):235–296

Cooper J, Sheets P (eds) (2012) Surviving sudden environmental change: lessons from archaeology. University of Colorado Press, Boulder

Crumley CL (ed) (1994) Historical ecology: cultural knowledge and changing landscapes. School of american research advanced seminar. Santa Fe: School of American Research

Crumley CL (2007) Historical ecology: integrated thinking at multiple temporal and spatial scales. The world system and the earth system: global socio-environmental change and sustainability since the Neolithic. Alf H, Carole C (eds), pp 15–28. Walnut Creek CA, Left Coast Press

Crumley CL (2018a) Historical ecology and the longue durée. In: Paul S, Jon M, Carole L, Crumley C, Carole L, Tommy L, Anna W (eds) Issues and concepts in historical ecology: the past and future of landscapes and regions, pp 13–40. Cambridge GB, Cambridge University Press

Crumley CL (2018b) New paths into the anthropocene: applying historical ecologies to the human future. The oxford handbook of historical ecology and applied archaeology, Isendahl C, Stump D (eds), pp 6–20. Oxford, Oxford University Press. (Online 2015, published)

d'Alpoim Guedes J, Crabtree SA, Bocinsky RK, Kohler TA (2016) Twenty-first century approaches to ancient problems: climate and society. Proc Natl Acad Sci 113(51):14483–14491

Dawdy SL (2009) Millennial Archaeology: Locating the Discipline in the Age of Insecurity/Doomsday Confessions. Archaeological Dialogues 16(2):131–142

Dawson T (2015) Eroding archaeology at the coast: How a global problem is being managed in Scotland, with examples from the Western Isles. in Journal of the North Atlantic: 2010 Hebridean Archaeology Forum. vol. Special Volume 9, Eagle Hill Publications, Steuben, Maine United States, pp. 83–98. https://doi.org/10.3721/037.002.sp905

Dawson TC, Nimura C, Lopez-Romero E, Daire M-Y (eds) (2017) Public archaeology and climate change. Oxbow, Oxford

Diamond J (1997) Guns, germs and steel. Vintage Classics, London

Diamond J (2005) Collapse: how societies choose to fail or survive. Penguin, London

Diamond J, Robinson JA (2010) Natural Experiments of History. Harvard University Press

Dietz T, Ostrom E, Stern PC (2003) The struggle to govern the commons. Science 302(5652):1907–1912

Donovan A and Oppenheimer C (2011) The 2010 Eyjafjallajökull eruption and the reconstruction of geography. Geogr J 177(1):4–11

Donovan A and Oppenheimer C (2012) Governing the lithosphere: Insights from Eyjafjallajökull concerning the role of scientists in supporting decision-making on active volcanoes. J Geophys Res 117:B03214

Dow K, Berkhout F, Preston BL, Klein RJT, Midgley G, Shaw MR (2013) Limits to adaptation. Nat Clim Change 3(4):305–307

Dugmore A, Vésteinsson O (2012) Black sun, high flame, and flood: volcanic hazards in Iceland. Surviv Sudd Environ Chang 67–90

Dugmore AJ, Thomas HM, Richard S, Christian KM, Konrad S, Christian K (2013) 'Clumsy solutions' and 'Elegant failures': lessons on climate change adaptation from the settlement of the North Atlantic islands in: a changing environment for human security: transformative approaches to research, policy and action. Linda S, O'Brien K, Johanna W (eds) Chapter 38. Routledge, UK, London

Dunne JA, Maschner H, Betts MW, Huntly N, Russell R, Williams RJ, Wood SA (2016) The roles and impacts of human hunter-gatherers in North Pacific marine food webs. Sci Rep 6(1):1–9

Dunning T (2008) Improving causal inference: strengths and limitations of natural experiments. Polit Res Q 61(2):282–293

Dunning T (2012) Natural Experiments in the Social Sciences. Cambridge University Press, Cambridge

Engelhard GH, Thurstan RH, MacKenzie BR, Alleway HK, Bannister RCA, Cardinale M, Clarke MW, Currie JC, Fortibuoni T, Holm P, Holt SJ, Mazzoldi C, Pinnegar JK, Raicevich S, Volckaert FAM, Klein ES, Lescrauwaet A-K (2015) ICES meets marine historical ecology: placing the history of fish and fisheries in current policy context. ICES J Mar Sci. https://doi.org/10.1093/icesjms/fsv219

Erlandson JM, Braje TJ (2013) Archeology and the Anthropocene. Anthropocene 4:1–7

Feola G (2015) Societal transformation in response to global environmental change: a review of emerging concepts. Ambio 44(5):376–390

Flecknoe D, Wakefield BC, Simmons A (2018) Plagues & wars: the 'Spanish Flu' pandemic as a lesson from history. Med Confl Surviv 34(2):61–68. https://doi.org/10.1080/13623699.2018.1472892

Ford JD, Pearce T, McDowell G, Berrang-Ford L, Sayles JS, Belfer E (2018) Vulnerability and its discontents: the past, present, and future of climate change vulnerability research. Clim Change 151(2):189–203

Fuchs S, Heiss K, Hübl J (2007) Towards an empirical vulnerability function for use in debris flow risk assessment. Nat Hazards Earth 7(5):495–506

Geels FW (2012) A socio-technical analysis of low-carbon transitions: introducing the multi-level perspective into transport studies. J Transp Geogr 24:471–482

Global Center on Adaptation (2020) State and trends in adaptation report 2020. https://gca.org/wp-content/uploads/2021/03/GCA-State-and-Trends-Report-2020-Online-3.pdf. Last Accessed 11 Aug 2021

Haider LJ, Schlüter M, Folke C, Reyers B (2021) Rethinking resilience and development: a coevolutionary perspective. Ambio 1–9

Haldon J, Mordechai L, Newfield TP, Chase AF, Izdebski A, Guzowski P, Labuhn I, Roberts N (2018) History meets palaeoscience: Consilience and collaboration in studying past societal responses to environmental change. Proc Natl Acad Sci 115(13):3210–3218

Hambrecht G, Anderung C, Brewington S, Dugmore A, Edvardsson R, Feeley F, Gibbons K, Harrison R, Hicks M, Jackson R, Ólafsdóttir GÁ, Rockman M, Smiarowski K, Streeter R, Szabo V, McGovern T (2018) Archaeological sites as distributed long-term observing networks of the past (DONOP). Quatern Int. https://doi.org/10.1016/j.quaint.2018.04.016

Hartley LP (1953) The go-between. Hamish Hamilton

Hartman S, Ogilvie AEJ, Ingimundarson JH, Dugmore AJ, Hambrecht G, McGovern TH (2017) Medieval Iceland, Greenland and the new human condition: a case study in integrated environmental humanities. Glob Planet Chang 156(2017):123–139. https://doi.org/10.1016/j.gloplacha.2017.04.007

Hartman S (2015) Unpacking the black box: the need for integrated environmental humanities. Future earth blog. Future earth, 3 June 2015. Web. 14 May, 2021. https://futureearth.org/2015/06/03/unpacking-the-black-box-the-need-for-integrated-environmental-humanities-ieh/

Hartman S (2016) Revealing environmental memory: what the study of medieval literature can tell us about long-term environmental change. In: Biodiverse Nr, 2 2016 (online)

Hartman S (2020) Into the fray: a call for policy-engaged and actionable environmental humanities. Ecozon@ 11(2):187–199

Hegmon M (ed) (2017) The give and take of sustainability: archaeological and anthropological perspectives on tradeoffs. United Kingdom, Cambridge University Press

Hendrickson DC, Mcmaster HR (1997) Dereliction of duty: Johnson, Mcnamara, the joint chiefs of staff, and the lies that led to Vietnam. Foreign Aff 76:153

Hess GR (1994) The unending debate: historians and the Vietnam war. Dipl Hist 18(2):239–264. https://doi.org/10.1111/j.1467-7709.1994.tb00612.x

Hicks M, Einarsson Á, Anamthawat-Jónsson K, Edwald Á, Friðriksson A, Þórsson ÆÞ, McGovern TH (2016) Community and conservation: documenting millennial scale sustainable resource use at lake Mývatn Iceland. In: Isendahl C, Stump D (eds) Handbook of historical ecology and applied archaeology. Oxford University Press. https://doi.org/10.1093/oxfordhb/9780199672691.013.36

Hilding-Rydevik T, Jon M, Carina G (2018) Baselines and the shifting baselines syndrome-exploring frames of reference in nature conservation. In: Carole C, Tommy L, Anna W (2018) Issues and concepts in historical ecology: the past and future of landscapes and regions, pp 112–144. Cambridge UP

Hochrainer-Stigler S, Colon C, Boza G, Brännström Å, Linnerooth-Bayer J, Pflug G, Dieckmann U (2020) Measuring, modeling, and managing systemic risk: the missing aspect of human agency. J Risk Res 23(10):1301–1317

Hoffman SM, Oliver-Smith A, Button GV (2001) Catastrophe & culture: the anthropology of disaster. School of American Research Press

Holm P, Adamson J, Huang H, Kirdan L, Kitch S, McCalman I, Ogude J, Ronan M, Scott D, Thompson KO, Travis C, Wehner K (2015) Humanities for the environment—a manifesto for research and action. Humanities 4(4):977–992

Hudson MJ, Aoyama M, Hoover KC, Uchiyama J (2012) Prospects and challenges for an archaeology of global climate change. Wiley Interdiscip Rev Clim Change 3:313–328

Hulme M (2011) Meet the humanities. Nat Clim Change 1(4):177–179

Huntington HP, Hamilton LC, Nicholson C, Brunner R, Lynch A, Ogilvie AEJ, Voinov A (2007) Towards understanding the human dimensions of the rapidly changing Arctic system: insights and approaches from five HARC projects. Reg Environ Change 7(4):173–186

Hynes W, Trump B, Love P, Linkov I (2020) Bouncing forward: a resilience approach to dealing with COVID-19 and future systemic shocks. Environ Syst Decis 40:174–184

Hynes W, Trump BD, Kirman A, Haldane A, Linkov I (2022) Systemic resilience in economics. Nature Physics 18(4):381–384

ICOMOS (2019) The future of our pasts: engaging cultural heritage in climate action. ICOMOS Climate Change and Heritage Working Group

Inkpen R (2016) 'Riskscapes' as a heuristic tool for understanding environmental risks: the Eyjafjallajokull volcanic ash cloud of April 2010. Risk Management 18(1):47–63

IPCC (2014) Climate change 2014: synthesis report. Contribution of working groups I, II and III to the fifth assessment report of the intergovernmental panel on climate change (Core Writing Team, Pachauri RK, Meyer LA (eds)), p 151. IPCC, Geneva, Switzerland

IRGC (2018) IRGC guidelines for the governance of systemic risks. International Risk Governance Center (IRGC)

IUCN (1980) World conservation strategy: living resource conservation for sustainable development. International Union for Conservation of Nature and Natural Resources (IUCN-UNEP-WWF)

Iuga A, Anna W, Bogdan I, Monica S, Haakan T (2018) Rural communities and traditional ecological knowledge. In: Carole C, Tommy L, Anna W (eds) Issues and concepts in historical ecology: the past and future of landscapes and regions, pp 84–112. Cambridge UP

Izdebski A, Holmgren K, Weiberg E, Stocker SR, Buentgen U, Florenzano A, Gogou A, Leroy SA, Luterbacher J, Martrat B, Masi A (2016) Realising consilience: how better communication between archaeologists, historians and natural scientists can transform the study of past climate change in the Mediterranean. Quat Sci Rev, 136:5–22

Jackson R, Dugmore A, Riede F (2017) Towards a new social contract for archaeology and climate change adaptation. Archaeol Rev Camb 32(2):197–221

Jackson RC, Dugmore AJ, Riede F (2018) Rediscovering lessons of adaptation from the past. Glob Environ Chang 52:58–65

Jin AS, Trump BD, Golan M, Hynes W, Young M, Linkov I (2021) Building resilience will require compromise on efficiency. Nature Energy 6(11):997–999

Keiler M, Sailer R, Jörg P, Weber C, Fuchs S, Zischg A, Sauermoser S (2006) Avalanche risk assessment–a multi-temporal approach, results from Galtür, Austria. Nat Hazards Earth Syst Sci 6(4):637–651

Kintigh, K, Altschul, J, Beaudry, M, Drennan, R, Kinzig, A, Kohler, T, Limp, W, Maschner, H, Michener, W, Pauketat, T, Wright, H and Zeder, M. (2014a). Grand Challenges for Archaeology. Proceedings of the National Academy of Sciences 111:879–880

Kintigh KW, Altschul JH, Beaudry MC, Drennan RD, Kinzig AP, Kohler TA, Limp WF, Maschner HDG, Michener WK, Pauketat TR, Peregrine P, Sabloff JA, Wilkinson TJ, Wrights HT, Zeder MA (2014b) Gran challenges for archaeology. Am Antiq 79(1):5–24

Kohler TA (2012) Social evolution, hazards, and resilience: some concluding thoughts. In Cooper J, Sheets P (eds) Surviving sudden environmental change: answers from archaeology. Boulder, University of Colorado Press

Kohler TA, Rockman M (2020) The IPCC: a primer for archaeologists. Am Antiq 85(4):627–651

Kohler TA, Smith ME (2018) Ten thousand years of inequality. Arizona University Press

Kohler TA, Bocinsky RK, Cockburn TA, Robinson CE, Kane AE (2012) Modelling Prehispanic Puebo societies in their ecosystems. Ecol Model 241:30–41

Khong K, Yuen F (1990) Analogies at war: Korea, Munich, Dien Bien Phu, and the Vietnam decisions of 1965. Princeton

Kuklicke C, Demeritt D (2016) Adaptive and risk-based approaches to climate change and the management of uncertainty and institutional risk: the case of future flooding in England. Glob Environ Chang 37:56–68

Kwok R (2017) Historical data: hidden in the past. Nature 549(7672):419–421

Lawson IT, Swindles GT, Plunkett G, Greenberg D (2012) The spatial distribution of Holocene cryptotephras in northwest Europe since 7 ka: implications for understanding ash fall events from Icelandic eruptions. Quat Sci Rev 41:57–66

Levy JS (2008) Counterfactuals and case studies. In: Box-Steffensmeier JM, Brady HE, Collier D (eds) The Oxford handbook of political methodology. Oxford, Oxford University Press

Linkov I, Trump B, Kiker G (2022a) Diversity and inclusiveness are necessary components of resilient international teams. HumIties Soc Sci Commun 9(1):1–5

Linkov I, Trump B, Trump J, Pescaroli G, Mavrodieva A, Panda A (2022b) Stress-test the resilience of critical infrastructure. Nature 603(7902): 578–578

Loyn D (2009) In Afghanistan: two hundred years of british, Russian, and American Occupation. Macmillan Press NY

Lund KA, Benediktsson K (2011) Inhabiting a risky Earth: the Eyjafjallajökull eruption in 2010 and its impacts. Anthropol Today 27(1):6–9

Mahony M, Hulme M (2012) The colour of risk: exploration of the IPCC's "Burning Embers" diagram. Spontaneous Gener J Hist Philos Sci 6(1):75–89

Mahoney E, Golan M, Kurth M, Trump BD, Linkov I. (2022). Resilience-by-Design and Resilience-by-Intervention in supply chains for remote and indigenous communities. Nature Communications 13(1): 1–5

Marks RB (2015) The origins of the modern world: a global and environmental narrative form the fifteenth to the twenty-first century. Rowman and Littlefield, London

Mazzorana B, Hübl J, Fuchs S (2009) Improving risk assessment by defining consistent and reliable system scenarios. Nat Hazards Earth Syst Sci 9(1):145–159

McAnany PA, Yoffee N (2010) Questioning collapse: human resilience, ecological vulnerability, and the aftermath of empire. Cambridge University Press, Cambridge

Meadows, Donnella H., Meadows, Dennis L., Randers, Jørgen and Behrens, William W. III (1972) The Limits to Growth: A report for the Club of Rome's project The Predicament of Mankind. Potomac Associates – Universe Books.

Meyer WJ, Crumley CL (2011) Historical ecology: using what works to cross the divide. Atlantic Europe in the first millennium BC: crossing the divide. In: Tom M, Lois A (eds), pp 109–134. Oxford, Oxford University Press

Middleton G (2017) Understanding collapse: ancient history and modern myths. Cambridge University Press, Cambridge

Middleton GD (2012) Nothing lasts forever: environmental discourses on the collapse of past societies. J Archaeol Res 20(3):257–307

Middleton GD (2017) The show must go on: collapse, resilience, and transformation in 21st-century archaeology. Rev Anthropol 46(2–3):78–105

Moen J, Keskitalo ECH (2010) Interlocking panarchies in multi-use boreal forests in Sweden. Ecol Soc 15(3):17

Moxnes JF, Christophersen OA (2008) The Spanish flu as a worst case scenario? Microb Ecol Health Dis 20(1):1–26. https://doi.org/10.1080/08910600701699067

Murphy JT, Crumley C (eds) (2021) If the past teaches, what does the future learn? Ancient urban regions and the durable future. Research in urbanism series, delft school of architecture. TU Delft OPEN Publishing

Nelson MC, Hegmon M, Kintigh KW, Kinzig AP, Nelson BA, Anderies JM, Abbott DA, Spielmann KA, Ingram SE, Peeples MA, Kulow S, Strawhacker CA, Meegan CA (2012) Long-term vulnerability and resilience: three examples from archaeological study in the Southwestern United States and Northern Mexico. In: Cooper J, Sheets P (eds) Surviving sudden environmental change: answers from archaeology. University of Colorado Press, Boulder

Nelson MC, Hegmon M, Kulow S, Schollmeyer KG (2006) Archaeology and ecological perspectives on reorganization: a case study from the Mimbres region of the US Southwest. Am Antiq 71(3):403–432

Nelson MC, Ingram SE, Dugmore AJ, Streeter R, Peeples MA, McGovern TH, Hegmon M, Spielmann KA, Simpson IA, Strawhacker C, Comeau LE, Torvinen A, Madsen CK, Hambrecht G,

Smiarowski K (2016) Climate changes, vulnerabilities, and food security. Proc Natl Acad Sci 113(2):298–303

Nelson MG, Kinzig AP, Arneborg J, Streeter R, Ingram SE (2017) 'Vulnerability to Food Insecurity: tradeoffs and Their Consequences'. In: Hegmon M (ed) The Give and Take of Sustainability: Archaeological and Anthropological Perspectives on Tradeoffs. Cambridge University Press, Cambridge

Nixon R (2011) Slow violence and the environmentalism of the poor. Harvard University Press, Cambridge, Mass

NOAA (2022) Carbon dioxide now more than 50% higher than pre-industrial levels. National Oceanic and Atmospheric Administration website. https://www.noaa.gov/news-release/carbon-dioxide-now-more-than-50-higher-than-pre-industrial-levels. 3 June 2022. Accessed 7 June 2022

November V (2008) Spatiality of risk. Environ Plan A 40(7):1523–1527

Nowotny H, Scott PB, Gibbons MT (2013) Re-thinking science: knowledge and the public in an age of uncertainty. Polity Press, Cambridge

O'Brien K (2018) Is the 1.5 C target possible? Exploring the three spheres of transformation. Curr Opin Environ Sustain 31:153–160

O'Brien K, Eriksen S, Nygaard LP, Schjolden ANE (2007) Why different interpretations of vulnerability matter in climate change discourses. Climate policy 7(1):73–88

Ogilvie AE (2010) Historical climatology, climatic change, and implications for climate science in the twenty-first century. Climatic Change 100(1):33–47

Oliver-Smith A, Hoffman SM, Hoffman S (eds) (2020) The angry earth: disaster in anthropological perspective. Routledge

Palma-Oliveira JM, Trump BD, Wood MD, Linkov I (2018) Community-driven hypothesis testing: a solution for the tragedy of the anticommons. Risk Anal 38(3):620–634

Popper K (1956) The poverty of historicism. Routledge, London

Redman CL (2005) Resilience theory in archaeology. Am Anthropol 107:70–77

Richer S, Stump D, Marchant R (2019) Archaeology has no Relevance. Internet Archaeol 53. https://doi.org/10.11141/ia.53.2

Riede F (2006) The Scandinavian connection: the roots of darwinian archaeology in 19th-century Scandinavian archaeology. Bull Hist Archaeol 16(1)

Riede F (2014a) Towards a science of past disasters. Natural Hazards 71(1):335–362

Riede F (2014b) Climate models: use in archaeology. Nature 513(7518):315

Riede F (2017) Past-Forwarding Ancient Calamities: Pathways for Making Archaeology Relevant in Disaster Risk Reduction Research, Humanities, vol. 6, no. 4, 79, pp. 1–25. https://doi.org/10.3390/h6040079

Riede F, Andersen P, Price N (2016) Does archaeology need an ethical promise? World Archaeol 48(4):466–481

Riede F, Jackson RC (2020) Do deep-time disasters hold lessons for contemporary understandings of resilience and vulnerability? The case of the laacher see volcanic eruption. In: Sheets P, Riede F (eds) Going forward by looking back: archaeological Perspectives on socio-ecological crisis, response, and collapse. Berghahn Books, New York

Rockman M (2003) 'Knowledge and learning in the archaeology of colonization'. In: Rockman M, Steele J. (eds) Colonization of Unfamiliar Landscapes: The archaeology of adaptation. Routledge, London

Rockman M (2012) 'The Necessary Roles of Archaeology in Climate Change Mitigation and Adaptation'. In: Rockman M, Flatman J (eds) Archaeology in Society. Springer, New York

Rockman M, Hritz C (2020) Expanding use of archaeology in climate change response by changing its social environment. Proc Natl Acad Sci 117(15):8295–8302. https://doi.org/10.1073/pnas.1914213117

Rolett B, Diamond J (2004) Environmental predictors of pre-European deforestation on Pacific islands. Nature 431(7007):443–446

Rounsevell MD, Metzger MJ (2010) Developing qualitative scenario storylines for environmental change assessment. Wiley Interdiscip Rev Clim Chang 1(4):606–619

Ruddiman WF (2013) The anthropocene. Annu Rev Earth Planet Sci 41:45–68

Ruddiman WF (2003) The anthropogenic greenhouse era began thousands of years ago. Clim Change 61:261–293

Ruddiman WF (2007) The early anthropogenic hypothesis: Challenges and responses. Rev Geophys 45(4). https://doi.org/10.1029/2006RG000207

Sachs J (2020) The Ages of Globalization. Columbia University Press, New York

Samuelson W, Zeckhauser R (1988) Status quo bias in decision making. J Risk Uncertain 1(1):7–59

Sigurðardóttir R, Newton AJ, Hicks MT, Dugmore AJ, Hreinsson V, Ogilvie AE, Júlíusson ÁD, Einarsson Á, Hartman S, Simpson IA, Vésteinsson O (2019) Trolls, water, time, and community: resource management in the Mývatn District of Northeast Iceland. In: Ludomir L, Thomas M (eds) Global perspectives on long term community resource management, pp 77–101. Springer Co, NY

Sinclair P, Moen J, Crumley CL (2018) Historical ecology and the longue durée. Crumley CL, Tommy L, Anna W (eds) Issues and concepts in historical ecology: the past and future of landscapes and regions, pp 13–40. Cambridge, UK, Cambridge University Press

Smith BD and Zeder MA (2013) The onset of the anthropocene. Anthropocene 4:8–13

Smith K (2013) Environmental hazards: assessing risk and reducing disaster. Routledge, London

Smith ML (2014) Citizen science in archaeology. Am Antiq 749–762

Spielmann K, Peeples MA, Glowacki DM, Dugmore A (2016) Early warning signals of social transformation: a case study from the US Southwest. PLoS One 11(10):1–18

Stefanovic IL (2000) Safeguarding our common future: rethinking sustainable development. New York University Press, New York

Steffen PJC (2003) How long have we been in the Anthropocene era? Clim Change 61(3):251

Steffen W, Broadgate W, Deutsch L, Gaffney O, Ludwig C (2015) The trajectory of the Anthropocene: the great acceleration. Anthropocene Rev 2(1):81–98

Steffen W, Grinevald J, Crutzen P, McNeill J (2011) The anthropocene: conceptual and historical perspectives. Philos Trans R Soc A Math Phys Eng Sci 369(1938):842–867

Stump D (2013) On applied archaeology, indigenous knowledge, and the usable past. Curr Anthropol 54:268–298

Stump D, Richer S (2017) 'Terraces are good, but sediment traps are better'. AAREA Policy Brief No. 1, University of York

Tainter JA (1988) The collapse of complex societies. Cambridge University Press, Cambridge

Tainter J (2003) A framework for sustainability. World Futur J New Parad Res 59(3–4):213–223

Thaler R and Sustein C (2008) Nudge: improving decisions about health, wealth, and happiness. Penguin, London

Trigger B (2006) A history of archaeological thought. Cambridge University Press, Cambridge

UN SDG (2021) Sustainable development goals decade of action. https://www.un.org/sustainabled evelopment/decade-of-action/. Last Accessed 17 May 21. (Online)

UNESCO (2020a) Cities, climate and culture: the urban research agenda in the upcoming IPCC Co-sponsored expert meeting on culture, heritage and climate change. https://whc.unesco.org/ en/events/1547/. Last Accessed 20 May 21. (Online)

UNESCO (2020b) Proposal for the establishment of BRIDGES as a MOST sustainability science coalition. https://unesdoc.unesco.org/ark:/48223/pf0000372656. Last Accessed 20 May 21. (Online)

Urry J (2003) Global complexity. Polity, London

Urry J (2007) Mobility. Polity, London

Waldman M (2013) System failure: the underlying causes of US policy-making errors in Afghanistan. Int Aff 89(4):825–843. https://doi.org/10.1111/1468-2346.12047

Walker B, Salt D (2012) Resilience thinking: sustaining ecosystems and people in a changing world. Island Press, London

Wallerstein IM (2004) World systems analysis: an introduction. Duke University Press, Durham

Webster CS (2005) The nuclear power industry as an alternative analogy for safety in anaesthesia and a novel approach for the conceptualisation of safety goals. Anaesthesia 60:1115–1122

Wigley TML, Ingram MJ, Farmer G (eds) (1981) Climate and history. Studies in past climates and their impacts on man, Cambridge University Press, Cambridge

Wise RM, Fazey I, Smith MS, Park SE, Eakin HC, Van Garderen ERMA, Campbell B (2014) Reconceptualising adaptation to climate change as part of pathways of change and response. Glob Environ Chang 28:325–336

WCED (1987) Our common future. World Commission on Environment and Development. New York and Oxford, Oxford UP

Yudkowsky E (2008) Cognitive biases potentially affecting judgement of global risks. In: Bostrom N, Milan C (eds) Global catastrophic risks, pp 91–120. Oxford, UP

Climate Change

Geoengineering and the Middle Ages: Lessons from Medieval Volcanic Eruptions for the Anthropocene

Martin Bauch

Abstract The existential challenge of mitigating anthropogenic climate change encouraged serious discussions on geoengineering approaches. One of them, Solar Radiation Management (SRM), would mean inserting aerosols into the atmosphere, thus imitating and perpetuating the cooling effects of large volcanic events, such as the 1815 Tambora eruption. However, artificially inserting sulphur aerosols into the atmosphere is connected with considerable uncertainties. One of them, pointed out by several climate scientists, is the different effects on temperature and precipitation in different parts of the globe. These are not the only ones, though. As the largest volcanic eruptions have taken place during the medieval times (ca 500–1500 CE), historical research can reveal further uncertainties in dating these eruptions and their connected socio-environmental effects, and hence on the actual climate and social impacts we might expect from SRM. A combination of humanist and scientific research on past volcanic eruptions therefore has the potential to produce a more precise understanding of past volcanic eruptions and their climatic consequences. As long as we do not acquire a consistent multi-disciplinary perspective on past volcanic eruptions, extreme caution should be taken before investing in geoengineering measures that include the artificial injection of sulphur aerosols in the atmosphere.

Keywords Volcanic eruptions · Geoengineering · Medieval history · Climate modelling · Sulphur injections

Geoengineering and the Anthropocene

The late chemist and Nobel Laureate Paul Crutzen (1933–2021) is well-known for having coined the term 'Anthropocene' in 2002, or at least for initiating the term's unhampered global success (Crutzen 2002). Once confined to the spheres of academia, the term has, over the course of the last decade, become a familiar part of the

M. Bauch (✉)
Leibniz Institute for the History and Culture of Eastern Europe, University of Leipzig, Leipzig, Germany
e-mail: martin.bauch@leibniz-gwzo.de

© The Author(s) 2022   111
A. Izdebski et al. (eds.), *Perspectives on Public Policy in Societal-Environmental Crises*, Risk, Systems and Decisions, https://doi.org/10.1007/978-3-030-94137-6_8

global discourse on environmental change and humans' role in the ongoing climate crisis (Ellis 2018), both in the public sphere and academic disciplines, amongst them history. But Crutzen did not stop at formulating a concept to describe how humanity could now exert influence as a geological force, altering hemispherical material flows. Just four years later, Crutzen published another short essay on the potential of implementing a technological solution to ongoing global warming (Crutzen 2006): 'Albedo Enhancement by Stratospheric Sulphur Injections'. The subtitle went on to claim that this might be a solution to our society's dilemma in climate policy, but the article also meant to break a taboo by suggesting that humans could inject stratospheric aerosols—preferably sulphur-based—using balloons or equivalent technical means into the higher strata of the atmosphere to counteract anthropogenic climate change (Robock 2014).

Volcanic Cooling and Solar Radiation Management

The basic idea behind this is simple and yet fascinating (Rampino et al. 1988): Large volcanic eruptions eject all kinds of dust and chemicals into the atmosphere, which reach even into the lower levels of the stratosphere to an altitude of up to 10 kms. While the relatively heavy volcanic ash falls down within a few weeks, the lighter material remains in the highest parts of the troposphere or even the lower parts of the stratosphere. Sulphur dioxide in particular reacts here with water to form tiny droplets of sulphuric acid. These very small droplets or solid particles called aerosols get caught up in moving air masses and are thus transported around the globe (Mather and Pyle 2015): a thin, yet effective, aerosol layer then deflects a considerable fraction of solar radiation before it can reach the ground. Any larger sulphur injections have the potential to influence global climate for up to three years. After this time period, the remaining aerosol particles sink to the ground and no longer influence global or regional climate (Oppenheimer and Donovan 2015; Timmreck 2018: 11–12).

The best-known historical case study for such a volcanic event (Schmidt and Robock 2015) was the eruption of Mount Tambora in Indonesia in 1815, which caused global temperatures to plunge and precipitation to increase in 1816 (and to a somewhat lesser extent in following years). This phenomenon has suggestively been described as 'the year without a summer', and as it is a particularly popular topic in the geosciences (Zeilinga de and Sanders 2002; Francis and Oppenheimer 2004), it is reasonable to assume a Tambora-like situation also came to the mind of researchers when thinking about the mitigation of anthropogenic climate change.

Crutzen's initiative from 2006 has ignited an ongoing public and scientific debate on the feasibility of using geoengineeering to mitigate climate change (Hamilton 2013; www.geoengineeringmonitor.org). A global overview of research projects on this so-called Solar Radiation Management (Boettcher et al. 2017) reveals an obvious concentration of these in anglophone countries such as the US, the UK and Australia (https://map.geoengineeringmonitor.org). Of these 23 projects, only seven propose the controlled injection of sulphur dioxide into the stratosphere to cool ground temperatures in an attempt to mimic the natural cooling effects of a big volcanic eruption. Notably, only three have actually conducted experiments on how to release sulphates, while the other four projects have relied on climate modelling to simulate the effects of artificial aerosol injections. Other projects have followed different paths of Solar Radiation Management: Some have preferred the injection of water droplets rather than sulphate aerosols, aiming to create brighter clouds that would deflect sunlight before it even reaches the ground. Other approaches aim to thin cirrus clouds to allow more heat to escape into space. And there are more ideas out there, such as enhancing ocean surface albedo by adding stable, nondispersive foam or microbubbles to the sea. Even more audacious studies have explored ideas such as clearing large areas of boreal forests in Russia to provoke a cooling effect, as snow-covered plains deflect sunlight more efficiently than snow-covered trees. So, stratospheric aerosol injection is just a small fraction of current research on solar radiation management, and, of course, a key concern is that it is not addressing the underlying increase in atmospheric greenhouse gases at all. In the political realm, it is criticized as being influenced by lobbyists supported by big corporations such as Shell and ExxonMobil with an urgent and vital interest in technology-based fixes to anthropogenic global warming. There are further concerns about and criticisms of the long-term governance of such aerosol injection projects and their possible weaponization (Costick and Ludlow 2020: 47–50).

All Kinds of Uncertainties

There are a number of climate- or climate-model-related uncertainties associated with stratospheric aerosol injection and its potential impact on human societies.

First, climate models are a crucial tool for understanding both future and past climatic changes and the role of climate drivers such as large volcanic eruptions, as these eruptions affect all components of the Earth System (Timmreck 2012; 2018: 10). Yet it is important to remember that modelling depends largely on instrumental and satellite measurements after the 1991 Pinatubo eruption, the best-researched and only large volcanic eruption with a considerable climatic impact in the twentieth century (Timmreck 2012: 548–549). Yet climate models still have difficulty reliably reproducing the well-known impact of the Pinatubo eruption, probably because many influencing factors on the actual volcanic impact such as the El Niño-Southern Oscillation are still poorly understood (Ibid.: 550–551; Cole-Dai 2010: 831).

Climate models are also key to estimating how artificial manipulation such as the injection of sulphate aerosols might influence global, regional and local climates. In particular, the finer scales down to regional and local weather impacts are hardly understood. Furthermore, climate reconstruction studies have focused on the changes in temperature after major volcanic eruptions while paying far less attention to the impacts on global and regional precipitation. A lot depends on the region where the volcanic eruption takes place: high-latitude eruptions in the Northern hemisphere (Kravitz and Robock 2011; Oman et al. 2005) change precipitation—for example, by weakening tropical monsoons—in a different way to tropical eruptions. Sulphate aerosol injections might produce winners and losers, or at least a rather unevenly distributed modification of climatic conditions over several years (Wegmann et al. 2014; Swingedouw et al. 2017).

This danger of uneven climate change—probably connected to a post-eruption effect of winter warming across the Northern parts of Eurasia (Zanchettin et al. 2013; Zambri et al. 2017)—has been clearly documented for 1816, the 'year without a summer', for while most of the continent faced severe cooling and continuous precipitation, Eastern Europe was hardly affected. Grain harvests failed in most parts of Europe (Post 1977), as an impressive memorial site, the so called Thanksgiving Box in the church of St. Michael in Schwäbisch Hall in southwest Germany demonstrates: four very small loaves of bread remember the famine year 1816, while the first wheat ears of the 1817 harvest were also preserved (Fig. 1). But summer temperatures were normal or even above average in Eastern Europe: Russian and Ukrainian grain producers profited from bountiful harvests that they sold to their starving neighbours at a considerable profit in 1816 and the following years. In fact, the rise of the port city of Odessa on the Black Sea (Herlihy 1986) has been associated with this economic success story after the Tambora eruption.

Fig. 1 'Erntedankkasten' (=Thanksgiving Box) in St. Michael, Schwäbisch Hall, Germany, presenting small bread loaves and wheat ears from 1816 and 1817. Picture: Klaus Graf, CC-BY SA 3.0, Wikimedia Commons

Beyond these uncertainties, previous research has focused on another crucial point: What we can learn from history is the irresolvable danger that, even after artificial sulphur injections in the atmosphere, assuming these could be controlled and targeted precisely, additional natural eruptions might change the overall-picture and drive global climatic conditions in highly problematic states as it happened, because of all-natural repeating eruptions, as was the case, for instance, in the sixth and seventh centuries (Costick and Ludlow 2020: 92–94).

The Role of Medieval History: More Precision

Medieval history has a special task in this context, as the greatest explosive volcanic eruptions of the past two millennia happened in the medieval period (Sigl et al. 2015; Toohey and Sigl 2017; Costick and Ludlow 2020; see also Fig. 2). First, a double eruption in the sixth century that triggered the so-called Late Antique Little Ice Age; then, eruption of the Samalas volcano in Indonesia in the mid-1250s; and, finally another, yet unidentified eruption event in the mid-fifteenth century, which was known as the Kuwae eruption until new research rightfully challenged the validity of this identification (Németh et al. 2007).

How do we know about these past volcanic eruptions? Without a single exception, and in stark contrast to the famous eruptions of the nineteenth century such as Tambora and Krakatoa (Simking and Fiske 1983; Winchester 2003), we do not have any written eyewitness accounts of medieval volcanic eruptions on a scale that could influence global climate. Narrative texts from the medieval period originating from the area of modern Indonesia do not provide concise chronologies, and other areas, such as premodern Melanesia, did not produce written records but relied on an oral tradition of genealogies that likewise do not provide reliable dating of reported

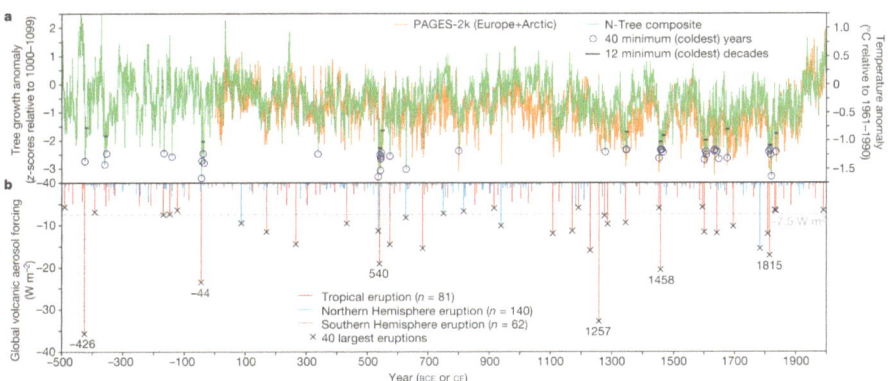

Fig. 2 Global volcanic aerosol forcing and Northern Hemisphere temperature variations for the past 2500 years. *Source* Sigl et al. 2015, Fig. 3

events. Even if we had more detailed information about the dates of volcanic eruptions in the premodern period, we could hardly estimate their influence on climate
from such accounts, for such deduction requires information regarding the amount
of sulphur injected into the atmosphere, something a medieval observer could hardly
quantify with naked eye.

The key to determining the magnitude of the sulphuric injection caused by a
specific eruption can be found in the polar ice shields (Cole-Dai 2010). The aerosols
of sulphuric acid fall in precipitation within a maximum of three years, most already
in the year after the eruption. The annual snowfall layers in polar regions, mainly
Greenland and Antarctica, are compressed into layers of ice, each containing distinctive chemical and physical traces, from Sahara dust to sulphuric acid. Scientists count
these layers backwards and measure these traces—for example, sulphuric acidity—
in each layer to create a chronology that provides us with a good record of past
volcanic eruptions that ejected large quantities of sulphur into the atmosphere and
hence, most probably, influenced global climate (Cole-Dai 2010).

There are, however, some inherent dating problems with this method: Snowfall
deposits and thus the amount of sulphuric acid in one specific ice core might vary
from place to place, so the only reliable information comes from cross-referencing
multiple ice cores to identify widespread sulphur peaks. Furthermore, researchers
cannot see the annual layers in ice cores that go back several hundred years, so
they have to measure seasonal variations of dust or chemicals in these ice cores to
identify annual layers. Eruption dates acquired using these methods may be off by two
to five years (Cole-Dai 2010: 828–829). So, it comes as no surprise that researchers
look for so-called stratigraphical markers: well-known volcanic events that should
have left a considerable peak of sulphur in all ice cores. These events are used to
synchronize peaks in different ice cores and validate their specific chronologies. For
the nineteenth century, the Krakatoa and Tambora events are precisely dated and
well documented in historical sources, but when it comes to the largest eruptions
of the last two millennia—we have mentioned them before: The so-called Kuwae
event in the fifteenth century, Samalas in the mid-thirteenth century, and a volcanic
double-event in the sixth century—historical evidence is much scarcer and yet even
more important.

Dating Uncertainty and Historical Sources

This is not to suggest that written sources can always be taken at face value, as
there are obviously pitfalls. Let me demonstrate this with regard to what is widely
accepted as the largest volcanic event of the last two millennia, the Samalas eruption
that is now dated to 1257, although it was once dated to 1258 or even 1259. A key
argument for dating the event to 1257 was advanced by Guillet et al. (2017) with
reference to historical sources: First, he evaluated chronicles from the whole of the
thirteenth century and counted the references to bad weather. In his results, 1258
stood out in an astonishing way, as we would expect after the most sulphur-rich

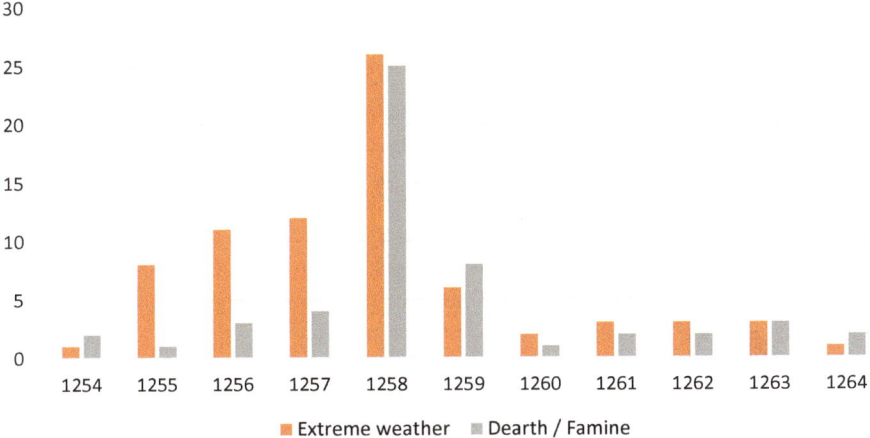

Fig. 3 Number of weather reports and accounts of dearth/famine per year. *Source* Bauch (2020)

eruption of the last two millennia. But there is a problem: While Guillet carefully collected all the relevant material from European chronicles for 1258, for the other years of the thirteenth century he relied on the important yet incomplete collection of Pierre Alexandre (1987), which focused exclusively on the territories of Continental Europe. Alexandre's collection does not include the British Isles, but the bulk of information on bad weather comes from English chroniclers. Excluding this body of sources for all years except 1258 is thus distorting the evidence from narrative sources to some extent. My alternative count (Bauch 2020) still shows a peak in 1258, but a period of extraordinarily bad weather set in as early as 1256 (Bauch 2020).

Furthermore, Guillet's argument relied on a quote from a German chronicle that referred to the year 1258 as '*munkeliar*', a strange vernacular term, the meaning of which is very unclear (Guillet et al. 2017, Supplementary Material, S2). Guillet followed faithfully the interpretation of the nineteenth-century editors of this chronicle, who interpreted *munkeliar* to mean a year of fog and darkness, which could refer to the visual effects of an unusually dense aerosol layer that year. With the Tambora eruption in mind and thinking of the year without a summer in 1816 this sounds very reasonable. However, a closer look at the quote in question (Bauch 2020: 220) reveals that the chronicler is not talking about weather or the sun's intensity but rather about the low quality of food, especially wine. Etymological dictionaries of wine-growing in Germany confirm that the verb *munkeln* means that wine tastes mouldy, which, of course, could be the result of the undisputed bad weather in 1258. The key question here is whether or not the bad weather started before 1258 and whether we can find more convincing descriptions of a volcanic aerosol layer. I argue elsewhere that we can indeed find such descriptions and probably should redate the eruption to 1256 or even 1255 (Bauch 2020).

What is the importance of all this for current discussions on geoengineering with solar radiation management? As we have seen before, modelling is a key feature when simulating the impact of past volcanic eruptions, and these climate models are based upon presumably precise data on eruption dates, sulphur injections, and subsequent temperature reductions. In the Samalas case, though, the maximum drop in global temperatures of the Northern hemisphere is detected in 1259 (Guillet et al. 2017: 126), while the geographical distribution of cooling in the northern hemisphere was rather uneven. If, in fact, there was a delay of two or even three years between an eruption and the related hemispherical cooling, this strengthens previous doubts about the strength and uniformity of the cooling effect of sulphur injections in the atmosphere (Timmreck et al. 2009). We should remain cautious about the predictive reliability of model simulations when the basic facts on the largest volcanic eruptions in the past, such as dates, sulphur loads and associated cooling, remain so unclear.

Let me provide another example from the Middle Ages: the second-largest volcanic eruption of the past two millennia has been identified in ice cores for the mid-fifteenth century (Fig. 2). It was long associated with the submarine caldera of Kuwae in Vanuatu (Gao et al. 2006) and had been dated to 1452. Rather baseless speculation (Pang 1993) even connected it with major historical events such as the fall of Constantinople in 1453. Recent volcanological research has precluded the association of Kuwae with the mid-fifteenth century event (Németh et al. 2007). Tree-ring and ice-core research has suggested there might have been two eruptions in close temporal proximity (Sigl et al. 2013; Cole-Dai et al. 2013): Probably one around 1453, and another one, according to ice-core analysis in 1458, while tree-ring research indicates a drop in temperature in 1453 and 1466 and hence an eruption in 1465 (Esper et al. 2017; Stoffel et al. 2015; Wilson et al. 2017). This paradox (Fig. 4)

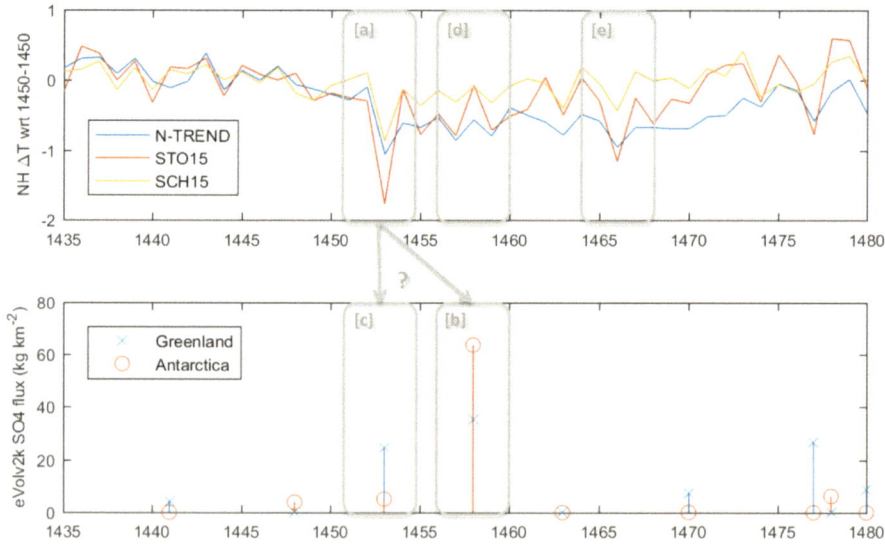

Fig. 4 Tree-ring-based reconstructed NH temperature anomalies (top) and average volcanic sulphate flux to Greenland and Antarctica (bottom), 1435–1480 Data from Toohey and Sigl 2017; Wilson et al. 2016; Stoffel et al. 2015; Schneider et al. (2015). Figure courtesy of Matthew Toohey

remains unsolved, especially as my analysis of historical sources (Bauch 2017) also points to 1465, with dramatic descriptions of a volcanic dust veil all across Europe in September 1464, while the climatic impact in 1453 is less well documented.

In a nutshell: The contradictory findings of ice-core research, dendroclimatology and history have not yet been reconciled. In other words, for none of the two of the largest volcanic eruptions of the past two millennia do we have consolidated evidence on eruption dates and hence on the connection between temperature decrease, precipitation and volcanic aerosols. What we should expect is, most of the time, a smooth fit of historical descriptions and dendroclimatological data, as both are annually resolved, yet often of different geographic origins and hence may not reflect the same local situations. Ice-core analysis is improving in the field of dating certainty, and yet still has its inherent uncertainties. But for any sulphur aerosol injection projects, we need a reliable attribution of specific eruptions and their reconstructed sulphur injections into the atmosphere. The work of traditional medievalists with narrative sources is crucial to the determination of best possible chronologies of past volcanic eruptions. Unless we can reasonably synchronize historical facts, ice-core data and tree-ring reconstructions, we should not assume that we have understood the complex climatic processes that followed large volcanic eruptions in the past two millennia.

Clarifying the Conditions of a Possible Future: Let Frankenstein Sleep

Let us return to the best-researched of all sulphur-rich volcanic eruptions of the past: Tambora. Many cultural phenomena have been associated with the experience of the year without a summer in 1816 (Wood 2015; Behringer 2015). Not all of these hypotheses are very well founded; however, one deserves our attention: When a group of young English writers, both men and women, decided to stay in their houses on Lake Geneva because of the 'End of the World' weather they were facing daily in this gloomy summer of 1816 that included 130 days of rain between April and September (Wood 2015: 1-11). One of them, 18-year-old Mary Godwin, better known by her later husband's family name as Mary Shelley, began drafting a story that eventually entered the literary canon: *Frankenstein*. It is not a mere horror novel but a parable for the modern age about how humans underestimate the risk of their technological ambitions, only to be chased and finally overwhelmed by a monster of their own creation. I hope I have demonstrated how medieval history might help us re-evaluate epistemological uncertainties within the whole approach of solar radiation management via artificial aerosol injections.

Historians also respond to the concerns of their own times. If some medievalists turn their attention now to questions of past climate change, natural disasters or pandemics of the past, they are not merely following another ephemeral fashion. But they do what their predecessors have done before—they try to find answers for questions their contemporary societies have: What can we still 'learn' from history for pressing matters such as climate change and adaptation or preventive measures, especially from such a remote time such as the Middle Ages? A possibility that was the indirect line of argumentation in this contribution is the classical 'orientation knowledge' (Koselleck 1989a, b) that no longer naively assumes a direct learning from past events and yet stresses that historiography can unveil the 'conditions of a possible future' (Koselleck 1989a, b: 157). In our context of volcanic eruptions and adaptive measures to a changing future climate, this means that we can hardly (Bethke et al. 2017) interpret societal reactions to past volcanic eruptions as guidelines for future political decision-making. But we can take full advantage of the epistemological possibilities that a combined approach of different disciplines, with medieval studies one of them, provides us a better understanding of climatic impacts of historical eruptions and their potential to design projects of geoengineering. History can help us to make sure that Frankenstein was the only monster created by a human mind in the dimmed sunlight of a global sulphate aerosol layer.

References

Alexandre P (1987) Le climat en Europe au Moyen Âge. Contribution à l'histoire des variations climatiques de 1000 à 1425, d'après les sources narratives de l'Europe occidentale. Paris: Editions de l'École des Hautes Études en Sciences sociales

Bauch M (2017) The day the sun turned blue. A volcanic eruption in the early 1460s and its possible climatic impact—a natural disaster perceived globally in the late middle ages? In: Schenk GJ (ed) Historical disaster experiences. Towards a comparative and transcultural history of disasters across Asia and Europe, pp 107–138. Heidelberg, Springer

Bauch M (2020) Chronology and impact of a global moment in the thirteenth century: the Samalas eruption revisited. In: Kiss A, Pribyl K (eds) The dance of death in late medieval and renaissance Europe. Environmental stress, mortality and social response, pp 214–232. Abingdon, Oxon, NY, Routledge

Behringer W (2015) Tambora und das Jahr ohne Sommer. Wie ein Vulkan die Welt in die Krise stürzte. Beck, München

Bethke I, Outten S, Ottera OH, Hawkins E, Wagner S, Sigl M, Thorne P (2017) Potential volcanic impacts on future climate variability. Nat Clim Chang 7:799–805

Boettcher M (2017) Solar radiation management. IASS fact sheet, p 2. https://doi.org/10.2312/iass. 2017.018

Cole-Dai J, Ferris DG, Lanciki AL, Savarino J, Thiemens MH, McConnell JR (2013) Two likely stratospheric volcanic eruptions in the 1450s C.E. found in a bipolar, subannually dated 800 year ice core record. J Geophys Res Atmos 118(14):7459–7466. https://doi.org/10.1002/jgrd.50587

Cole-Dai J (2010) Volcanoes and climate. Wires Clim Change 1(6):824–839. https://doi.org/10. 1002/wcc.76

Crutzen PJ (2002) Geology of mankind. Nature 415(6867):23. https://doi.org/10.1038/415023a

Crutzen PJ (2006) Albedo enhancement by stratospheric sulphur injections: a contribution to resolve a policy dilemma? Clim Change 77(3):211–220. https://doi.org/10.1007/s10584-006-9101-y

Ellis EC (2018) Anthropocene. a very short introduction. Oxford University Press, Oxford

Esper J, Büntgen U, Hartl-Meier C, Oppenheimer C, Schneider L (2017) Northern Hemisphere temperature anomalies during the 1450s period of ambiguous volcanic forcing. Bull Volcanol 79(6):41. https://doi.org/10.1007/s00445-017-1125-9

Francis P, Oppenheimer C (2004) Volcanoes. Oxford, Oxford University Press

Gao C, Robock A, Self S, Witter JB, Steffenson JP, Clausen HB, Siggaard-Andersen M-L, Johnsen S, Mayewski PA, Ammann C (2006) The 1452 or 1453 A.D. Kuwae eruption signal derived from multiple ice core records. Greatest volcanic sulphate event of the past 700 years. J Geophys Res 11(D12). https://doi.org/10.1029/2005JD006710

Guillet S, Corona C, Stoffel M, Khodri M, Lavigne F, Ortega P, Eckert N, Dkengne Sielenou P, Daux V, Churakova OV, Davi N, Edouard J-L, Zhang Y, Luckman BH, Myglan VS, Guiot J, Beniston M, Masson-Delmotte V, Oppenheimer C (2017) Climate response to the Samalas volcanic eruption in 1257 revealed by proxy records. Nat Geosci 10:123EP. https://doi.org/10.1038/ngeo2875

Hamilton C (2013) Earthmasters: the dawn of the age of climate engineering. Yale University Press

Herlihy P (1986) Odessa: a history, 1794–1914. Harvard Univ. Press, Cambridge/Mass.

Koselleck R (1989a) 'Darstellung, Ereignis und Struktur. Über die Auflösung des Topos im Horizont neuzeitlich bewegter Geschichte' in: R. Koselleck, ed. Vergangene Zukunft. Zur Semantik geschichtlicher Zeiten. Frankfurt/Main: Suhrkamp: 144–157

Koselleck R (1989b) 'Historia Magistra Vitae. Über die Auflösung des Topos im Horizont neuzeitlich bewegter Geschichte' in R. Koselleck, ed. Vergangene Zukunft. Zur Semantik geschichtlicher Zeiten. Frankfurt/Main: Suhrkamp: 38–66

Kostick C, Ludlow F (2020) Medieval History, Explosive Volcanism, and the Geoengineering Debate. In : Jones C, Kostick C and Oschema K (eds) Making the Medieval Relevant. How Medieval Studies Contribute to Improving our Understanding of the Present, pp 46–98, Berlin: De Gruyter https://doi.org/10.1515/9783110546316-003

Kravitz B, Robock A (2011) Climate effects of high-latitude volcanic eruptions. Role of the time of year. J Geophys Res 116(D1). https://doi.org/10.1029/2010JD014448

Mather TA, Pyle DM (2015) Volcanic emissions: short-term perturbations, long-term consequences and global environmental change. In Schmidt A, Fristad KE, Elkins-Tanton LT (eds) Volcanism and global environmental change, pp 208–227. Cambridge, Cambridge University Press

Németh K, Cronin SJ, White JD (2007) Kuwae caldera and climate confusion. Open Geol J 1:7–11

Oman L, Robock A, Stenchikov, G., Schmidt, G.A., Ruedy, R. 2005. 'Climatic response to high-latitude volcanic eruptions' Journal of Geophysical Research 110. https://doi.org/10.1029/200 4JD005487

Oppenheimer C, Donovan A (2015) On the nature and consequences of super-eruptions. In: Schmidt A, Fristad KE, Elkins-Tanton LT (eds) Volcanism and global environmental change, pp 16–29. Cambridge, Cambridge University Press

Pang KD (1993) Climatic impact of the mid-fifteenth century Kuwae caldera formation, as reconstructed from historical and proxy data fall meeting. Am Geophys Union 106

Post JD (1977) The last great subsistence crisis in the western world. The Johns Hopskins University Press, London, Baltimore

Rampino MR, Self S, Stothers RB (1988) Volcanic winters. Annu Rev Earth Planet Sci 16:73–99

Robock A (2014) Stratospheric aerosol geoengineering. In: Hester R, Harrison RM (eds) Geoengineering of the climate system. Cambridge University Press, Cambridge, pp 162–185

Schmidt A, Fristad KE, Elkins-Tanton LT (eds) (2015) Volcanism and global environmental change. Cambridge, Cambridge University Press

Schmidt A, Robock A (2015) Volcanism, the atmosphere and climate through time. In: Schmidt A, Fristad KE, Elkins-Tanton LT (eds) Volcanism and global environmental change. Cambridge University Press, Cambridge, pp 195–207

Schneider L, Smerdon JE, Büntgen U, Wilson R, Myglan VS, Kirdyanov AV, Esper J (2015) Revising midlatitude summer temperatures back to A.D. 600 based on a wood density network. Geophys Res Lett 42(11):4556–4562. https://doi.org/10.1002/2015GL063956

Sigl M, McConnell JR, Layman L, Maselli O, McGwire K, Pasteris D, Dahl-Jensen D, Steffensen JP, Vinther B, Edwards R, Mulvaney R, Kipfstuhl S (2013) A new bipolar ice core record of volcanism from WAIS divide and NEEM and implications for climate forcing of the last 2000 years. J Geophys Res Atmos 118(3):1151–1169. https://doi.org/10.1029/2012JD018603

Sigl M, Winstrup M, McConnell J et al (2015) Timing and climate forcing of volcanic eruptions for the past 2,500 years. Nature 523(7562):543–549

Simkin T, Fiske RS (1983) Krakatau 1883. The volcanic eruption and its effects, Washington

Stoffel M, Khodri M, Corona C, Guillet S, Poulain V, Bekki S et al (2015) Estimates of volcanic-induced cooling in the Northern Hemisphere over the past 1,500 years. Nature Geosci 8(10):784–788. https://doi.org/10.1038/ngeo2526

Swingedouw D, Mignot J, Ortega P, et al (2017) Impact of explosive volcanic eruptions on the main climate variability modes. Glob Planet Change 150:24–45

Timmreck C (2012) Modelling the effects of large volcanic eruptions. WIREs Clim Change. https://doi.org/10.1002/wcc.192

Timmreck C (2018) Climatic effects of large volcanic eruptions. Habilitation thesis, Universität Hamburg, Hamburg. https://doi.org/10.17617/2.2566000

Timmreck C, Lorenz SJ, Crowley TJ et al (2009) Limited temperature response to the very large AD 1258 volcanic eruption. Geophys Res Lett 36:1–5

Toohey M, Sigl M (2017) Volcanic stratospheric sulphur injections and aerosol optical depth from 500 BCE to 1900 CE. Earth Syst Sci Data 9(2):S.809–831. https://doi.org/10.5194/essd-9-809-2017

Wegmann M, Brönniman S, et al (2014) Volcanic influence on European summer precipitation through monsoons: possible cause for 'Years without a Summer'? J Clim 27:3683–3691

Wilson R, et al (2016) Last millennium northern hemisphere summer temperatures from tree rings: Part I: the long term context. Quat Sci Rev 134:1–18. https://doi.org/10.1016/j.quascirev.2015.12.005

Winchester S (2003) Krakatoa: the day the world exploded 27 August 1883. London/New York

Wood GDA (2015) Tambora: the eruption that changed the world. Princeton University, NJ

Zambri B, LeGrande AN, Robock A, Slawinska J (2017) Northern Hemisphere winter warming and summer monsoon reduction after volcanic eruptions over the last millennium. J Geophys Res Atmos 122(15):7971–7989

Zanchettin D, Timmreck C, Bothe O, et al (2013) Delayed winter warming: a robust decadal response to strong tropical volcanic eruptions? Geophys Res Lett 40(1):204–209. https://doi.org/10.1029/2012GL054403

Zeilinga de BJ, Sanders DT (2002) Volcanoes in human history. The far-reaching effects of major eruptions. Princeton/Oxford

A Perfect Tsunami? El Nino, War and Resilience on Aceh, Sumatra

Emmanuel Kreike

I would like to thank John Haldon, the editors, and the Environmental History and Climate Change and History Research Initiative for inviting me to participate in the seminar series that led to this publication. Thanks also to Tsering Wangyal Shawa, Geographic Information Systems and Map Librarian, Lewis Science Library, Princeton University, who expertly and generously advised and assisted me with the ArcGIS analysis.

Abstract The history of Aceh, Indonesia highlights societies' resilience and vulnerability in the face of natural and human-made disasters. A multi-scalar, qualitative and quantitative analysis of land use changes in nineteenth century Greater Aceh by using GIS analysis, highlights that processes may play out differently at the system and subsystem levels. At the system's meso and micro levels, the episodic and the structural violence of war, climate anomalies, and tsunamis wiped out entire communities and families of people, animals, and plants while at the macro scale Aceh society showed remarkable resilience. Greater Aceh's case also suggests that the impact of war through population displacement and the destruction of such environmental infrastructure as homes, villages, orchards, and irrigated fields while less immediately and directly destructive than such episodic events as the devastating 2004 tsunami, nevertheless may have a comparable impact because the events are more sustained and cumulative over a timeframe of years and decades.

Keywords War- environmental aspects · Global environmental change · Climate change and resilience/collapse · Human ecology · Natural and human-made disasters · Post-crisis reconstruction

E. Kreike (✉)
History Department, Princeton University, 132 Dickinson Hall, Princeton, NJ, NJ 08542, USA
e-mail: kreike@princeton.edu

A. Izdebski et al. (eds.), *Perspectives on Public Policy in Societal-Environmental Crises*, Risk, Systems and Decisions, https://doi.org/10.1007/978-3-030-94137-6_9

Introduction

The history of Greater Aceh (modern Aceh) on the northern tip of Sumatra, Indonesia highlights societies' resilience and vulnerability in the face of natural and human-made disasters. A comparative perspective using a more recent, more data-rich case study provides detailed insights into the processes involved that may shed light on the dynamics involved in other cases discussed in the book for which the data are scarcer. The chapter offers a multi-scalar, qualitative *and* quantitative analysis of land use changes in nineteenth century Greater Aceh by using GIS analysis. It sheds light on key issues relating to spatial and temporal scales of analysis. What and who collapsed or proved resilient and to what extent are the qualifications of collapse and resilience dependent on the temporality of the analytical framework? What does a focus on the abstract level of "systems" reveal or hide? What is "the system:" a polity represented by a state elite and state infrastructure or an ecosystem or agro-ecological system? How do processes at the meso-level communities of people, animals, and plants factor? What about the microlevel of households and individuals? Indigenous Americans experienced genocide and ecocide, and demographic and societal collapse between 1492 and the 1880s. Yet, the Huron, Iroquois, Sioux and many other survived as tribes (Kreike 2021: 59–96, 137–172, 279–317). The history of Greater Aceh raises the same questions. Greater Aceh was the core of the Aceh Sultanate, which was a major political and economic power in Southeast Asia from the sixteenth century to the late nineteenth century. In 1873, the Dutch invaded the sultanate, triggering a war that dragged on for decades causing destruction, displacement, and massive loss of life (Van 't Veer 1969; Reid 1969; Stolwijk 2016; Hagen 2018: 438–480; Kreike 2021: 318–357). The humanitarian and economic costs of the war initiated a public debate in the Netherlands, with a former colonial officer accusing the Dutch government of genocide in Aceh (Wekker 1907; Kreike 2021: 318–320, 355–357). A century later, the Netherlands' colonial wars in Indonesia and elsewhere are once again under scrutiny because of the intense violence and destruction that accompanied them (Luttikhuis and Moses 2014; Enthoven et al. 2013).

Monsoons and El Niño Southern Oscillation (ENSO)

Greater Aceh, located on the northwestern tip of Sumatra (Fig. 1), is subject to the monsoons, with a wet season in December-February and a dry season in June–August. Irrigated cultivation has marked the densely populated Aceh River Valley lowlands since the sixteenth century and many households also had dryland fields in the foothills and mountains surrounding the floodplain. In addition to the staple paddy rice crop, farmers cultivated such highly valuable export crops as pepper, betel nuts and betel leaves. Heavy rains that are exacerbated by wetter ENSO La Niña events cause flooding in Greater Aceh (Fig. 2) while onsets of El Niño are associated with droughts (Ilhamsyah et al. 2019). Flooding can severely damage buildings, fields,

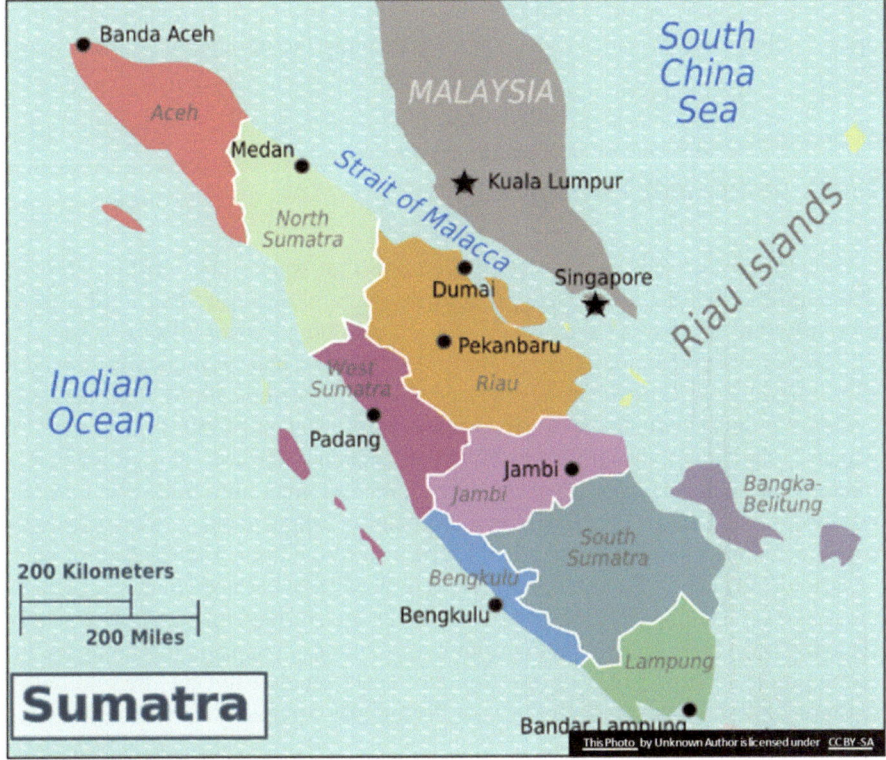

Fig. 1 Sumatra with Aceh (*source* Wikimedia commons/creative commons)

Fig. 2 The flooded village of Meulaboh south of Banda Aceh, the capital of Aceh in November 2014 (*source* Getty images/christian science monitor)

and crops not only in the low-lying floodplain of the Aceh River, but also in the upland valleys and mountainous slopes surrounding the floodplain (Azmeri and Isa 2018).

Earthquakes and Tsunamis

Greater Aceh also lies in an earthquake zone that caused the terrible Christmas Tsunami in 2004 that killed over 100,000 people and displaced half a million with mortality rates in some of the coastal districts climbing to over 20% (Fig. 3).

The sea waters inundated areas up to 4 km inland and permanently moved the shore of Aceh 1.5 km inland (Doocy et al. 2007; Borrero 2005). Archaeological evidence points to equally devastating tsunamis in 1394 and 1450 (Fig. 4). The 1394 tsunami wiped out the small settlements that existed in the coastal region where the modern city of Banda Aceh is located but left the thriving Lamri trade port unscathed because it was located on higher ground behind a tongue of land. During the 2004 tsunami, Lamri's location once again saved it from the impact of the wall of water.

In the 16th century, however, Lamri was replaced as a key node in the Silk Road by the new Aceh sultanate (Fig. 5) that arose in the area of modern Banda Aceh despite it having been hit twice by a tsunami in half a century. It appears that refugees from Samudra/Pasai further east on the Sumatra coast recolonized the area around the modern capital of Banda Aceh after the Portuguese occupied their town (Daly et al. 2019; Meltzner et al. 2004).

In 2004, massive aid from the Indonesian government and international emergency and development assistance sustained the survivors and allowed the region to

Fig. 3 Flooding caused by 2004 Tsunami (*source* Wikimedia commons/US Navy)

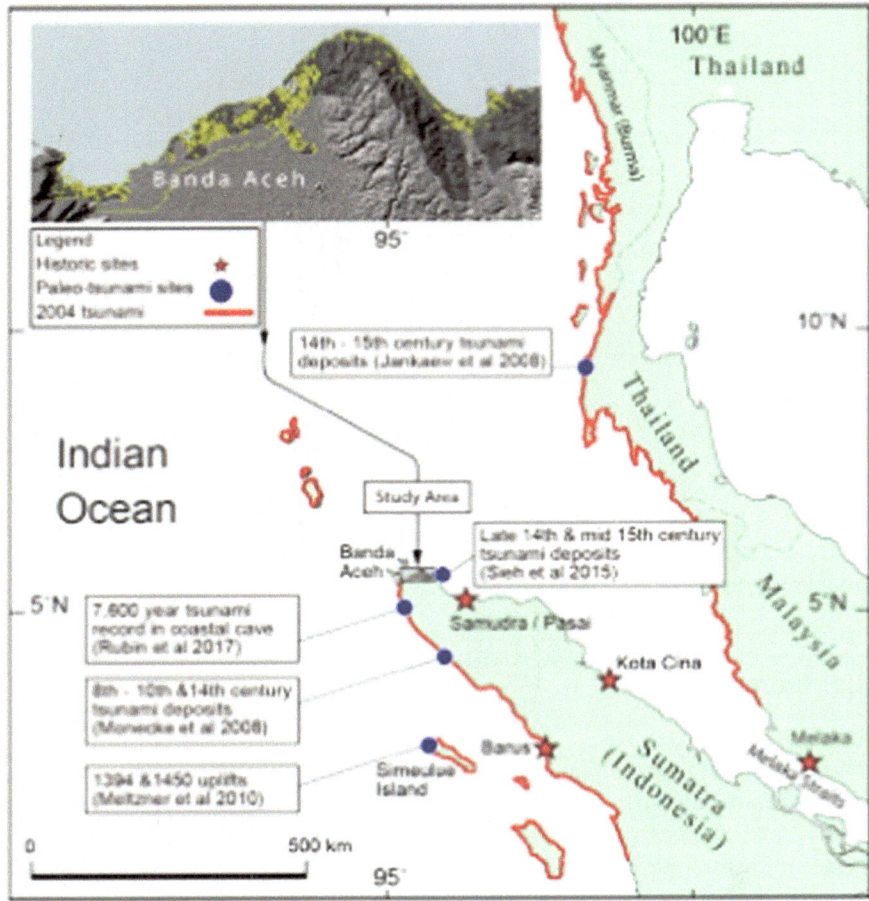

Fig. 4 Map showing pre-historical and historical record tsunami impacts Aceh (*source* Daly et al. 2019)

be rebuild. Research on the post-2004 recovery suggests that restoring individual and family level livelihoods were at least as significant for recovery as the reconstruction of system-level physical and institutional (i.e., state) infrastructure. Restoring livelihoods (including shelter) was especially important to rebuilding resilience for those who were displaced from their homes by the disaster (Sina et al. 2019).

Natural Disasters and War

Other contributors to this volume emphasize that the phenomena of collapse and resilience are multi-causal: the Horsemen of the Apocalypse seldom ride alone. The colonial war of conquest in the second half of the nineteenth century is the main focus.

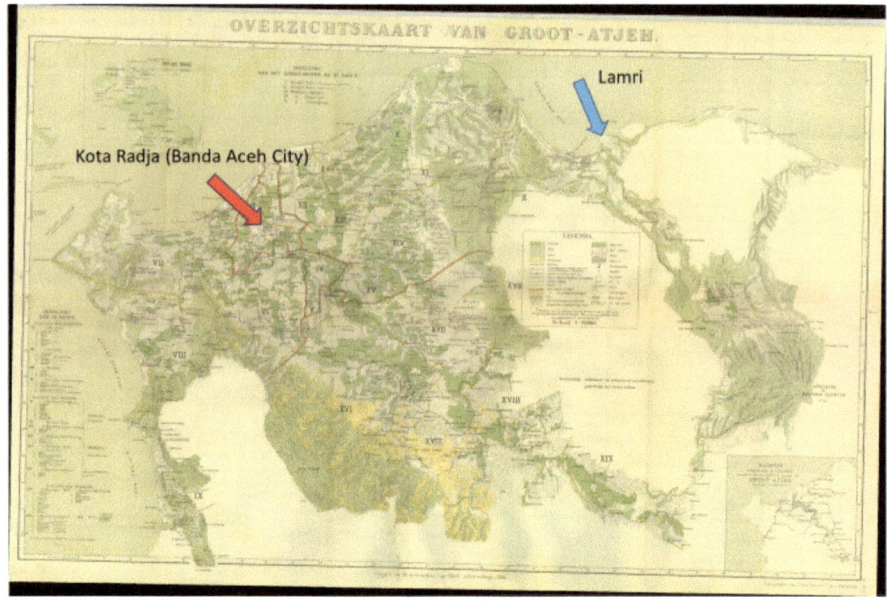

Fig. 5 Dutch late-nineteenth century map of Aceh Sultanate with arrows indicating Lamri and Banda Aceh City (*source* base map E.B. Kielstra, Beschrijving van den Atjeh-Oorlog, 3 vols. (Den Haag: Van Cleef, 1883–1885))

It lasted well into the twentieth century, causing not only direct destruction through scorched earth but also by triggering massive population displacement, tearing people from villages, homes, granaries, water sources, seed supplies and other key environmental infrastructure and exposing people and animals to heat, cold, hunger, thirst, epidemic disease, and death. Seasonal and cyclical droughts and floods in the Aceh River Valley and the surrounding mountainous forests in the monsoon climate exacerbated the impact of war, disease, and famine. Tens of thousands were displaced and thousands perished (Hagen 2018: 479–480; Reid 1969:187–188).

The Dutch invaded Aceh in 1873 (Fig. 6). The first invasion was a failure, with heavy casualties due to the fighting and disease and the Dutch withdrew before the full onset of the monsoon that made it impossible to effectively deploy their men and material.

A new invasion force sent from the Netherlands reached Aceh in December 1873 and occupied the palace of the Aceh Sultan in January 1874, in the middle of the rainy season. Dutch campaign reports on the war are replete with complaints of how flooded and swampy terrain handicapped their operations, with heavier losses due to disease than to actual combat (Van't Veer 1969: 49–81). The last three decades of the nineteenth century saw an unprecedented clustering of ENSO events that coincided with the Dutch conquest wars. El Niño events occurred in 1873 and 1874, 1876, 1877, 1878, 1881, 1887, 1888, 1889, 1891, 1895, 1896, and 1897. The hotter

Fig. 6 The village of Glumpang (Gloempang) burning while under Dutch attack in 1873 (*source* J.A. Kruijt. Atjeh en de Atjehers: Twee Jaren Blokkade (Leiden: Gualth. Kolff, 1877))

and drier conditions associated with El Niño years made the terrain more passable for men, horses, carts, and cannon. La Niña events associated with cooler and wetter conditions occurred in 1876, 1878, 1879, 1880, 1886, 1887, 1889, 1890, 1892, 1893, and 1894 and may have exacerbated inundations in Aceh. Analysis from early twenty-first century ENSO events suggests that western Sumatra and Greater Aceh may be less affected by ENSO effects than other parts of Indonesia and that anomalies cluster in the dry season months of June–August. An extreme double ENSO El Niño that occurred in 1877–1878 likely had a substantial impact. The relative temperature anomalies of the 1877–1878 event were comparable to the events in 1982–1983 and 1991, which caused heavy padi rice losses and severe bush fires across Indonesia. Moreover, high-temperature dry ENSO El Niño periods are usually followed by cooler and wetter years (Harger 1995; As-syakur et al. 2014; Davis 2001: 271, table 8.8; Giese and Ray 2011: C2, tables 1 and 2). In Aceh, the 1878–1879 El Niño may have favored Dutch operations, literally fanning the flames of scorched earth. The 1878–1879 El Niño may also have depressed irrigated and dryland rice production in regions that were not directly subject to the hostilities. The cooler and wetter years that followed may have intensified flooding and encouraged the invasion of *alang alang* grasses in the rice fields, and bamboo and brushwood in village gardens and dryland fields; these invasive species negatively impacted agricultural yields. ENSO events thus were likely to have been a significant factor during the late 1870s and the early 1880s, exacerbating the impact of the war and population displacement. The severest ENSO events of the late nineteenth century coincided with the peak of the Dutch scorched earth campaign and its immediate aftermath, when many villages lay abandoned, along with their gardens, plantations, orchards and irrigated rice fields (Fig. 7).

Fig. 7 Damaged Aceh houses (*source* National Archive, The Hague, The Netherlands (henceforth NL-HaNA) 2.20.46_850_15)

1873–1880: Scorched Earth and Population Displacement

From 1878 to 1880, the Dutch invaders intensified their attempts to conquer the territory, advancing up the Aceh valley inland while relying heavily on scorched earth: any village that resisted or was found abandoned was burned to the ground. In many cases, the inhabitants narrowly escaped with the little they could carry and their cattle.

The Dutch colonial soldiers captured or destroyed the smallstock and the rice stores left behind. The Dutch timed their campaigns with the onset of the rice harvest season in March–May when the rains ended, and the irrigated rice fields were drained, making them passable for the soldiers and their heavy equipment. By 1880, the Aceh Valley had been subdued. But even the general who had led the second invasion in 1873 concluded that the price of Dutch success had been too heavy: 500 villages had been reduced to ashes and 30,000 Acehnese had perished in the fighting or had fallen victim to starvation or disease, including smallpox and cholera (Kreike, 2021: 318–342).

1880–1884 War, Resilience, and Recovery

Many villagers had fled the Aceh Valley and those who returned found themselves homeless and short of labor, food, seed, and water buffaloes (Fig. 8) to prepare the rice fields (Kreike 2021: 337–338, 341–342).

Fig. 8 Waterbuffalo in rice field (*source* NL-HaNA 2.20.46_850_29)

The May–June 1880 padi-rice harvest was poor, many rice fields remained abandoned. Returned villagers only managed to clear, plant, and harvest the fields closest to their homes. Meanwhile, violence continued to hamper a return to normal life. Dutch patrols only had control over the countryside by day. By night, rebels extracted food and war taxes from the villagers at will and punished anyone suspected of collaborating with the Dutch invaders. Having temporarily abandoned their blanket scorched earth strategy for a policy of "exemplary punishment" during the early 1880s, Dutch security forces now limited themselves to the selective and occasional burning of single villages accused of actively supporting the rebels (Kreike 2021: 342–350).

The Concentrated Line

In 1884, due to the continued high military and economic costs of the occupation of Aceh and the failure of its civilian-led pacification campaign, the Dutch government radically changed its policies. Instead of trying to control the entire territory, it retreated to the far west of Greater Aceh. The so-called "Concentrated Line" consisted of a 5–6 km deep territory that extended from the coast inland, basically the lower

Aceh River Delta, and included the capital Kota Radja (modern Banda Aceh) and the Oleh Leh Port. All territory beyond the concentrated line was abandoned by the Dutch military and administration. The line consisted of a number of larger and smaller forts connected to one another and to Kota Radja and Oleh Leh by all-weather roads and a narrow-gauge railroad. The roads and the railroad were elevated to ensure that they were above the monsoon flood levels of the Aceh River. Despite having placed drains where the elevated roads and railroad track crossed smaller and larger rivers, the roads and railroads effectively functioned as dikes, increasing the incidence of flooding upstream and causing water shortages downstream.

Dutch soldiers and laborers also created and maintained a one-kilometer free fire zone beyond the line of forts that faced inland along the entire length of the defensive line. Forced laborers razed homes, trees, graves and entire villages to create free fields of fire for the Dutch artillery installed in the forts. The defensive line, the free fire zone and the areas directly beyond transformed into a deserted no man's land. Dutch patrols had standing orders to shoot to kill anyone crossing the defensive line without permission. The Dutch also responded to shots fired from beyond the territory under their control, and indeed, anything seen as a provocation by shelling villages beyond the defensive line in retaliation. In 1891, a mere six years after the completion of the defensive line in 1885, a broad swath of territory beyond the one-km-wide free fire zone lay entirely abandoned as villagers and farmers fled their homes and fields.

The red line on the 1891 Dutch military map (Fig. 9) highlights the defensive line with the railroad connecting the forts. The green color indicates abandoned villages and the blue color marks abandoned and inundated rice fields. Population flight led to a decline in the repair and maintenance of the infrastructure of dikes, dams, drains, and sluices in the padi-rice fields, turning them into swamps or weed deserts. Clustered ENSO events increased the incidence of flooding and water shortages and the Dutch road and railroad dikes reshaped and impeded pre-existing drainage and irrigation patterns in the delta (Kreike 2021: 350-355). Not all flooding was incidental. Hostile villagers closed dams, sluices and drains to flood the rice fields during the off-season in order to hinder the Dutch advance. The flooding forced the Dutch soldiers to advance very cautiously over the warren of narrow slippery dikes and rickety bridges towards the villages (Fig. 10) that were surrounded by bamboo stakes and fences and palisades (Kreike 2021: 328, 331). It is somewhat ironic that the Dutch reports increasingly portrayed the conflict as being fought in swamps and jungle, that is, as wilderness warfare, although the swamps and the jungle-like flora they encountered in the Aceh Valley were actually the outcome of their own destructive war tactics.

Measuring Resilience at the Systemic Scale Through GIS

Dutch military surveyors accompanied the troops during the fighting and used trees and mosques as markers to draw maps that contain detailed information about villages, vegetation, and land use. An 1885 map and an 1891 update were separately georeferenced in GIS using up to 100 points to tie features on each map to the same features identified on the modern satellite image (Fig. 11).

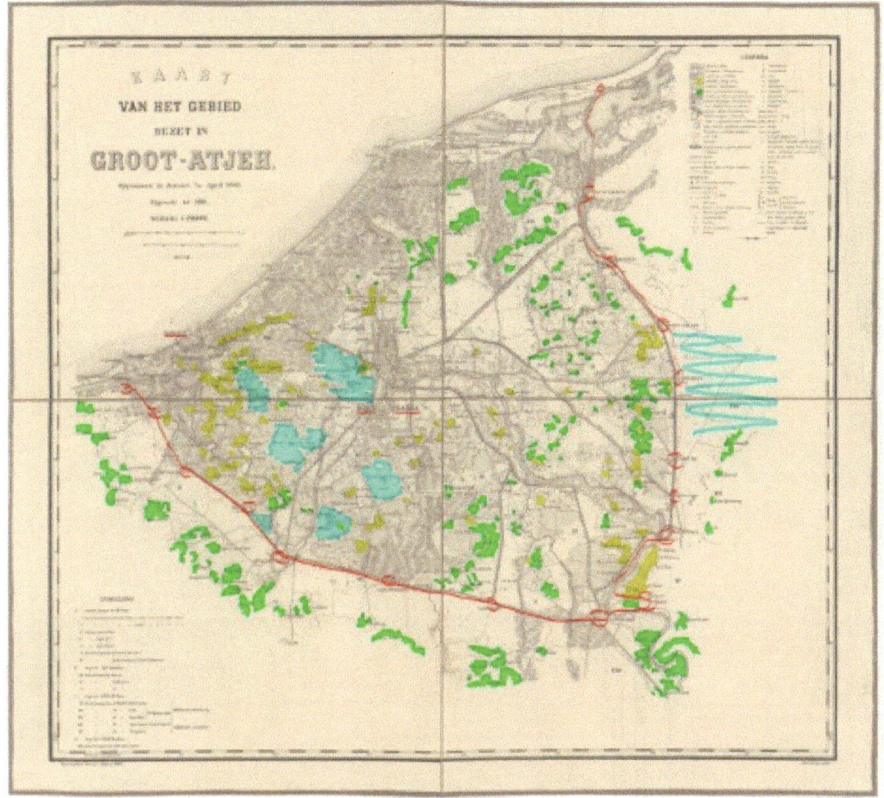

Fig. 9 1891 Dutch military map of Aceh with color mark up in Paint software (*source* base map: NL-HaNA, 4. Ministerie van Kolonien (henceforth MIKO), 1414, 1891)

Next, villages, padi rice fields, and other land use features on the georeferenced maps were marked with polygons. The polygons allowed the surface totals for each feature to be calculated and facilitated assessing the land use changes between 1885 and 1891.

Many villages and rice fields already had been abandoned before 1885 because of the fighting since 1873. The polygons in the georeferenced 1885 map (Fig. 12) mark different types of land use or abandonment, including previously abandoned village lands (coded as *Previouslyabandoned* in the legenda and marked by the blue-lined transparent polygons); actively cultivated ricefields (*Ricefield*: red filled polygons); partially cultivated ricefields (*Ricefieldpartiallycultivated* in dark pink); uncultivated ricefields (*Ricefielduncultivated* in light pink); and inhabited villages (*Village* in dark green).

The same categories and colors were used for the analysis of the 1891 map (Fig. 13). One complication is that the land use information provided on the two maps was not always consistent between the two maps, making the comparison challenging. Not all village spaces and rice fields indicated on the 1885 map were

Fig. 10 Dutch officers on a narrow dike with a bridge in the background (*source* NL-HaNA 2.20.46_850_17)

Fig. 11 Dutch 19th-century military map of Aceh superimposed on satellite image for georeferencing (*source* map: NL-HaNA, 4 MIKO, 1414, 1891)

marked in the same way as on the 1891 map. For example, some of the abandoned village sites on the 1885 map were identified as bush vegetation on the 1891 map. Similarly, the locations of abandoned padi-rice fields on the 1885 map appeared as swamps or grasslands on the 1891 map, without any reference to their previous

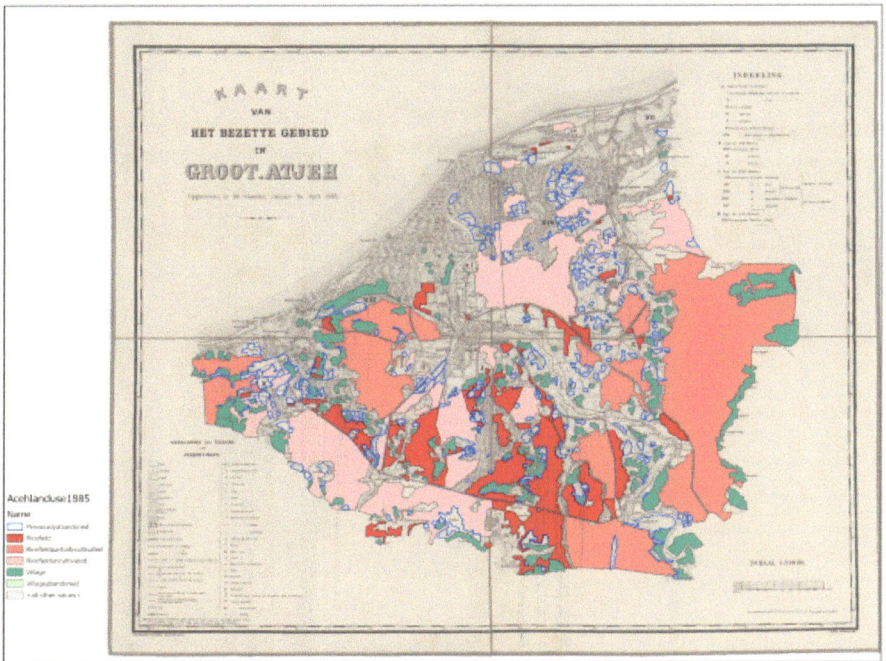

Fig. 12 Georeferenced map: Aceh land use 1885 (*source* base map: NL-HaNA, 4 MIKO, 1239, 1885)

classification on the 1885 map. To measure the total size of the village surfaces abandoned between 1885 and 1891, for example, I marked the village surfaces that were marked as abandoned *before* 1885 with blue-lined transparent polygons in the 1885 map to highlight that they are different from the villages on the 1891 map that could be identified as having been newly abandoned between 1885 and 1891 (coded as *Villageabandoned* and marked in light green).

To better visualize the land use changes, I separated the changes in the size of village spaces from the changes in the sizes of irrigated rice fields in the following images and displayed the changes in each category on a single map. The first image (Fig. 14) shows the occupied and inhabited village spaces in 1885 (in dark blue), with the village spaces still inhabited in 1891 (in orange superimposed on the blue 1885 polygons) within the Dutch-occupied Concentrated Line. The image indicates that most of the villages were entirely or partially abandoned. The total surface of inhabited village space shrunk from 880 ha in 1885 to 267 ha in 1891, a loss of 520 ha of village lands (homes, gardens, orchards) that constituted a 60 percent reduction in 6 years. The abandoned village lands transformed into dense thickets and bush. Indeed, on the 1891 map, many areas that had been identified as abandoned village lands on the 1885 maps were now relabeled as forest and wilderness.

The next image of the 1891 map (Fig. 15) includes the category of abandoned villages (coded *Villageabandoned1891* in yellow) and adds two new sets of data: the

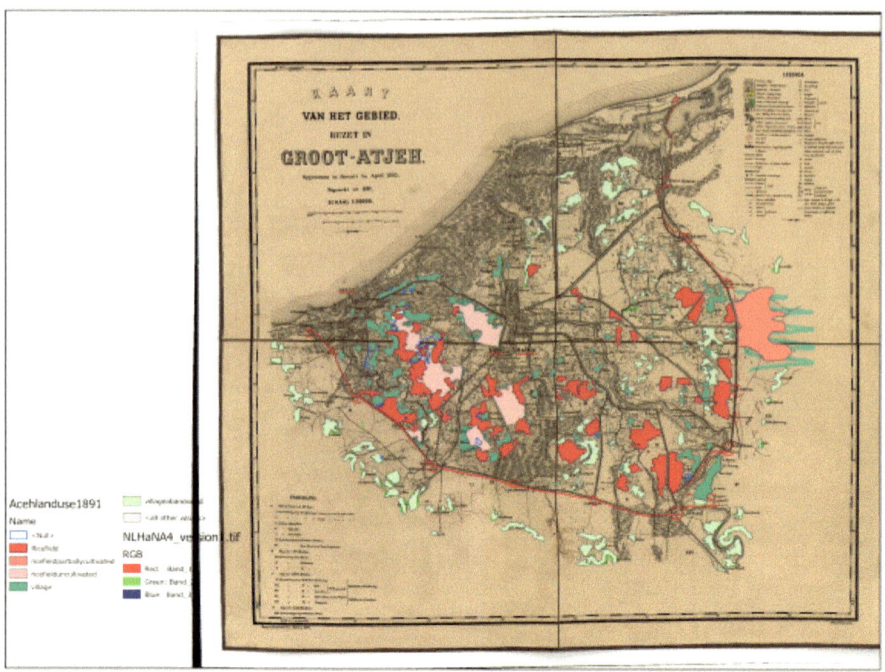

Fig. 13 Georeferenced map: Aceh land use 1891 (*source* base map: NL-HaNA, 4 MIKO, 1414, 1891)

extent of the abandonment of 1885 village sites by 1891 (in orange); and projected village spaces that appear to have been abandoned before 1885 (in blue).

The next image (Fig. 16) compares cultivated irrigated rice fields in 1885 and 1891. The image shows that the surface area of irrigated padi rice fields shrunk from 984 ha in 1885 (*Ricefield1885* in light green) to 572 ha in 1891 (coded *Ricefield1891* in dark green), a reduction of 45%. On the maps, abandoned padi rice fields often are marked either as overgrown by *alang alang* grass (*Imperata cylindrica*) or as wild sugar cane (*Sacharum spontaneum*) infested swamplands. In particular, *alang alang* is known for its invasive properties. *Alang alang* invasion in irrigated rice fields is a major problem especially if fields are cleared too late in the season to plant the rice. Two-and-half years after the 2004 Tsunami and despite a massive influx of aid and machine-assisted clearing of the rice fields, rice production had not resumed because of *alang alang* invasion, flooded fields and saltwater intrusion. (Thorburn 2009).

The land use data suggest that more than half of the villages and almost half of the irrigated rice fields in Banda Aceh were abandoned between 1885 and 1891, while many, if not most, villages and ricefields already had been abandoned before 1885. Yet, as a political, societal, and environmental system and measured at a time scale of give or take 20 years (from the Dutch invasion of 1873 to1891), Aceh seemed to be surprisingly resilient. Despite the use of total war and scorched earth, the Dutch retreated within the Concentrated Line in 1885, leaving the Aceh Sultan and the rebels

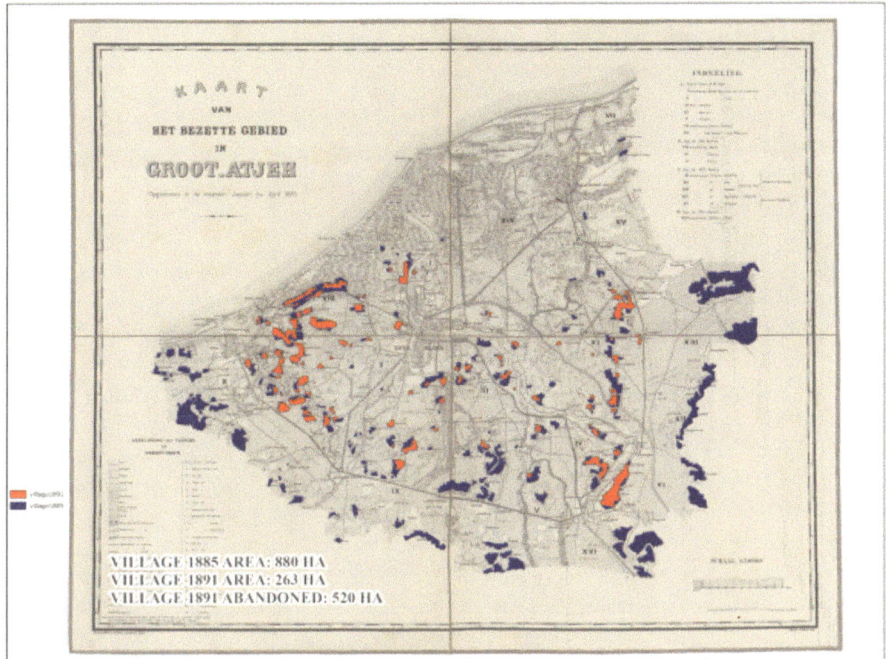

Fig. 14 Abandoned villages 1891: 520 hectares of village lands abandoned (*source* base map: NL-HaNA, 4 MIKO, 1414, 1891)

in control of the remainder of the territory. Although dramatically reduced in size, the preserved villages and rice fields continued to sustain the remaining population. If, however, the year 1880 is selected to assess Aceh at the system level (using a time scale of less than a decade: 1873–1880), collapse seems the more appropriate description with the sultanate defeated, over 500 villages burned, and the population exposed to the monsoons, drought, famine, and disease.

Sub-Systemic Resilience and Collapse: A Village View in Lamara

Data on the village of Lamara (Fig. 17) allow for a sub-systemic assessment of resilience and collapse, providing insights into the dynamics at a village meso-level that also point to differentiation at the micro-level of households. The violence and the displacement prevented clearing, cultivating, and weeding fields, gardens, orchards, and plantations at the village level. Unchecked weed growth and denser undergrowth during the wetter La Niña years resulted in an increased availability of fuel and consequently tinderbox conditions during the dry season. The extra heat that marked

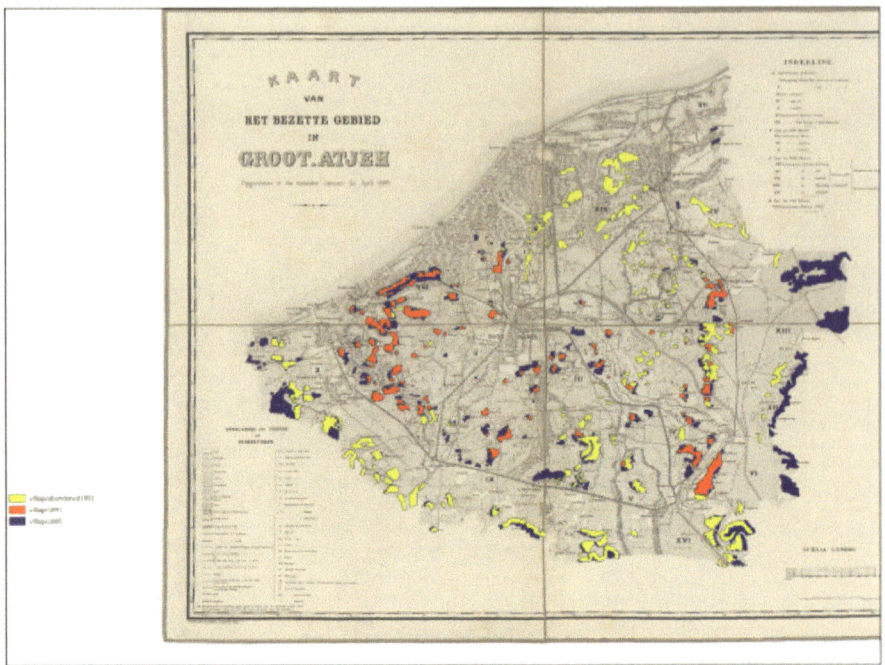

Fig. 15 Villageabandoned1891 (*source* base map: NL-HaNA, 4 MIKO, 1414, 1891)

El Niño events rendered scorched earth and fires more destructive. In the dry season month of June 1880, an accident with a cooking fire caused a blaze in Lamara that destroyed most of the 30 newly rebuilt houses.

Fortunately, the fire occurred during the day, when the villagers were engaged in the rice fields and most livestock was in the pastures. One man, a cow, and a water buffalo suffered burns and six goats perished. The ease with which the fire spread may have been related to an abundance of weedy undergrowth in the village that only recently had been resettled in what was a La Niña year. Seven months earlier, in 1879, Dutch soldiers had burned Lamara to the ground after they found the population had abandoned it. Returning refugees had rebuilt part of the village and a new mosque. The house frames consisted of betel nut palm poles (*Areca catechu* or *pinang*), with bamboo sides, and a roof covered with mangrove palm leaves (*Nypa fruticans* or *atap*), all locally available resources. Only seven homes and the mosque survived the June 1880 fire. In November of the same year, with irrigated rice cultivation under way, the Lamara men organized themselves in shifts of 25 individuals, taking turns to work in the fields and in the village to rebuild their homes once again (Kreike 2021: 338, 340). The mosque is indicated by a symbol indicated by the yellow arrow on the Dutch map (Fig. 18).

In 1882, some of Lamara's inhabitants left the village to heed the mosque Iman's advice to construct a new village called Manga in the mountains surrounding the Aceh Valley. Colonial workers had destroyed a section of the village's irrigation

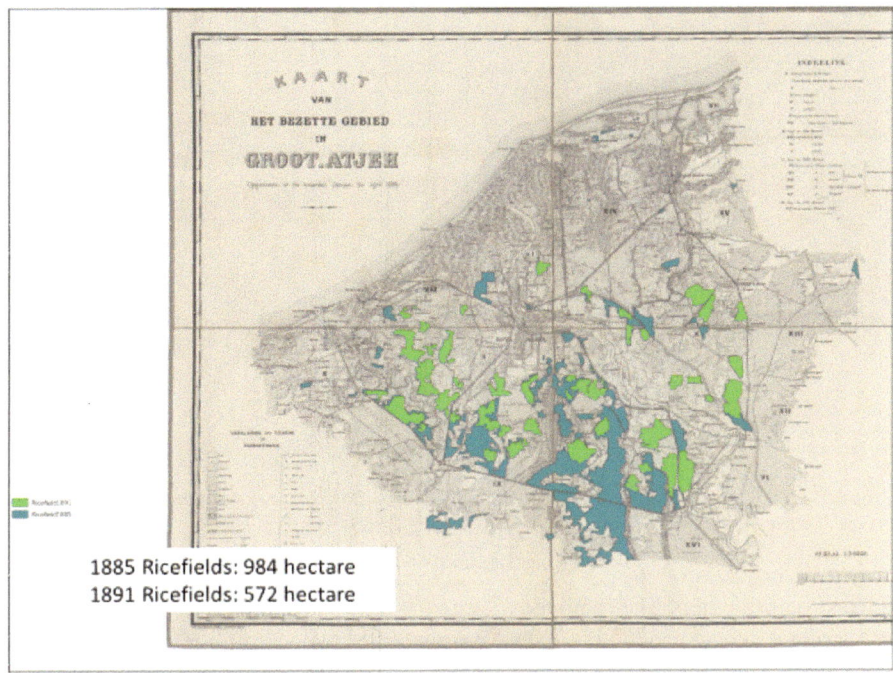

Fig. 16 Rice fields 1885 and 1891 compared (*source* base map: NL-HaNA, 4 MIKO, 1414, 1891)

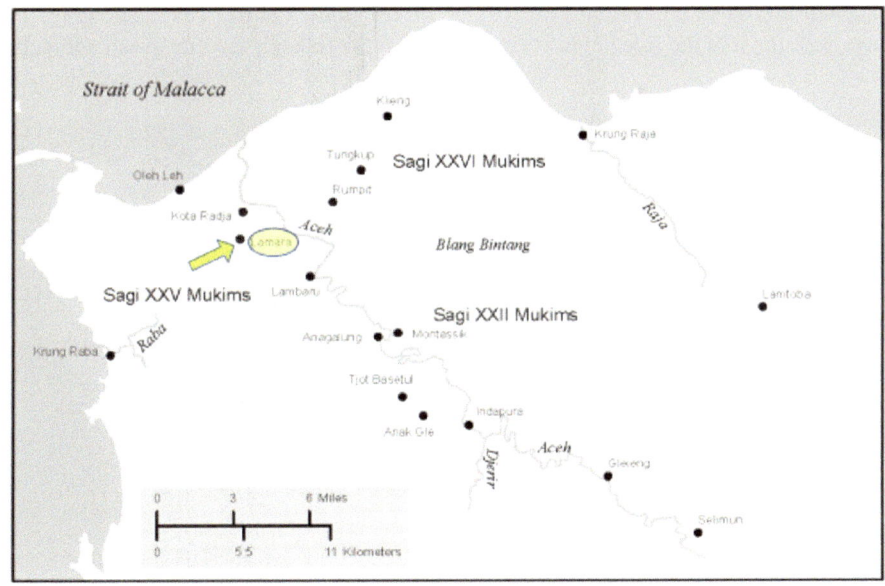

Fig. 17 Greater Aceh with Lamara

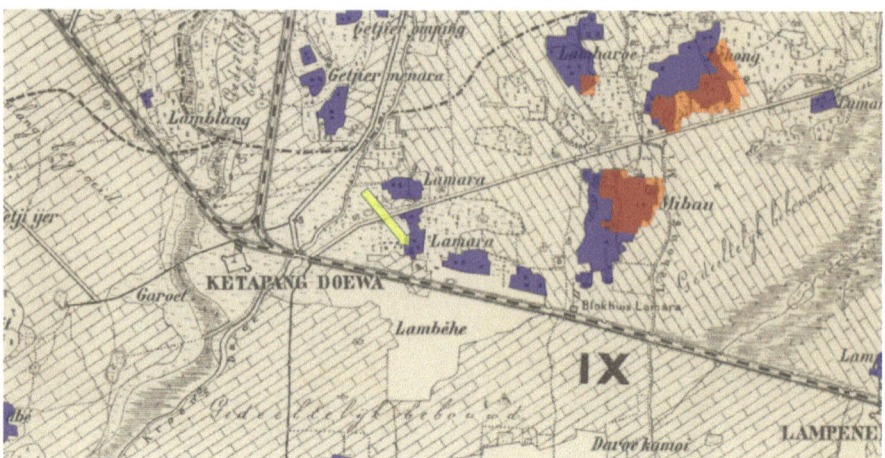

Fig. 18 Dutch map with Lamara mosque indicated by yellow arrow, identified on the 1885 map as in use. (*source* base map: NL-HaNA, 4 MIKO, 1239, 1885)

infrastructure when they removed a large quantity of gravel from the local river to construct a new elevated all-weather road nearby. As a result, the river water could no longer reach the ditches to irrigate the Lamara rice fields leading its owners to migrate (Kreike 2021: 342). On the 1885 map, the mosque is still in use and parts of the village are still inhabited (marked in purple). Extensive cultivated rice fields (the hatched areas) extend around the village on all sides. On the 1891 map (Fig. 19), however, most of the rice fields are uncultivated (the transparent bluegreen unhatched

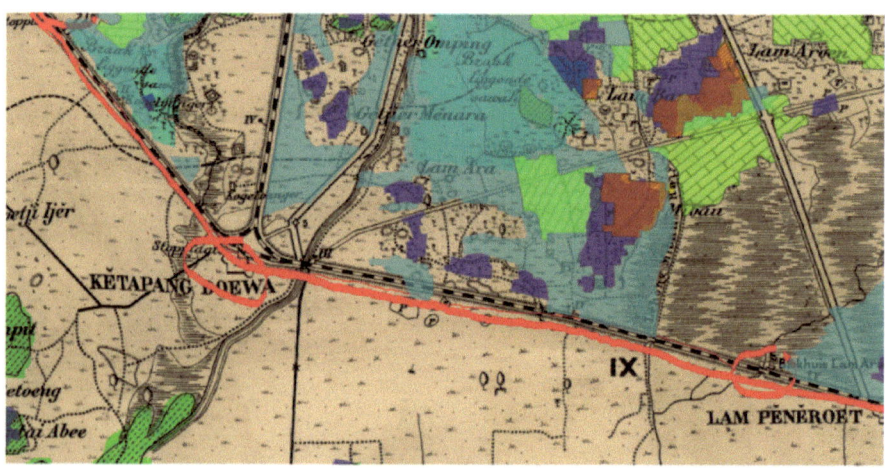

Fig. 19 1891 Lamara with abandoned ricefields (*source* base map: NL-HaNA, 4 MIKO, 1414, 1891)

areas) and only greatly reduced patches to the North remain (the transparent light green hatched areas).

The Village Was Entirely Abandoned in 1891

For comparison, the purple polygons marking the 1885 inhabited village patches are superimposed on the 1891 map. The close-up of the 1891 map (Fig. 20) highlights the symbol for Lamara's mosque which was rebuilt in 1880 (indicated by the yellow arrow) and in ruins 11 years later.

The 1891 map marks the padi fields north of the railroad (the interrupted black line) as uncultivated and the fields to the south of the railroad are labeled as *alang alang* grasslands. Thus, the Lamara community, which in 1880 had proved resilient by having constructed 30 new homes and a mosque only seven months after the Dutch had torched their village, had once again abandoned its home by 1891. The prospect of peace after the end of the 1879 Dutch scorched earth campaign had enticed part of Lamara's population to return from their places of refuge and to rebuild homes and their mosque. Neither the 1880 accidental blaze nor the damage to their irrigation system by Dutch roadbuilders in 1882 had broken their spirit. The 1885 map shows that the village continued to be partly inhabited and that its adjacent rice fields were under cultivation.

By 1891, however, the village community and its environmental infrastructure had collapsed. The village grounds were overgrown by bush and trees, the mosque in ruins, and the padi ricefields suffocated by invasive grasses. Lamara was located just inside the Dutch Concentrated Line that had been constructed in 1884 and 1885.

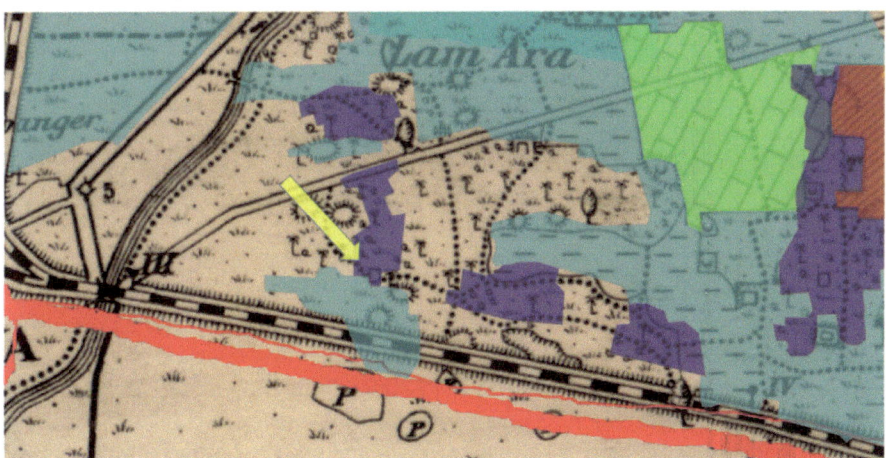

Fig. 20 Detail of Fig. 19 highlighting abandoned mosque, indicated still as a structure on 1891 map but not identified as active. (*source* base map: NL-HaNA, 4 MIKO, 1414, 1891)

The dike for the narrow-gauge railroad line cut through its fields just south of the village. The railroad is visible right next to the red marker line that indicates the Dutch defenses; the red marker circles highlight the position of two nearby Dutch forts. Because the 1885 map depicts the village as partially inhabited, the mosque in operation, and the rice fields as under cultivation, it seems that Lamara's returnees did not immediately give up. But the continued violence along the border made lives and livelihoods very precarious. The 1880s damage inflicted upon the village's hydraulic infrastructure was compounded by the construction of the railroad on the southern edge of the village and a new road to the west of Lamara. Both the railroad and the road, which ran across new dikes that rose above the estimated highest flood levels, crossed and inhibited the drainage lines that the irrigated padi cultivation depended on. The levees for the road and the railroad likely led to the removal of even larger quantities of gravel from the local river, which lowered its water level below the level of the padi rice fields, and thereby incapacitating the gravity-forced irrigation system. Significantly, in other villages, the railroad and roads functioned as dikes, transforming rice fields into swamps because they interfered with drainage. The maps show Lamara's rice fields as *alang alang* drylands and deprived of water. Thus, although Lamara's society and environment demonstrated remarkable resilience during the early 1880s, when the village literally rose from the ashes after a decade of war, population displacement, scorched earth, epidemic disease, famine, continued insecurity, and the construction of Dutch defenses that affected local land use, proved fatal and by the 1890s led to the total collapse of the village's society and environment.

Conclusion

Violence in Greater Aceh continued until the Second World War, when it was occupied by Japan until the end of the war. When Indonesia's nationalist leaders demanded independence, the Netherlands dispatched a large army, but the Dutch failed to reoccupy Greater Aceh. After a bloody war from 1945 to 1949, the Dutch government accepted Indonesia's independence. War returned to Greater Aceh in the 1970s as Aceh guerillas resisted Indonesian rule in another drawn out insurgency that displaced tens of thousands. Peace returned only after the deadly 2004 tsunami.

   Plagued by a long history of devastating and deadly tsunamis, climate anomalies, and wars, Greater Aceh's society and the environmental infrastructure that it shaped and depended on demonstrated both its vulnerability to collapse and its remarkable resilience. Paradoxically, the region at times simultaneously revealed its vulnerability and resilience, as recurred during the 1879–1891 Dutch invasion of the territory. A detailed analysis deploying ARC-GIS methodology to analyze Dutch contemporary military maps in conjunction with Dutch documentary records facilitates identifying the impact of natural and human-made disasters at different temporal and spatial scales, highlighting that processes may play out differently at the system and subsystem levels. Moreover, these processes may not be synchronized, with different outcomes at different spatial and temporal scales. At the meso-level of the village, Lamara's society and environmental infrastructure proved highly resilient by the early 1880s, despite the violence of war and scorched earth, the impact of a cluster of ENSO effects that rendered the village a tinderbox, and the destruction wrought upon its irrigation system by the construction of colonial military infrastructure. Yet barely a decade later, the population of Lamara had entirely abandoned the area and the village's environmental infrastructure was overgrown by weeds and bush. At a macro systems level, however, Aceh's society and environmental infrastructure in more ways than one displayed remarkable resilience from the sixteenth century to the present, despite a series of episodic and structural natural and human-made challenges that, moreover, reinforced one another. How meaningful then, is it to speak of the resilience of such an abstract system as "Greater Aceh" or the "Aceh Sultanate" if much of the population is displaced or killed as occurred during the 1871–1900 Dutch war of conquest or the 2004 tsunami, and the environmental infrastructure that the society created and that sustained it is destroyed? At the system's meso and micro levels, the episodic and the structural violence of war, climate anomalies, and tsunamis wiped out entire communities and families of people, animals, and plants. Moreover, to what extent does the use of the term "collapse" negate the strength and creativity of the survivors of war and tsunamis who managed to rebuild workable societies and environments, with or without aid from the outside?

The Greater Aceh case also suggests that the impact of war and insecurity, with the attendant population displacement and the destruction and deterioration of such environmental infrastructure as homes, villages, orchards, and irrigated fields while less immediately and directly destructive than the 1394 and 2004 tsunamis, nevertheless may have a comparable impact because the events are more sustained and cumulative over a timeframe of years and decades. In addition, war and population displacement may intensify and in turn be exacerbated by climate anomalies or changes, such as the clustered ENSO events that coincided with the late nineteenth century Dutch invasion. Although the war and ENSO events dominated for two decades to bring about what the 1394 and 2004 tsunamis accomplished in mere hours, for many the outcome was the same: death, devastation, displacement of the traumatic survivors, exposing them to hunger, disease, and more death.

References

As-syakur AR, Adnyana IWS, Mahendra MS, Arthana IW, Merit IN, Kasa IW, Ekayanti NW, Nuarsa IW, Sunarta IN (2014) Observations of spatial patterns in the rainfall response to ENSO and IOD over Indonesia using TRMM multisatellite precipitation analysis (TMPA). Int J Climatol 34(15):3825–3839

Azmeri A, Isa AH (2018) An analysis of physical vulnerability to flash floods in the small mountainous watershed of Aceh Besar Regency Aceh province, Indonesia. Jàmbá J Disaster Risk Stud 10(1):550

Borrero JC (2005) Field data and satellite imagery of Tsunami effects in Banda Aceh. Science 308(5728):1596

Daly P, Sieh K, Seng TY, Edwards ME, Parnell AC, Ardiansyah FRM, Ismail N, Majewski J (2019) Archaeological evidence that a late 14th-century tsunami devastated the coast of northern Sumatra and redirected history. PNAS 116(24):11679–11686

Davis M (2001) Late victorian holocausts: El Niño, famines, and the making of the third world. Verso, London

Doocy S, Gorokhovich Y, Burnham G, Balk D, Robinson C (2007) Tsunami mortality estimates and vulnerability mapping in Aceh, Indonesia. Am J Public Health 97(Suppl 1):S146–S151

Enthoven V, Den Heijer H, Jordaan H (eds) (2013) Geweld in de West: Een Militaire Geschiedenis van de Nederlandse Atlantische Wereld. Brill, Leiden

Giese BS, Ray S (2011) El Niño variability in simple ocean data assimilation (SODA), 1871–2008. AGU J Geophys Res Oceans 116:C02024, tables 1 and 2

Hagen AP (2018) Koloniale oorlogen in Indonesie: Vijf eeuwen verzet tegen vreemde overheersing. De Arbeiderspers, Amsterdam

Harger JRE (1995) Air-temperature variations and ENSO effects on Indonesia, the Philippines, and El Salvador: ENSO patterns and changes from 1866–1993. Atmos Environ 29(16):1919–1942

Ilhamsyah Y, Farhan A, Irham M, Setiawar I, Haditiar Y, Irwandi (2019) Greater Aceh, Indonesia enters climate change: climate on extreme ENSO 2015–2016. IOP Conf Ser Earth Environ Sci 273:012002

Kreike E (2021) Scorched earth: Environmental warfare as a crime against humanity and nature. Princeton University Press, Princeton

Luttikhuis B, Moses AD, (eds) (2014) Colonial counterinsurgency and mass violence: The Dutch Empire in Indonesia. Routledge Milton Park, Abington

Meltzner AJ, Sieh K, Chiang H.-W, Shen C.-C, Suwargadi BW, Natawidjaja DH, Philibosian BE, Briggs RW, Galetzka J (2010) Coral evidence for earthquake recurrence and an A.D. 1390–1455 cluster at the south end of the 2004 Aceh-Andaman rupture. J Geophys Res 115:B10402

Reid A (1969) The contest for North Sumatra: Atjeh, the Netherlands and Britain, 1858–1898. Oxford University Press, London

Sina D, Yan C-R, Wilkinson S, Potangaroa R (2019) A conceptual framework for measuring livelihood resilience: relocation experience from Aceh Indonesia. World Dev 117:253–265

Stolwijk A (2016) Atjeh. Het Verhaal van de Bloedigste Strijd uit de Nederlandse Koloniale Geschiedenis. Prometheus, Amsterdam

Thorburn C (2009) Livelihood recovery in the wake of the tsunami in Aceh. Bull Indones Econ Stud 45(1):85–105

Van 't Veer P (1969) De Atjeh-Oorlog. De Arbeiderspers, Amsterdam

Wekker (1907) Hoe beschaafd Nederland in de twintigste eeuw vrede en orde schept op Atjeh. Avondpostdrukkerij, 's-Gravenhage

Social Responses to Climate Change in a Politically Decentralized Context: A Case Study from East African History

William Fitzsimons

Abstract Over the past 3,000 years, speakers of the Ateker family of languages in East Africa chose various strategies to respond to periods of climate change including the end of the African Humid Period and the Medieval Climate Anomaly. Some Ateker people made wholesale changes to food production, adopting transhumant pastoralism or shifting staple crops, while others migrated to wetter lands. All borrowed new economic and social idea from neighbors. These climate-induced changes in turn had profound social and political ramifications marked by an investment in resilient systems for decentralizing power, such as age-classes and neighborhood congresses. By integrating evidence from historical linguistics and oral traditions with paleoclimatological data, this paper explores how a group of stateless societies responded to climate change. It also considers whether these cases complicate concepts such as "collapse" and "resilience" that are derived from analyses of mostly state-centric climate histories.

Keywords Climate change · East Africa · Historical linguistics · Ateker · Medieval climate anomaly

Introduction: Collapse, Resilience, and the Centralized State in Historical Climatology

The specter of global warming has, over the past couple of decades, amplified climate scientists' and historians' interest in one another's research. While some historians have looked anew at the long-term climate impact of historical epochs such as the agricultural and industrial revolutions while others have examined how past climate change was a shaping factor of earlier human events (Chakrabarty 2009; White 2011, 2017; Degroot 2018). At the same time, climate scientists increasingly recognize that any effective response to global warming must take into account sociological factors. Climate researchers are turning to the historical record to better understand how

W. Fitzsimons (✉)
Columbia College of Chicago, Chicago, IL, USA
e-mail: wfitzsimons@colum.edu

© The Author(s) 2022
A. Izdebski et al. (eds.), *Perspectives on Public Policy in Societal-Environmental Crises*,
Risk, Systems and Decisions, https://doi.org/10.1007/978-3-030-94137-6_10

reconstructed paleoclimatic conditions affected people in real life, and seek to "learn lessons from history" while designing global or local responses to current anthropogenic climate change predictions (Costanza et al. 2007; Gelorini and Verschuren 2012).

Two narrative devices—"collapse" and "resilience"—have come to prominence at this intersection of climatology and history (Butzer and Endfield 2012; Walker and Cooper 2011). Recently, humanistic scholars have questioned the utility of these concepts, pointing out that they are both overly simplistic and carry an implicit normative judgment in favor the status quo (Haldon et al. 2020; Cote and Nightingale 2012: 476; Endfield 2014; Izdebski 2018). Following decades of critical scholarship "from the bottom," historians and others are poised to ask just whose "resilience" we are speaking of, and whether the "collapse" of elite power structures proven ill-equipped to respond to climate change is truly to be mourned? Consider a classic example of climate-induced historical collapse: the Old Kingdom of ancient Egypt, c. 2180 BCE. After the Sixth Dynasty perished through a mixture of toxic elite politics and failing annual Nile floods, "[w]ealth was dispersed to new centers, with economic growth, artistic and cultural change, and a shift to a different style of social complexity. Some of the elite were deeply disturbed by the course of events..." (Butzer 2012: 3633). Historians cognizant of social inequity and wary of privileging elite narratives of history might readily ask, whose "collapse" was this, really?

Left mostly unchallenged in these critiques is another way that collapse/resilience narratives are skewed: they focus on centralized states, leaving aside the majority of past human experience (White et al. 2018; Lieberman and Gordon 2018; Scott 2009: 3). This over-focus on centralized states, which usually entails a parallel focus on sedentary farming or industrial economies, impoverishes the "lessons" that climate scientists can learn from past examples of collapse or resilience (Cosmo et al. 2018). Non-centralized political systems found across much of the premodern non-western world represent some of humanity's most flexible forms of social organization, and have proven resilient in the face of large-scale challenges, including climate change (McIntosh 1999: 1–31; Mattalia et al. 2018). Typical hallmarks of decentralized and mobile societies are distributed risk, inter-group alliance building, and flexible boundaries—in other words, factors that can contribute to resiliency. Historical study contributes improves decision-making not by providing a "how to" manual or even a roadmap from the past, but by generating a reservoir of informative experiences beyond what any single person could accrue in a lifetime—a point perhaps best articulated by military historians (Luvaas 1982; Griess 1988: 37). By overlooking past examples of non-centralized societies in our search for historical examples, we are missing a critical piece of the human-climate story, and thereby limiting the usefulness of historical study for policy makers.

The habit of visualizing "collapse" and resilience" through the lens of the state impedes a more expansive understanding of these twinned concepts that centers ordinary people rather than their governments. To be sure, this tendency reflects biases inherent in research produced within the framework of today's nation-state system. Yet, one of the biggest hurdles to incorporating decentralized societies into climate history is more mundane: the documentary records available to climate historians are almost exclusively produced within sedentary centralized states (Brönnimann et al. 2018: 30; Haldon et al. 2018: 3212; Carey 2012: 236). Recent innovative work combining archaeology and documentary evidence with paleoclimate data has pushed the study of climate history to include rural areas on the margins of states (Roberts 2018). Nevertheless, times and places with little archaeological research or lying well outside the purview of centralized states remain almost entirely absent from climate historiography. Even when climate proxies such as sedimented pollen indexing innovative agricultural activity can provide some indication of broad economic responses to climate change in undocumented regions, historians still lack an "insider perspective" that is necessary to understand the social implications of such shifts (Butzer and Endfield 2012: 3629; Rosen and Rivera-Collazo 2012).

This paper uses the methodology of historical linguistics—a largely untapped source of climate history—to explore the *longue durée* history of social, political, and economic responses to climate change among the politically decentralized Ateker-speaking populations in the grasslands of east Africa (for examples of similar methodological convergence, see Ehret 2011; de Luna 2016; Bostoen et al. 2015; Friedrich 1970). The earliest Ateker speech community emerged in the aftermath of widespread southward migration between 1500 and 500 BCE spurred by the desiccation of the eastern Sahara desert after the end of the African Humid Period (AHP). This culture remained linguistically contiguous for more than a millennium, during which time the Ateker people adapted to a new variable environment in part by borrowing a slew of economic and cultural practices from neighboring groups. Much later, aridity during the semi-global Medieval Climate Anomaly (MCA) c. 900 to 1250 CE spurred some Ateker-speakers to migrate further south, while others endured in their homeland by adopting new cattle herding techniques. After the return of rains c. 1250 CE, Ateker-speakers drew on these arid-period innovations to rapidly expand across a wide area of today's Uganda-Kenya-Ethiopia-South Sudan borderlands. Throughout all this history, Ateker-speakers never established centralized political institutions. Examining this history shows how linguistics can be combined with paleoclimatic data to yield a fuller picture of human-climate interaction in understudied historical settings, while also shedding light on non-centralized responses to climate change.

500 BCE—900 CE: Resilience Through Interaction

The first significant climate event to shape Ateker history actually occurred before the Proto Ateker language was spoken (Fitzsimons 2020). Around 5,000 years ago, seasonal rainfall in the Nilotic Sudan began shifting south, inaugurating the end of the African Humid Period (AHP) and desiccation of the Sahara (Shanahan et al. 2015; Berke et al. 2012; Kröpelin et al. 2008). Over the following two millennia, southward-retreating rains were followed by diverse Sudanese populations, many of whom eventually settled along today's Uganda-South Sudan border, forming a

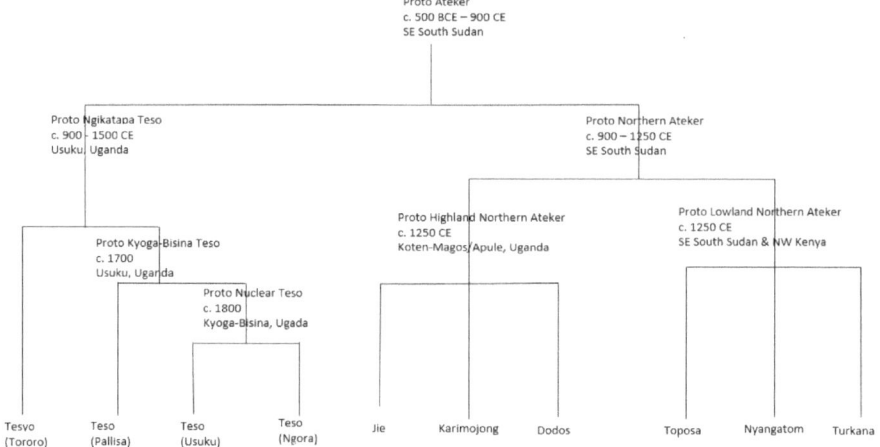

Fig. 1 Linguistic tree of Ateker language family, including postulated homelands and dates

culturally heterogenous zone (David 1982).[1] There, by about 500 BCE, Proto Ateker emerged as a distinct language spoken in the mountainous mixture of grasslands and acacia forest on this zone's eastern edge. The early Ateker lived in close proximity to a range of diverse protolanguage communities, including especially those of the Southwest Surmic, Lwo, and Rub language families (Ehret 1982). Proto Ateker culture and language was first formed, therefore, in a context of migration, cultural contact, and re-settlement in response to long-term climate change. This language community would remain basically intact until c. 900 CE (Fig. 1), demonstrating a marked resilience in the face of climate change.

Having left behind a flat savanna with intermittent swamps and rivers, Ateker-speakers settled in an area of ecological diversity marked by hot low-lying plains tapering off into scrub land to the drier north and cooler forested mountains reaching over 6,000 feet ASL in elevation to the south. When Proto-Ateker speakers first occupied this land, they were already familiar with cultivating finger millet (*Eleusine coracana*) and keeping cattle, goats, and chickens. Archaeological excavations demonstrate the antiquity of livestock keeping in the region, and genetic studies point to the domestication of finger millet in East Africa by this period. We can further pinpoint Ateker participation through linguistics (Kay et al. 2019; Marshall and Hildebrand 2002; Dida et al. 2008) because reconstructed words for "finger millet" *-kima, "goat" *-kine, "cow" *-kiteng and "chicken" *-(ko)-kor, were all inherited from Proto Eastern Nilotic (the * symbol marks a reconstructed protoword; evidence underlying all linguistic reconstructions can be found in Fitzsimons 2020). From their arrival, Proto Ateker-speakers were probably the most skilled herders in their neighborhood, so they found little reason to borrow foreign words to describe sophisticated herding techniques. Proto Ateker-speakers inherited a robust herding lexicon,

---

[1] The term "heterogenous zone" to describe this area is my own.

including items such as *-woro "cow dung," *-dong "to castrate by pounding," *-gelem "to castrate by cutting," *-mong "ox," *-dak "to graze (of livestock)," and *-kyok- "to herd livestock." Newly innovated words, including *-manangit "very young calf," *-kori "he-goat," *-dongot "cow-bell," *-lepit "milking can," *-kere-t "gourd for churning milk," *-doot "salt lick," *-bela "herding stick," *-gum- "to bleed cattle," *-coto "cow urine," and *-tub(w)a "watering trough" were all derived from native Eastern Nilotic sources, while only *-masanik "bull" was borrowed, probably from Southwest Surmic. Globally, livestock have always served as an important hedge against drought and crop failure, and a robust herding culture undoubtedly contributed to Ateker resilience. Throughout this period, livestock of all kinds continued to be prized by Ateker-speakers, who coined a generic term *-bar-en to mean "livestock (not species-specific)," derived from an earlier word meaning simply "wealth."

In the realm of cereal agriculture, Proto Ateker-speakers had more to learn from their new neighbors. Though they could still grow the finger millet inherited from their Eastern Nilotic ancestors, experience with periodic drought appears to have spurred cereal crop diversification. Drought-resistant sorghum (*Sorghum bicolor*) had long been domesticated in the wider region and Proto Ateker speakers paid renewed attention to the crop, borrowing *-momw- from an unknown source language to name it specifically (Winchell et al. 2017). This lexical innovation accompanied a wider revision in the Proto Ateker lexicon for producing cereal-based foodstuffs. About half of these new terms had Eastern Nilotic etymologies, such as *-kut "digging stick," *-kir(i)ya- "grinding stone," and *-tap "bread." Others were borrowed from an array of neighbors, including *-pyet "to winnow" from Proto Lwo, and *edula "granary" from the Moru-Madi language family. Proto Ateker speakers adapted to uneven climatic conditions by growing a more drought-resistant secondary staple and developing new techniques for processing and storing grains, both with the input of neighboring groups.

Food collection–hunting, fishing, gathering–underwent an even more far-ranging transformation during the Proto Ateker period. Ateker speakers learned from already-established communities how to exploit natural resources in their new mountainous homeland. From the Rub, Proto Ateker speakers borrowed the verb *-pok- "to ensare in a trap," as well as the local name *-jeje for the honeyguide bird (*Indicator indicator*) that led followers to rich beehives. Hunting was also part of the early Ateker economy. The root *-riɣ, borrowed from the nearby Proto Nuer-Dinka language with the meaning "to encircle and compress" formed the Proto-Ateker verb *-ri(g)(k)a "to hunt in a group," as well as the noun for the *-ri(g)(k)ak "hunting party" that used this tactic. Finally, Proto Ateker-speakers were ecumenical in borrowing fishing technologies, adopting *-kol- "fish" from Rub, *-biti "small fishing spear" from Lwo, and *-golo "fish-hook" from Southwest Surmic.

New subsistence practices supporting life in a variable environment were buttressed by new technology. Most significantly, the settlement of Proto Ateker speakers in their new homeland marked the start of their participation in East Africa's Pastoral Iron Age (Ehret 1982). They did not smelt iron themselves, but instead acquired iron implements from their neighbors. The iron spear haft (*-morok), iron hoe (*-melek), and iron fish-hook (*-golo), were all borrowed from Southwest Surmic, while the iron axe (*-jep) used to clear forests was borrowed from Lwo. This period also saw improvements in transportation. The donkey, called *-sigiria, was borrowed from Southwest Surmic, and was affixed with a pack-saddle, *saaja-t, to stabilize loads. Other innovations included a gourd bottle, *-tuo, for carrying water on journeys, and new word for a head-carrying pad, *ikit, innovated from a Rub borrowing.

The key to successful farming and pasturing two thousand years ago in East Africa was rainfall, and Proto Ateker-speakers would have paid much attention to precipitation. Here, greater elevation changes in the Proto Ateker community's new home played a crucial role (Jackson 1956). In a region where temperatures in low-lying plains routinely exceed 100 degrees Fahrenheit, moist air coming from the Indian Ocean became super-heated and rose rapidly after encountering physical barriers such as the region's plateaus and mountain complexes. As moist rising air began to cool after reaching elevations between 15,000 and 20,000 feet ASL, the tumultuous blend of falling icy precipitation and rising hot air created static charges that frequently caused powerful thunderstorms and heavy precipitation around mountains. It was, perhaps, after witnessing this entirely new weather pattern that the Proto Ateker language underwent a semantic shift in which the Proto Eastern Nilotic root *-kudyu "rain" (pronounced *-kuju in Proto Ateker) came to denote the "sky" as a conceptually distinct entity, while a new word for the noun "rain" was formed by attaching a verb prefix /aki-/ to the inherited root *-ru "to water plants/animals." Whereas the Proto Eastern Nilotic language had described rain as a noun without agency—something that just happened—Proto Ateker-speakers spoke of rain, *akiru, as a transitive verb performed by a distinct actor, *-kuju "the sky," with the implied intention of watering plants and animals. Although a fully developed theory of *akuj* as the "High God" most likely did not enter Ateker philosophy until a later period, the Proto Ateker language community had already trained their thoughts upwards. To name phenomena such as *-top "Venus," *-kacer "star," and *-gir "thunder," they borrowed words from their Rub neighbors. Terms for unfamiliar high-elevation water sources, such as *-bur "mountainous water pool," and *ecoa "rocky ground spring," were also innovated in Proto Ateker, although these words' origins are unknown.

Over more than a millennium, Proto Ateker-speakers slowly refined their economic and social practices to achieve their own ends. The primary Proto Ateker sociopolitical group was the exogamous clan, called *ateker, of which there were probably no more than two dozen. Livestock were preeminent in Proto Ateker culture, and cattle belonging to each clan were marked with special brands, called *-macar. Hospitality revolved around meat-feasting, with the word *-pej "visitor" deriving from an earlier root meaning "to roast meat." Animal sacrifice, in the meantime, was seen as essential for both mourning rituals, called *-puny, and for harnessing spiritual forces to particular ends through a rite called *-sub-an. These and countless other Proto Ateker practices and values were well-adapted to their local environment, while economic diversification and trade created what was in all likelihood a prosperous and growing community.

900 to 1400—"collapse" and Transformation

Rainfall in the Proto Ateker homeland between the end of the AHP and c. 900 CE, though marked by decadal variations and intermittent drought, was overall sufficient to support the Proto Ateker way of life. Subsistence diversification and intercultural exchange generated enough flexibility in the local socio-ecological system to enable the resilience and longevity of the Proto Ateker language community. However, during a three-hundred-and-fifty-year period beginning c. 900 CE and corresponding with the semi-global MCA, excessively dry conditions caused a permanent rupture to this equilibrium (Lüning et al. 2018; Lüning et al. 2017). Finger millet cultivation became nearly impossible and sorghum cultivation challenging, while water sources for livestock were fewer and farther away.

Water scarcity creates a challenge for the reconstruction of local climate histories, because the best tropical palaeoclimatological sources are found at the bottom of standing lakes, which are not common in the Proto Ateker homeland (Gasse 2000). Bodies of standing water contain piled lake sediments that hold a sequenced history of climatic detritus which can be analyzed to determine the nature of past environments. Types of pollen or the photosynthetic material embedded in stromatoliths found in datable sedimentary layers can index varying climatic conditions at the time of their stratigraphic formation, as can the chemical compositions of lake sediment soils. Throughout the Northern Ateker world, there is only one standing water body—Lake Turkana—that paleoclimatologists have closely studied, and it stands 150 miles away from the Proto Ateker homeland (Morrissey and Scholz 2014; Garcin et al. 2012). This lake can be considered a proxy for the regional environment (Bloszies and Forman 2015). Halfmann et al. measured ratios of fine-grained carbonite isotopes in Lake Turkana sediments as a proxy for water levels, with the assumption that lower ratios would indicate higher overall water volumes. Using this evidence, they conclude that Lake Turkana experienced overall low water levels in the period c. 900–1100 CE (Halfman et al. 1994: 94). Looking throughout the wider region affected by the same monsoon-derived seasonal rains, Verschuren et al. combine evidence from Lake Naivasha in Kenya and Lake Tanganyika in Tanzania with Halfman's study from Lake Turkana to suggest a regional period of aridity on either side of c. 1000 CE that stretched from Tanzania to Northern Kenya (Verschuren et al. 2000). A final unique source for climatic conditions is the "Nilometer" on Rhoda island in Cairo, which has been used to document annual Nile flood levels since the seventh century CE. Because the lower Nile is partially fed by rainfall in northern Uganda, Nile flood levels can be considered an imprecise proxy for climate conditions in the Ateker region. The Rhoda Nilometer matches paleoclimatological studies, indicating periodic severe droughts amid overall low water levels from CE 900–1250 CE (Hassan 2007; Kondrashov et al. 2005; Putter et al. 1998).

A convergence of evidence indicates regional drought in eastern Africa during the MCA at the end of the first millennium CE. This period corresponds with language-based estimates for the divergence of Proto Ateker into Proto Northern Ateker (PNA) and Proto Teso. The story that emerges is one where Proto Ateker speakers faced an existential question at the onset of this arid period: should they migrate south towards wetter climes to maintain an economy based on finger millet and localized livestock-rearing, or undergo a fundamental transformation towards subsistence practices better suited to reduced rainfall? In the end both choices were made, leading to the bifurcation of the Proto Ateker language into two distinct sub-groups. Those who remained behind formed the PNA. They adapted to dry conditions through the use of herding technique called transhumant pastoralism in which households split during the annual dry season so that young adults can take herds in search of far-out grazing lands while elders watch over permanent homes and young children. In contrast, the Proto Teso, remembered in oral traditions as the *Ngikatapa* or "finger millet bread people," moved south and ultimately settled in wetter eastern Uganda (Lamphear 1976). There, they established finger millet farms and joined other migrating families in loose neighborhood associations. Both groups embarked upon paths that fundamentally transformed their social and political lifeways.

Lexical reconstructions of selected flora help track the local impact of the MCA climate shift on the PNA. The most economically significant of these innovations was *erau, for "pearl millet," borrowed from Lwo speakers. Pearl millet (*Pennisetum glaucum*) is a global cereal crop that was first domesticated in West Africa by the third millennium BCE, though it was not significant in Proto Ateker agriculture before 900 CE (Manning et al. 2011). Pearl millet has the advantage of being farmable in areas receiving an average of 400–650 mm per annum of rainfall, as opposed to finger millet's minimum requirement of 600 mm (Dida and Devos 2006). Pearl millet supplemented the cultivation of sorghum, which requires 300–700 mm of rain per annum (Singh and Lohithawa 2006).

Increased aridity is also captured in the PNA arboreal lexicon. When a local climate dries, thirstier tree and shrub species die out except in areas near consistent water sources such as rivers, and they give way to new "pioneer" species with lesser rainfall requirements. We can see evidence of this process playing out in the linguistic record. All newly innovated PNA words for flora refer to species with minimum annual requirements of 200–450 mm and some with upper limits of 800 mm.[2] These include *Commiphora africana* or *ekadeli (250–800 *mm*), *Hyphaene compressa* or *-kVngol (300–900 *mm*), *Salvadora persica* or *esiokon (300–800 mm), and the fodder grass *Bothriochloa insculpta* or *elet (450–1500 mm).

Words describing the dispersed cattle herding techniques that enabled PNA-speakers to adapt to arid conditions can be dated to this same period. These include the annual assembly inaugurating cattle movement, called *-ud-akin and derived from Proto Ateker *-ud "to prod, to push together," and the act of blessing to "release" cattle, called *wos from a root meaning "to let go." PNA-speakers also innovated the term *-bor for dry season camps from the root "to depart, to break away," and directionally specific verbs *-ram "to drive cattle toward" and *-twar "to drive cattle away" in order to describe the annual to-and-fro. Finally, they adopted new objects to assist with long-distance herding, including *-ku-wos(i) "gourd for carrying cow

---

[2] For higher rainfall in higher elevations in South Sudan, see J. K. Jackson, "The Vegetation of the Imatong Mountains, Sudan," *Journal of Ecology*, 44, 2 (1956), 343.

urine" (used for cleaning), *ku-tam "leather sack for carrying butter," and *-gec "stirring stick (used for mixing milk and blood)."

Greater spatial separation challenged inherited political ideals emphasizing "gathering together" while also causing anxiety among elders, who gave up control over life-giving cattle herds for half of each year (Lamphear 1983: 114). The most significant political innovation during this period was an age-class system, called *asapan, which effectively addressed both concerns (Fitzsimons 2020: 157–210). With clans and lineages thinly spread across an arid landscape, the formation of overarching social classes based on age provided a substitute ideology of group-making that could bind together fragmented settlements. The particular structure of *asapan, which granted political power to the generation of the "fathers" over that of the "sons," helped quell elders' anxieties about the growing economic responsibility of the youth while also providing young men with a predictable pathway to authority in due time. Women, who had traditionally exercised power through clan structures, were initially disenfranchised by this new system, but later adopted parallel initiation institutions taking the form of "singing groups" (Gourlay 1970; Fitzsimons 2020: 240).

Turning south, the Proto Teso who migrated south to continue cultivating finger millet met climate change with an entirely different set of responses, also altering the older patriclan system of the Proto Ateker. As families and friends migrated together in small pioneer groups, the approximately two dozen over-arching Ateker patriclans lost their practical coherence (Webster et al. 1973). These migrants continuously invested in the invention and re-invention, naming and re-naming, of new smaller and more localized "sub-clans," leading to the proliferation of well over one thousand Teso sub-clans by the eighteenth century. When they reached new lands, migrants encountered new peoples, many of whom spoke versions of Lwo and Rub languages. The general pattern of settlement, then, was one of small groups of families and friends living in diverse neighborhoods comprised of both numerous Teso-speaking sub-clans and entirely foreign cultures. The Proto Teso language community eventually came to fully occupy the area of eastern Uganda between the lakes Bisina and Kyoga, the heart of today's Teso sub-region.

To coordinate territorial activities between unrelated small-scale settlements, Proto Teso-speakers innovated a political assembly called *etem, derived from the Proto Ateker word meaning "hearth." These assemblies proved effective for integrating non-Teso communities, including especially members of the Lwo-speaking Inomu and Ikomolo clans, which became thoroughly "Teso-ized" during this period (Cohen 1972: 144–145; Cohen 1988: 67–68). As they integrated these Lwo communities, Teso-speakers also learned about their new landscape from them, borrowing words for two important fodder grasses dominant in the region—Guinea Grass (*Panicum maximum*), called *edinyo, and Elephant Grass (*Pennisetum purpureum*), called *egada.

Within the *etem* institution, Proto Teso-speakers took care to maintain a balance of power between local exogamous sub-clans, preventing any single sub-clan from becoming too dominant. Assembly speakers, called *airabis from the root word *rab "to speak," were elected according to meritocratic principles, and the position of speaker was not inherited from father to son. Structural equality between sub-clans was also pursued through strict protocols related to marriage exchanges. To prevent one sub-clan from becoming indebted to another through marriage, bridewealth exchanges were converted into a years-long process called *-yit, from the Proto Ateker root meaning "to drip slowly." Marriage transactions involved a series of reciprocal gift-giving calculated so that families united through marriage would end up with a clean balance sheet between them (Nagashima 1981).

The practice of migration within a sedentary farming economy also spurred new social norms at the individual level, with an increased expectation that young families should build their own separate homes. A new word for "marriage," *-many, was derived from an earlier verb meaning simply "to dwell together," while a married man was called *-duk-okina, literally "one who has built a home." That new homesteads should have been separate from one's extended family is implicit in an innovated word for "sibling," *inac, derived from the root "to avoid, to pass by." Notions of prosperity also followed greater investment in cereal agriculture. Whereas the Proto Ateker had only conceptualized wealth in terms of livestock, cereal-growing Proto Teso-speakers innovated a corollary term, *amion, meaning "wealth in crops" specifically (Stephens 2018).

By the end of the thirteenth century, rainfall in East Africa returned to pre-MWP levels. Although the aridity that had stimulated transformations among Ateker-speakers disappeared, the fundamental social changes it had wrought were permanent. Political institutions such as *asapan* age-classes and *etem* assemblies, initially created to ensure the survival of Ateker communities during hard times, provided a foundation for the rapid expansion of Ateker-speakers across a large swath of the region once rains returned. The flexible and decentralized strategies that enabled resilience during a climate crisis also proved effective for constituting political communities in times of relative plenty. By 1800, linguistic descendants of the PNA and Proto Teso peoples dominated more than 100,000 square miles of grasslands in Uganda, South Sudan, Kenya, and Ethiopia, while the *asapan* system was adopted by neighboring groups throughout the region.

Conclusion

Reconstructing the *longue durée* socio-ecological history of a decentralized society such as the Ateker offers new insights into the themes of collapse and resilience which have long been at the heart of historical climate research. To begin with, this focus offers a new set of metrics by which such concepts can be measured. While state-centric historical climatology tends to be framed by the start- and end-dates of particular dynasties or government systems, these neat dividing lines are unavailable for undocumented and decentralized societies, and they can be misleading as climate trends do not adhere to political histories. Protolanguage longevity can serve as a useful alternative proxy, because lack of linguistic divergences may indicate an absence of major disruptions to other long-term social factors. In a climatic regime characterized by decadal variability, such as that experienced by the early Ateker, linguistic longevity implies a certain degree of socio-ecological resilience. In the Ateker case, resilience was produced by a flexible subsistence system comprised of internal innovations and foreign borrowings.

The bifurcation of Proto Ateker as a result of climate-induced migrations beginning c. 900 CE provides a useful example of the complex relationships scholars have recently noted between collapse, resilience, and transformation. Traditional narratives pose collapse and resilience as opposing outcomes, but analysis of stateless histories reveal this as too simple of a dichotomy. There is no obvious standard to judge whether the more "resilient" Ateker speakers were those who migrated south to maintain their basic subsistence practices, or those who adopted transhumant pastoralism to remain in their homeland. At the level of the society, if not the individual, both strategies seem to have been successful, since descendant communities of both groups increased in population and territory through the rest of the millennium. But they did so by either undertaking a wholesale transformation of their economic and political systems, or leaving their homeland entirely. The ultimate value of bringing examples from decentralized societies into the field climate history may be that, by removing the deceptively straight-forward phenomenon of the state from consideration, these very categories themselves are shown to be insufficient.

So what are the "lessons," if any, that climate researchers might take away from this Ateker case study? Two come to mind. The first is the value of building bridges to foreign communities–welcoming new ideas and new people. The unusual longevity of Proto Ateker language and culture in a marginal environment surely would not have been possible without borrowing subsistence practices and material technologies from a diverse range of neighbors. Beyond these practical borrowings, however, Proto Ateker speakers also adopted foreign ways to speak about topics ranging from the cosmos and ritual practice to trade and the geographical features of their homeland. After c. 900 CE the *asapan age-class system enabled the Proto Northern Ateker to forge cohesive political communities from multiple far-flung lineages, while the *etem assemblies did the same for Proto Teso-speakers.

The second is that a community's survival during periods of climate change may require fundamental transformations to basic lifeways and shared ideals. Ateker speakers endured the arid period of 900–1250 CE—and thrived thereafter—because they were willing to take bold collective steps in response to environmental pressure. For the Proto Teso, this meant leaving the homeland they had known for over a millennium and finding their way among new people living in a new place. For the PNA, this meant giving up a lifestyle of sedentary farming and revising family practices to cope with months of geographical separation.

For a world facing an anthropogenic climate change crisis today, these "lessons" are not directly applicable, of course. No one would suggest we should all become transhumant pastoralists. However, the general themes of openness to foreign people and their ideas, and a willingness to undergo fundamental economic and socio-political transformation do seem relevant today. In the end, the greatest value that we can glean from examples of historical responses to climate change is to build a richer repertoire of narratives about our shared transhistorical human condition of living in societies that are inextricable from their natural environment. Focusing only on examples from centralized states impoverishes that repertoire, and threatens to stifle the collective creative imagination that must be at the center of any effective response to global warming.

References

Berke M, Johnson TC, Werne JP, Schouten S, Sinninghe Damsté JS (2012) A mid-Holocene thermal maximum at the end of the African Humid Period. Earth Planet Sci Lett 351–352:95–104

Bloszies C, Forman S (2015) 'Potential relation between equatorial sea surface temperatures and historic water level variability in Lake Turkana Kenya. J Hydrol 520:489–501

Bostoen K, Clist B, Doumenge V, Grollemund R, Hombert J, Muluwa JK, Maley J (2015) Middle to late holocene paleoclimate change and the early bantu expansion in the rain forests of Western Central Africa. Curr Anthropol 56(3):354–384

Brönnimann S, Pfister C, White S (2018) Archives of nature and archives of societies. In: White S, Pfister C, Mauelshagen F (eds) The Palgrave handbook of climate history. Palgrave, London, pp 27–36

Butzer K, Endfield G (2012) Critical perspectives on historical collapse. PNAS 109(10):3628–3631

Butzer K (2012) Collapse, Environment, and Society. In: Proceedings of the National Academy of Sciences (PNAS), 109(10):2633

Carey M (2012) Climate and history: a critical review of historical climatology and climate change historiography. Wires Clim Change 3(3):233–249

Chakrabarty D (2009) The climate of history: four theses. Crit Inq 35:197–222

Cohen DW (1972) The historical tradition of Busoga: Mukama and Kintu. Clarendon, Oxford

Cohen DW (1988) The cultural topography of a Bantu Borderland: Busoga, 1500–1800. J Afric Hist 29(1):57–79

Costanza R, Graumlich L, Steffen W, Crumley C, Dearing J, Hibbard K, Leemans R, Redman C, Schimel D (2007) Sustainability or collapse: what can we learn from integrating the history of humans and the rest of nature. Ambio 36(7):522–527

Cote M, Nightingale A (2012) Resilience thinking meets social theory: situating social change in socio-ecological systems (SES) research. Prog Hum Geogr 36(4):475–489

David N (1982) The BIEA Southern Sudan expedition of 1979: interpretation of the archaeological data', In: J Mack, P Robertshaw, (eds.) Culture history in the Southern Sudan: archaeology, linguistics and ethnohistory. British Institute in East Africa, Nairobi, pp 49–57

De Luna K (2016) Collecting food, cultivating people: subsistence and society in Central Africa. Yale University Press, New Haven

De Putter T, Loutre M, Wansard G (1998) Decadal periodicities of Nile River historical discharge (A.D. 622–1470) and climatic implications. Geophys Res Lett 25(16):3193–3196

Degroot D (2018) Frigid golden age: climate change, the little ice age, and the Dutch Republic, 1560–1720. Cambridge University Press, New York

Di Cosmo N, Hessl A, Leland C, Byambasuren O, Tian H, Nachin B, Pederson N, Andreu-Hayles L, Cook ER (2018) Environmental stress and steppe nomads: rethinking the history of the Uyghur Empire (744–840) with Paleoclimate Data. J Interdiscip Hist 48(4):439–463

Dida MM, Devos KM (2006) Finger Millet. In: Kole C (ed) Genome mapping and molecular breeding in plants, vol 1. Cereals and Millets. Springer, New York, pp 333–344

Dida M, Wanyera N, Harrison Dunn ML, Bennetzen JL, Devos KM (2008) Population structure and diversity in finger millet (Eleusine coracana) germplasm. Tropical Plant Biology 1:131–141

Ehret C (2011) History and the testimony of language. University of California Press, Berkeley

Ehret C (1982) Population movement and culture contact in the Southern Sudan, c. 3000 BC to AD 1000: a preliminary linguistic overview. In J Mack, P Robertshaw, (eds.) Culture history in the Southern Sudan: archaeology, linguistics and ethnohistory. British Institute in East Africa, Nairobi, pp 19–39

Endfield G (2014) Exploring particularity: vulnerability, resilience, and memory in climate change discourses. Environ Hist 19:303–310

Fitzsimons W (2020) Distributed power: climate change, elderhood, and republicanism in the grasslands of East Africa, c. 500 BCE to 1800 CE. Doctoral Dissertation, Northwestern University

Friedrich P (1970) Proto-Indo-European trees: the arboreal system of a prehistoric people. University of Chicago Press, Chicago

Garcin Y, Melnick D, Strecker M, Olago D, Tiercelin J (2012) East African mid-Holocene wet-dry transition recorded in paleo-shoreline of Lake Turkana, northern Kenya Rift. Earth Planet Sci Lett 331–332:322–334

Gasse F (2000) Hydrological changes in the African tropics since the Last Glacial Maximum. Quatern Sci Rev 19:189–211

Gelorini V, Verschuren D (2012) Historical climate-human-ecosystem interaction in East Africa: a review. Afr J Ecol 51:409–421

Gourlay KA (1970) Trees and anthills: songs of karimojong women's groups. Afr Music 4(4):114–121

Griess T (1988) A Perspective on Military History. In: Jessup J, Coakley R (eds) A guide to the study and use of military history. Center of Military History, Washington, DC, pp 25–41

Haldon J, Mordechai L, Newfield T, Chase A, Izdebski A, Guzowski P, Labuhn I, Roberts N (2018) History meet palaeoscience: consilience and collaboration in studying past societal responses to environmental change. PNAS 115(13):3210–3218

Haldon J, Chase A, Eastwood W, Medina-Elizalde M, Izdebski A, Ludlow F, Middleton G, Mordechai L, Nesbitt J, Turner BL (2020) Demystifying collapse: climate, environment and social agency in pre-modern societies. Millennium: Yearbook on the Culture and History of the First Millennium C.E., 17(1):1–33

Halfman J, Johnson TC, Finney B (1994) New AMS dates, stratigraphic correlations and decadal climatic cycles for the past 4 ka at Lake Turkana Kenya. Paleogeograp Paleoclimatol Paleoecol 111:83–98

Hassan F (2007) Extreme Nile floods and famines in Medieval Egypt (AD 930–1500) and their climatic implications. Quatern Int 173–174:101–112

Izdebski A (2018) The social burden of resilience: a historical perspective. Hum Ecol 46:291–303

Jackson JK (1956) The vegetation of the Imatong Mountains, Sudan. J Ecol 44(2):343

Kay DK, Lunn-Rockliffe S, Davies M (2019) The archaeology of South Sudan from c. 3000 BC to AD 1500. Azania Archaeol Res Africa 54(4):516–537

Kondrashov D, Feliks Y, Ghil M (2005) Oscillatory modes of extended Nile River records. Geophys Res Lett 32(10)

Kröpelin S, Verschuren D, Lézine AM, Eggermont H, Cocquyt C, Francus P, Cazet JP, Fagot M, Rumes B, Russel JM, Darius F, Conley DJ, Schuster M, von Suchodoletz H, Engstrom DR (2008) Climate-driven ecosystem succession in the Sahara: the past 6000 years. Science 320:765–768

Lamphear J (1976) The traditional history of the Jie of Uganda. Clarendon, Oxford

Lamphear J (1983) Some thoughts on the interpretation of oral traditions among the central paranilotes. In: R Vossen, M Bechaus-Gerst, (eds.) Nilotic studies: proceedings of the international symposium on language and history of the nilotic peoples, Cologne, January 4–6, 1982. Dietrich Reimer Verlag, Berlin, pp 110–146

Lieberman B, Gordon E (2018) Climate change in human history: prehistory to the present. Bloomsbury, New York

Lüning S, Gałka M, Danladi IB, Adagunodo TA, Vahrenholt F (2018) Hydroclimate in Africa during the medieval climate anomaly. Palaeogeogr Palaeoclimatol Palaeoecol 495:309–322

Lünning S, Gałka M, Vahrenholt F (2017) Warming and cooling: the medieval climate anomaly in Africa and Arabia. Paleoceanography 32:1219–1235

Luvaas J (1982) Military history: is it still practicable? Parameters 12(1):2–14

Manning K, Pelling R, Higham T, Schwenniger J (2011) 4500-year old domesticated pearl millet (*Pennisetum glaucum*) from the Tilemsi Valley, Mali: new insights into an alternative cereal domestication pathway. J Archaeol Sci 38(2):312–322

Marshall F, Hildebrand E (2002) Cattle before crops: the beginnings of food production in Africa. J World Prehist 16(2):99–143

Mattalia G, Volpato G, Corvo P, Pieroni A (2018) Interstitial but resilient: nomadic shepherds in piedmont (Northwest Italy) amidst spatial and social marginalization. Hum Ecol 46(5):747–757

McIntosh SK (1999) Beyond chiefdoms: pathways to complexity in Africa. Cambridge University Press, Cambridge, UK

Morrissey A, Scholz C (2014) Paleohydrology of Lake Turkana and its influence on the Nile river system. Palaeogeogr Palaeoclimatol Palaeoecol 403:88–100

Nagashima N (1981) Is the wife-giver superior? The affinal relationship among the Iteso of Kenya with special reference to Ivan Karp's proposition. In: N Nagashima, (ed.) Themes in socio-cultural ideas and behaviour among the six ethnic groups of Kenya. Hitotsubashi University, pp 45–68

Roberts N (2018) Not the end of the world? Post-classical decline and recovery in rural Anatolia. Hum Ecol 46:305–322

Rosen A, Rivera-Collazo I (2012) Climate change, adaptive cycles, and the persistence of foraging economies during the late Pleistocene/Holocene transition in the Levant. PNAS 109(10):3640–3645

Scott JC (2009) The art of the not being governed: an anarchist history of upland Southeast Asia. Yale University Press, New Haven

Shanahan TM, McKay NP, Hughen KA, Overpeck JT, Otto-Bliesner B, Heil CW, King J, Scholz CA, Peck J (2015) The time-transgressive termination of the African Humid Period. Nat Geosci 8:140–144

Singh H, Lohithawa C (2006) Sorghum. In: Kole C (ed) Genome mapping and molecular breeding in plants, vol 1. Cereals and Millets. Springer, New York, pp 257–302

Stephens R (2018) "Wealth", "Poverty" and the question of conceptual history in oral contexts: Uganda from c. 1000 CE. In: Fleish A, Stephens R (eds) Doing conceptual history in Africa. Berghahn Books, New York, pp 21–48

Verschuren D, Laird K, Cumming B (2000) Rainfall and drought in equatorial east Africa during the past 1,100 years. Nature 403:410–414

Walker J, Cooper M (2011) Genealogies of resilience: from systems ecology to the political economy of crisis adaptation. Secur Dialogue 42(2):143–160

Webster JB, Emudong CP, Okalany DH, Egimu-Okuda N (1973) The Iteso during the Asonya. East African Publishing House, Nairobi

White S (2011) The climate of rebellion in the early modern Ottoman Empire. Cambridge University Press, New York

White S (2017) A cold welcome: the little ice age and Europe's encounter with North America. Harvard University Press, Cambridge, MA

White S, Pfister C, Mauelshagen F (eds) (2018) The Palgrave handbook of climate history. Palgrave, London

Winchell F, Stevens C, Murphy C, Champion L, Fuller D (2017) Evidence for sorghum domestication in Fourth Millennium BC Eastern Sudan. Curr Anthropol 58(5):673–683

Resilience at the Edge: Strategies of Small-Scale Societies for Long-Term Sustainable Living in Dryland Environments

Arlene Rosen

Abstract Modern Western communities have much to learn from the ways in which small-scale societies have survived and even thrived while cycling through phases of profoundly shifting moist to dry environmental conditions. In doing so, these small communities display a resilience developed from thousands of years of being rooted in what Western Society considers 'marginal' environments. The most important of the solutions they developed are sustainably rooted in deep-time and identifiable in archaeological records. The ability to live sustainably in these kinds of challenging environments emerges from a profound and long-term reservoir of 'Traditional Ecological Knowledge' that includes a keen awareness of the interface between human needs and natural processes. Although these traditional solutions may not apply to massive complex systems that drive the survival of large cities as a whole, we can benefit a great deal from the study of these past societies to help generate ideas for smaller segments and sub-systems of larger cities, such as neighborhood collectives, urban gardening, water conservation methods, and others that will lead us towards a more sustainable existence on our planet through the use ground-up solutions.

Keywords Wetland exploitation · Foraging societies · Pastoralists · Resource storage · Diversification

Introduction

Complex societies have the potential to provide members with resilience to climate change through numerous interconnected social institutions and advanced technological innovations. Small-scale societies lack this overarching umbrella effect of a vast social and economic network and infrastructure. These smaller, more localized groups must rely on a number of "ground-up" mechanisms for resilience and ultimate survival when abruptly changing climates lead to shifts in environments that affect

A. Rosen (✉)
University of Texas, Austin, USA
e-mail: amrosen@austin.utexas.edu

© The Author(s) 2022   161
A. Izdebski et al. (eds.), *Perspectives on Public Policy in Societal-Environmental Crises,*
Risk, Systems and Decisions, https://doi.org/10.1007/978-3-030-94137-6_11

the plant and animal resources these groups rely on for subsistence and materials critical for maintaining their lifeways. This is especially true in semi-arid environments with plant and animal distributions that are highly sensitive to variations in rainfall and droughts. Our current situation of a steadily warming planet with the accompanying increase of drought years, fires, and decrease in rangelands across the globe, have led numerous geographers, anthropologists, and archaeologists to investigate the impacts of these adverse climatic changes on societies through time. Many have studied the ways in which small-scale societies living in present-day dryland environments use their deeply rooted systems of Traditional Ecological Knowledge (TEK) to exist in these challenging settings.

Modern Western communities have much to learn from the ways in which small-scale societies have survived and even thrived while cycling through phases of dramatic environmental changes from moist to dry conditions, and in doing so, display a resilience that can only come from thousands of years of being rooted in these environments considered marginal by more complex agricultural societies. Although our present-day technological societies are far more able to apply innovative scientific and engineering solutions to immediate situations of drought, decreasing crop yields, and increasing vulnerability to fires, what they lack are solutions for long-term sustainable maintenance of the landscape and agricultural resources. Adaptations to dryland environments by past societies can help provide insights to future sustainability in an ever-challenging environment of global warming. This may be especially true for some of the smaller, more independent social subsets within the larger more complex modern social systems, such as smallhold farmers, and urban gardeners. To gain insights into some of the challenges and innovations of adaptations by prehistoric small-scale societies to climate change and droughts in drylands, I will explore some of the commonalities of strategies for maximizing the predictability of critical resources in two dryland regions. One is the semi-arid zones of the Southern Levant, and the other is the Gobi Desert of southern Mongolia.

Modern researchers studying contemporary and ethnohistorical accounts of foragers in areas such as Australia and the Great Basin in the Western United States have identified the ways in which these groups respond to abrupt adverse climate change in the present and recent past. Archaeologists have built on these ethnographic observations using evidence from archaeological sites. Together these data can contribute a deep time perspective that shows there are patterns of adaptive strategies, which foragers implement in response to cycles of climatic degradation and amelioration over the course of thousands of years. These patterns suggest there are long-term memories of methods by which these small-scale societies respond to adverse environmental changes. These methods and solutions to survival pass down through the generations as TEK through mechanisms such as generational memory, stories of the recent past, and finally myths and legends with this information embedded within. These memory-messages may include the ways of exploiting plants and animals that thrive in harsh environments, including some poisonous plants and not-so-palatable animals, ways of finding sources of sub-surface water, and rituals that promote social bonding and resilient social institutions such as traditions of sharing. Some of this Traditional Knowledge and lifeways for thriving under adverse climatic conditions are recognizable in the archaeological record and can help us understand the longevity of a forager and early pastoral lifestyle in these very harsh and unpredictable dryland regions. More importantly, they contain methods for the long-term sustainable use of resources. Some of the most important of the strategies that are identifiable in the archaeological record include the following four factors:

(1)   *Water Resources*, including knowledge of the range of variability of unmanaged water sources, and/or the technology to manage these sources at varying levels of engineering skills.
(2)   *Strategies for diversification* of resources, diet, social institutions, and toolkits.
(3)   *Some form of low-level storage*, either a natural characteristic of a particular plant food, or technologies for caching or preserving food items in times of abundance.
(4)   *"Niche Construction"*—the ability to alter habitats for improved reliability of resource acquisition, at some level of action.

All four of these strategies rely heavily on a society's Traditional Ecological Knowledge, which can shift, or increase through time. For hunter-gatherers living in dryland regions, these four important elements of dryland survival are all facilitated by proximity to wetland environments. The wetlands are a key focal point, providing water and food resources, but also provide for increased diversity, natural storage, and a fertile environment for niche construction at varying levels of technology.

(1)   Water and desert adaptations

Watery places in desert environments typically become "landscape anchors" (Hammer 2014) and "persistent places" (Schlanger 1992; Olszewski and al-Nahar 2016; Maher 2019) which attract human groups over the course of millennia or throughout the lifespan of these water sources. These can be lakes, rivers, springs, marshy wetlands, ephemeral playa lakes, and human-made cisterns and wells. Water is a limiting factor in semi-arid environments and arguably one of the most critical resources for human settlement and subsistence. Settled agricultural societies deal with the negative effects of too little or too much water on the landscape in the form of droughts and floods respectively which destroy crops, agricultural infrastructures and lead to famine and human suffering. But the advantage for settled societies is their long-term experience and profound knowledge of their immediate environs. They have had time to adapt to and learn about the variability of the rainfall patterns leading to droughts or floods, assess risks, and respond accordingly. If cycles of droughts and floods impact these societies with relative regularity, then agricultural societies can prepare for such situations through technological and social planning. If drought or flood inflection points are unpredictable and severe occurrences, they can enhance the inherent weaknesses within a social and political system of organization. This contrasts with small-scale mobile populations which tend to have more resilient strategies of mobility, social flexibility for altering group size, and knowledge of a broader range of environmental zones across the landscape.

The control of water resources in dryland environments is an area of much research, discussion and debate among anthropological archaeologists concerned with developing agricultural societies and the rise of complex social systems (Mithen 2010; Hammer 2018). These debates go back to the 1930 and 1950s with Childe (1936), and Wittfogel (1957), including the proposition that the very process of building irrigation systems was the single most important factor leading to the rise of elite managers and eventually to stratified societies. Of course, this proposition led to much debate and eventual rejection of the idea in its simple and basic form (Adams 1960; Butzer 1996; Mithen 2010). This focus on the core area of settled agricultural societies and their relationships to water resources in prehistory has vastly over-shadowed the adaptations of smaller societies living on the peripheries of complex societies, and indeed their forager predecessors who ultimately developed resilient adaptations to water scarcity in their home ranges.

These seemingly diverse human social/ecological systems, early complex societies, marginal small-scale societies (foragers, pastoralists, and subsistence farmers living on the peripheries of contemporaneous complex societies), and preceding generations of bands of foragers, are closely related, either in the vertical trajectory of deep time with learned experiences of a landscape—or horizontally across a social/economic system where peripheral small-scale societies are tightly woven into a more complex core. Yet the use, exploitation, limitations, and impact of water sources in these small-scale societies thriving in dryland regions have received much less attention from archaeologists (Fuller and Qin 2009; Hammer 2018).

Another challenge to consider in the study of adaptations by small-scale societies to water resources in drylands is that water availability is a moving target at all temporal scales. Water sources vary seasonally with fluctuations in the amount and timing of rainfall events. Yearly averages rise and fall dramatically around a statistical mean, and total amounts can shift greatly over the course of decades and millennia in a secular or stochastic manner (Butzer 1982). A key element to consider then, is the "predictability" of rainfall and water sources. Human social/ecological systems are innovative and can adapt to the driest deserts on the planet if these environments are predictably dry. It is the series of unpredictable and unexpected weather events, causing sudden droughts and floods that throw social/economic systems off balance and subsequently exacerbate inherent weaknesses in the social and economic systems. This is true of both small-scale and complex societies.

Archaeological and ethnohistorical evidence points to various patterns of movement and exploitation of water resources among mobile foragers in dryland environments. Two of the most important are basin-centered wetlands, and linear valleys containing spring-fed perennial flow.

Basin Wetlands

It is well-known that wetlands are a key ecological zone for foragers (Janetski and Madsen 1990; Kelly 1990; Nicholas 1998; Ramsey 2016; Ramsey and Rosen 2016). In regions with broad basins, the wetlands form from the emergence of underground springs yielding perennial marshes, catchment basins with no outlet that accumulate standing water and form seasonal playa lakes which collect water in rainy seasons but have no outlet, or more permanent saline lakes. They also can occur around the margins of permanent lakes with perennial outlets.

Wetlands are replete with essential resources that provide materials for shelters, basketry, matting, and protein in the form of birds, fish, eggs, mammals, and reptiles that are attracted to the water sources. One of the most important nutrients provided by wetlands are the calories from starches derived from the roots rhizomes and shoots of cattails (*Typha* sp) and sedges (Cyperaceae), for example *Cyperus* sp. and *Scirpus* sp. (Hillman et al. 1989; Wollstonecroft et al. 2008). The contribution of roots and rhizomes supplied by wetland plants is a critical resource for foragers living in general dryland habitats. There are few other sources of these essential nutrients in semi-arid environments, especially in the interfluves. However, in contrast, protein from small mammals and reptiles is more readily available in the uplands and interfluves of these zones.

The water, food, and materials from wetlands provide a key focal point for foragers living in semi-arid regions especially in times of drought. Ramsey and Rosen (2016) have focused on these ecozones as an essential part of forager adaptations to dryland environments in the Southern Levant, describing these societies as being tethered to the wetlands, since they are often a vital component of their seasonal migrations. The tethering effect of the wetlands might lead to a radial pattern of movements among mobile foragers as they would be free to radiate out from the wetland focal points to hunt and collect other important resources. In years of abundant rainfall, migrations can extend further from the wetlands, in years of more sparse rainfall, populations would be restricted in the distances they could travel away from the water and its embedded resource base. Pastoralists had the added concern of pasturage for their flocks (Hammer 2018).

The configuration of basin centric wetlands existed for millennia in both the Southern Levant, and the Gobi Desert of southern Mongolia during the late Pleistocene through the middle Holocene. The Levantine corridor has been a conduit for hominins and *Homo sapiens* for close to two million years (Goren-Inbar and Saragusti 1996; Gaudzinski 2004). It is a narrow strip of land flanked by the Mediterranean Sea to the west, and the harsh deserts of the Arabian Plateau to the east. Throughout these time periods there have been many dramatic shifts in climate and water availability. In the Levant, wetlands have been geobotanical foci which attracted numerous groups of foragers through time. Wetlands around Lake Kineret (Sea of Galilee) in present-day Israel were home to the Kebaran Period foragers at the site of Ohalo II around 23,000 cal BP. These peoples lived in a semi-sedentary village on the lake shore during the coldest-driest episode of the Late Pleistocene, known by climatologists as the Last Glacial Maximum (LGM) (Nadel et al. 2004; Ramsey et al. 2017). This time period was marked by the contraction of the oak/pistachio forest, expansion of the grassland vegetation and frequent droughts (Rosen 2007).

The unique situation of Ohalo II as a waterlogged site, has led to phenomenal preservation of organic remains, giving archaeologist a full picture of how these Kebaran Period foragers not only existed, but thrived on the shores of these wetlands, exploiting fish, birds, small-grained grasses, wild cereals, and the rhizomes of sedges (Weiss et al. 2004a, b). Ohalo II demonstrates that the presence of the lake shore and accompanying wetland surrounding it was a key focal point contributing to the resilience of these foragers during a long phase of otherwise unpredictable resource distribution and availability.

Likewise, at the Natufian settlement at Eynan/Ain Mallaha, foragers were able to live year/round by the marshland of the Hula Lake and wetlands in modern-day northern Israel, while in other localities further south, Natufian foragers had to increase their mobility to satisfy their subsistence requirements (Goring-Morris, Hovers et al. 2009). During the Younger Dryas cool-dry climatic episode (ca. 12,900–11,700 BP) of the terminal Pleistocene, the Late Natufian foragers adapted to the dry conditions by broadening their use of multiple species of plants and animals including small-grained grasses, and a wide variety of small and medium-sized mammals (Stiner 2004; Weiss, Wetterstrom et al. 2004a, b; Munro and Atici 2009; Rosen and Rivera-Collazo 2012), compared with the Early Natufians of the preceding Bølling–Allerød warm-moist episode (ca. 14,700–12,900 BP). The Late Natufians increased their mobility and organized into smaller, more highly mobile groupings. They frequently visited springs spots, lake shores, seasonal playa lakes and wetlands (Ramsey et al. 2015, 2016; Maher et al. 2016), exploiting much the same types of resources as were targeted by the Kebaran Period peoples who lived on the shores of Lake Kineret (Galilee) during the very cold-dry Last Glacial Maximum (LGM), at 23,000 BP at the site of Ohalo II. Evidence from the Kebaran Period archaeological site of Ohalo II, show heavy exploitation of a wide range of wetland resources including fish, birds, reeds, rushes, and importantly both small-seeded grasses as well as wild cereals. The Late Natufian populations living through similar cold-dry conditions at sites such as Eynan/Mallaha on the shores of Lake Hula in modern-day northern Israel relied on a similar broad resource base of small/medium-sized mammals, fish, and small-seeded grasses as well as wild cereals.

It is worth noting the that the Early Natufian (ca. 15,100–13,000 cal BP) populations who lived during the warmer/wetter Bølling/Allerød Period may have differed from both the preceding Kebaran (ca. 23,000 BP) and the subsequent Late Natufians (13,000–11,700 cal BP) in their plant resource selection. This was a period when the oak woodlands were expanding, and the phytolith data suggests an increased reliance on woodland resources, including acorns and perhaps pistachios, and less on grass seeds, indicating a movement away from the broader, more diverse plant resource base.

Archaeologists have shown a similar pattern of reliance on basin wetlands in the Azraq Basin of northern Jordan (Ramsey et al. 2015). Here again, the seasonal expansion of the playa lake formed a key draw for Late Pleistocene Epipaleolithic foragers, allowing the exploitation of critical protein and starch resources throughout the late and terminal Pleistocene (Janz 2012).

In the desert-steppe zones of the southern Mongolian Gobi, Lisa Janz's research has shown that early Holocene foragers began intensive exploitation of a broader range of plant resources such as the 'underground storage organs' (UGS) of sedges, wild grass seeds, and small-fast "*r*-selected species" with high reproductive rates, as the large mammal populations diminished at the end of the Pleistocene. She attributes this to a more specialized intensification of land-use potential (Janz et al. 2017). This went together with increasing warmth and humidity, and the early development of wetlands, as they began to appear within swales of eroded Pleistocene lake beds, and other localities in the Gobi dryland regions (Feng et al. 2005; Janz et al. 2021; Rosen et al. n.d.). In her research, Janz defined three phases of terminal Pleistocene through late Holocene occupation, which are distinctly focused on the intensive use of wetlands and their resources in the semi-arid desert-steppe region (Janz 2012). Janz's terminology is outlined as:

(1) Oasis 1/Mesolithic (13,500–8000 cal BP), defined by sites showing the earliest use of the wetland environments and their adjacent ecozones in the vicinity of rivers or lakes
(2) Oasis 2/Neolithic (8000–5000 cal BP) with sites indicating the intensive exploitation of wetland oases, characterized by camp sites within dune-fields and marshlands, and notably a common occurrence of grinding stones, suggesting the exploitation of small-seeded grasses and other steppe plants, along with a wider range of microlithic tool types, also suggesting the possibility of composite tools for hunting small-fast prey, harpooning fish, and possible reaping activities associated with these wetlands.
(3) Oasis 3/Bronze Age or Eneolithic (5000–3000 cal BP) with evidence for even more intensive use of wetland habitats including larger numbers and types of ceramics, and bifacial flaking of projectile points, knives, and other small tools (Janz 2012).

The subsistence activities of the desert-steppe inhabitants during these three periods are distinctly different from the Big-Game hunting specialists which preceded the terminal Pleistocene, and the later Bronze Age pastoralists which succeeded these populations in the later Holocene. It is useful to think of the unique adaptations of the middle Holocene foragers in terms of 'Push' and 'Pull' factors. The distinction of the resilient adaptations of mid-Holocene foragers from the terminal Pleistocene Big Game hunters can be attributed to the Push of declining populations of large game animals due to the increasing trends of the Pleistocene Extinctions of the large mammals in Eurasia, which undoubtedly undermined the advantage of this strategy. The Pull factor would have been the increased rainfall from the strengthening of the East Asian Monsoons, raising ground water tables and forming ponds and small lakes across the landscape of the eastern Gobi. In southern Mongolia, the region would have been at the very northernmost edge of the East Asian Monsoonal systems, and thus still a semi-arid zone. Thus, these newly formed wetlands would have been an attractive draw to the hunters and foragers living in this region.

These middle Holocene foragers would have employed adaptive strategies that were not available to the later Holocene inhabitants who were presented with an ever increasingly dry environment. In the later Holocene the Gobi Desert experienced weakened East Asian Monsoons, which led to the disappearance of the lakes, marshlands, and small perennial streams in this region. It is also a period in which the inhabitants of the eastern Gobi adopted animal husbandry and a nomadic pastoral lifestyle (Honeychurch and Makarewicz 2016; Honeychurch 2017; Wright et al. 2019).

Linear Valleys

In addition to the basin-centric wetland focus of mid-Holocene adaptation, the Gobi Desert of southern Mongolia presents another configuration of wetland adaptation. This is a mobile movement along "hydrological corridors". These are a series of valleys, many trending north/south, that form belts of springs leading from the lusher steppe zones north towards central Mongolia, to the wetlands and marsh zones fringing the former Paleolake beds of the Mongolian Gobi Desert. Examples of sites along these kinds of corridors were studied by Holguin and Sternberg (2018), Holguín (2019) in the region of Ulaan Nuur, and by Rosen and others at the Ikh Nart Nature Reserve (Schneider et al. 2016; Rosen et al. 2019). The foragers living in these regions reduced their subsistence risk by taking advantage of the mosaic of micro-environments associated with these extensive linear palaeohydrological systems. These geomorphological features accommodated both the resource needs of these foragers and later hunter/pastoralists. They also facilitated mobility and movement throughout the Gobi in both east/west and north/south directions.

A second critical factor for survival in drylands, and especially in situations of increasing desertification is the capacity for diversification. This applies to resilient strategies for subsistence diversity, as well as the flexibility to alter social patterns and institutions.

(2) *Diversification*

The ability to diversify our diet, material culture, and social institutions is one of the most powerful implements in the survival toolkits of our genus *Homo*. We are a species that can consume a wide variety of foods, withstand many types of climatic conditions and ecological zones, move easily from habitat to habitat, and adapt our material and non-material culture accordingly.

When climatic conditions are favorable and rainfall and other resources are abundant, people living in dryland environments are free to target the foods that most appeal to them. They have the option to narrow their resource selection to high yield, highly nutritious, or large packets of foods, and specialize in the collection, hunting, and processing of a narrower range of resources. Examples of this are big game hunting, and heavy reliance on collection of nuts from groves of nut-bearing trees (Mason 1995; Abrams and Nowacki 2008). With a non-mobile food source, this can lead to increasing sedentary residential patterns as well as specialization in the use of that resource (Rosen and Rivera-Collazo 2012). These populations can afford to engage in high-risk, high profit endeavors. Some societies may choose to increase their risk and select for more "costly" resources, others may choose not to do so, but the option is there.

In periods of increasing drought, climatic degradation, and especially in situations when rainfall events become more stochastic and unpredictable over the long term, the most resilient strategy would be to diversity subsistence resources. This diversity is a approach referred to by many archaeologists as a "Broad Spectrum" strategy (Stiner and Munro 2002; Weiss et al. 2004a, b; Stutz et al. 2009; Janz 2016). It may take the form of exploiting more animal food sources that are small, fast, difficult to catch, and have less palatable meat with little fat content such as rabbits and other small mammals (Stiner and Munro 2002). Plant exploitation may take the form of concentrating more on small-seeded grasses (Weiss et al. 2004a, b; Rosen 2010), or plants that require more intensive labor in collection, processing, and cooking. Diversification of plant resources may also include plants which are rich in nutrients but require extensive processing to remove toxins which would otherwise be harmful to humans if consumed directly.

In the Near East, Rosen and Rivera-Collazo (2012) point to a cyclicity of narrow to broad spectrum targeting of plant resources depending upon the expansion or contraction of forest and grassland zones with changes in climate from cold/dry to warm/wet conditions. They suggest that the resilience of the adaptation lies in the ability of hunter-gatherer groups to switch their plant exploitation strategies from the narrower focus of gathering tree foods (including acorns, pistachios, almonds and carob) in times of warn/moist climatic amelioration, to the more broad spectrum exploitation of diverse resources (including small-grained grasses and wild cereals), and more intensive use of wetlands in the periods of harsher cool/dry climates when grassland zones expanded. But rather than a linear trajectory from one type of plant collection strategy to the other, Rosen and Rivera-Collazo (2012) suggest that the resilience lies in the ability of foragers to switch from one to the other (narrow to broad) and back again, in harmony with the distribution of plant communities in a cyclical fashion. This pattern only ceases when populations of foragers begin to increase in the Southern Levant during the early Holocene. This led to the inception of growing settlements of sedentary foragers, decreasing size of home ranges, and impeded mobility. It is only then, under the lower-risk and more predictable warm/wet conditions that the foragers narrowed their focus even further in the middle Holocene to the cultivation and domestication of wheat and barley. Interestingly, the cyclicity in plant collection strategies is not mirrored by species selection in hunting. Animal species targeted seem to follow a more linear pattern of narrow to broad throughout the late Pleistocene into the early Holocene (Rosen and Rivera-Collazo 2012).

In the Gobi, Holguin applies the work of Australian researchers who have robust theoretical perspectives on the use of desert landscapes as foragers arrive in these regions and begin to understand the cycles of abundance and fluctuations of resource microenvironments (Holguin 2019: 29–31). Holguin cites cases from Australian research, which refer to a sequence of the first foragers arriving in the desert during times of moister climate giving them the opportunity of developing a sophisticated ecological knowledge of the behavior of the plant and animal resources. With drying climate, these populations use this ecological understanding to broaden the types of resources they are able to exploit (Veth 1989; Hiscock and Wallis 2005; Smith et al. 2005).

Again, wetland environments are a source of high biodiversity, and attracts mobile foragers for its diversity in food resources, lending a large measure of increased resilience to droughts and otherwise drying climatic conditions and/or unpredictable rainfall events. The site of Ohalo II is a poster child for this kind of wetland contribution to a highly diverse diet in times of climatic degradation in the Levant at the LGM (Nadel et al. 2004; Weiss et al. 2004a, b).

(3)   *Storage and Caching*

Storage of resources is a key part of resilience in unpredictable environments. Storage technologies are critical for agricultural societies and will allow farmers to recover from at least one year of severe drought, and often two or even three years of persistent rainfall shortages and crop failures. Mobile foragers and early pastoralists will also use systems of storage and caching of food and other resources which can accommodate the group for recovery from unproductive drought years in their home ranges (Rowley-Conwy and Zvelebil 1989; Gerber et al. 2004; Bettinger 2009; Morgan 2012; Tushingham and Bettinger 2013). Tushingham and Bettinger (2013) have explored the concept of "Front-Loaded" and "Back-Loaded" storability of resources from the perspective of Optimal Foraging Theory, to explain the selection of food items from the view of energy output for collecting, storing, processing, and planning future return. "Front-loaded" resources are costly to acquire and store, but preparation for consumption is easy. "Back-loaded" resources are easy to acquire and store, but energy intensive to prepare. They note that back-loaded resources involve less risk because the energy expenditure needed to acquire and store them is lower, and therefore if the cache is lost or not needed, then there was little wasted effort. We can explore this concept from the perspective of wetland exploitation as well.

One great advantage of wetland localities in dryland environments is their high potential as natural caches for the storage of important high-value year-round resources (Ramsey and Rosen 2016). Some of the most important of these are the underground storage organs (USOs) of sedges such as club rush (*Bolboschoenus maritimus*), yellow nutsedge (*Cyperus esculentus*), and purple nutsedge (*Cyperus rotundus*) (Hillman et al. 1989; Wollstonecroft et al. 2008). These resources are available for collection as needed for most of the year, thus exploitable at a comparable level with low-risk "back-loaded" stored resources. Additionally, they are rich in caloric value, being the starch repository for the plants. As mentioned above, calories are hard to come by in dryland regions such as the Southern Levant and the Gobi where the animal foods are lean, and there is less access to other forms of calorie-rich plants such as nuts, cereals, or an abundance of small-grained grasses. Some desertic environments provide high-caloric succulents such as cactus and Agavaceae species as in the American Southwest, but in the Gobi and Levantine deserts these foodstuffs are less prevalent.

(4)   *Niche Construction*

The success of our *Homo sapiens* species, and indeed our genus *Homo* lies in our ability to innovate and survive in in a multitude of environments far from the tropical grasslands where we evolved. A substantial element of our survival is based on our propensity to shape our habitats to fit our subsistence and other needs. These abilities have allowed our species to inhabit the entire planet and thrive in niches well beyond our African homeland (Potts 2012). One essential requirement for all human societies is an understanding of how to predict and control water availability. This is especially true for dryland regions. Water can be a limiting factor inhibiting settlement and subsistence, and at times—in the cases of massive floods and severe droughts—"nature" may have the upper hand. But water is also a resource that human societies have controlled to a certain extent through varying degrees of technology and TEK.

The study of how societies have manipulated water resources for irrigation farming has a long history of investigation, with many authors drawing links to increasing social complexity and the political ecology of water management back to the earliest Neolithic farming communities. But small-scale mobile societies also manipulate water resources to intensify and secure food resources. Research into the ways mobile communities manage water can be challenging due to transitory evidence left on the landscape by these societies which leave behind only a faint environmental footprint. Yet such evidence does exit, and examples are also surmised by auxiliary data (Harrower 2016).

Reaching back to the more distant past, Harrower's (2016) excellent comprehensive study makes a strong case for water management and manipulation on the part of foragers and other small-scale societies worldwide. Fuller and Qin (2009) surmise that pre-agricultural foragers collecting the wild ancestral *Oryza sativa japonica* rice along the Yangtze River in China, would have needed to enhance the supplies of water to the wetland microenvironments in order to increase the productivity of that ancestral rice sub-species. They suggest that this took the form of burning and weeding out undesirable species to clear the waterways and allow more water to flow to the stands of wild rice and directing drainage so selected areas were flooded during the rainy season, but dry out when the rice grains had ripened. A more recent example of the enhancement of water resources among foraging societies comes from Australia. Lourandos (2010) writes about substantial water manipulation systems in southwestern Victoria which native Australian peoples excavated for the purpose of fishing for eels. Here the hunter-gatherer groups dug channels that extended hundreds of meters in length for the expansion of water courses and the intensification of fishing. Likewise, Campbell (1965) and Tindale (1977) discuss cases in which foragers in Australia manipulated water flowing in and out of wetlands to increase the extent of wild plant foods.

Mobile pastoralists are also adept at enhancing water supplies in their dryland home ranges, which is an adaptive strategy that extends far into antiquity and predates the advent of farming communities (Mithen 2010; Hammer 2018). Pokrandt (2014) maintains that in Jordan's south-eastern desert there is substantial evidence for water source enhancement among mobile pastoralists that dates back to the middle Holocene. This evidence includes well-building, channel-type watering systems, and rainwater harvesting techniques of capturing ponded water in depressions with small-stone dams.

Conclusions

This paper briefly touched on some of the effective and resilient strategies employed by small-scale societies in the past for thriving in dryland environments through sustainable use of their environments. I argue that the ability to live sustainably in these kinds of challenging environments emerges from a profound and long-term reservoir of Traditional Ecological Knowledge. These strategies are based in part on (1) the knowledge of the range of variability of unmanaged water sources, and/or the technology to manage these sources at varying levels of engineering skills, (2) Strategies for diversification of resources, diet, social institutions, and toolkits, (3) A system of storage, either a natural characteristic of a particular plant food, or technologies for caching or preserving food items in times of abundance. This is also enhanced by sharing and the accumulation of "social capital". Finally, (4) "Niche Construction" and the management of local environments for sustainable production of essential resources. For small-scale, mobile societies living in dryland regions in the past and present, these highly successful strategies are markedly enhanced by living near and exploiting wetland environments.

In our modern world consisting of highly technical and sometimes over-engineered solutions to environmental problems, it might seem that small-scale and mobile societies of the past have little to offer us in terms of solutions to global issues of abruptly changing climates, increasing desertification, and the politics of decreasing water resources. However, there is much to learn from such societies and their "ground-up" solutions which operated effectively and sustainably for generations and sometimes millennia. These adaptations, based in Traditional Ecological Knowledge and environmental understanding over deep time may translate well as modern-day small-scale community-based solutions. We can benefit a great deal from the study of these past societies to help generate ideas about neighborhood collectives, urban gardening, water conservation methods, and others that will lead us towards a more sustainable existence on our planet.

References

Abrams MD, Nowacki GJ (2008) Native Americans as active and passive promoters of mast and fruit trees in the eastern USA. Holocene 18(7):1123–1137

Adams RM (1960) Early civilizations, subsistence, and environment. In: Kraeling CH, Adams RM (eds) City invincible: a symposium on urbanization and cultural development in the ancient near East. University of Chicago Press, Chicago, pp 269–295

Bettinger RL (2009) Hunter-gatherer foraging: five simple models. Eliot Werner Publications Inc., Clinton Corners, NY

Butzer KW (1996) Irrigation, raised fields and state management: Wittfogel Redux? (book review). Antiquity 70:200–204

Butzer KW (1982) Archaeology as human ecology: method and theory for a contextual approach. Cambridge: Cambridge University Press

Campbell AH (1965) Elementary food production by the australian aborigines. Mankind 6(5):206–211

Childe VG (1936) Man makes himself. Oxford University Press, Oxford

Feng ZD, Wang WG, Guo LL, Khosbayar P, Narantsetseg T, Jull AJT, An CB, Li XQ, Zhang HC, Ma YZ (2005) Lacustrine and eolian records of Holocene climate changes in the Mongolian Plateau: preliminary results. Quatern Int 136(1):25–32

Fuller DQ, Qin L (2009) Water management and labour in the origins and dispersal of Asian rice. World Archaeol 41(1):88–111

Gaudzinski S (2004) Subsistence patterns of early Pleistocene hominids in the Levant—taphonomic evidence from the 'Ubeidiya Formation (Israel).' J Archaeol Sci 31(1):65–75

Gerber LR, Reichman OJ, Roughgarden J (2004) Food hoarding: future value in optimal foraging decisions. Ecol Model 175(1):77–85

Goren-Inbar N, Saragusti I (1996) An Acheulian biface assemblage from Gesher Benot Ya'aqov, Israel: indications of African affinities. J Field Archaeol 23(1):15–30

Goring-Morris N, Hovers E, Belfer-Cohen A (2009) The dynamics of Pleistocene and early Holocene settlement patterns and human adaptations in the Levant: an overview. In: Shea JJ, Lieberman DE (eds) Transitions in prehistory: essays in honor of Ofer Bar-Yosef. Oxbow Books, London

Hammer E (2014) Local landscape organization of mobile pastoralists in southeastern Turkey. J Anthropol Archaeol 35:269–288

Hammer E (2018) Water management by mobile pastoralists in the middle East. In: Holt E (eds) Water and power in past societies, vol 7. State University of New York Press, New York, pp 63–88

Harrower MJ (2016) Water histories of ancient Yemen in global comparative perspective. In: Harrower MJ (ed) Water histories and spatial archaeology: ancient Yemen and the American West. Cambridge University Press, Cambridge, pp 51–83

Hillman GC, Madeyska E, Hather JG (1989). Wild plant foods and diet at Late Palaeolithic Wadi Kubbaniya (Upper Egypt): evidence from charred remains. In: Wendorf F, Schild R, Close A (eds) The prehistory of Wadi Kubbaniyavol: stratigraphy, subsistence and environment, vol 2. Southern Methodist University, Dallas, pp 162–242

Hiscock P, Wallis LA (2005) Pleistocene settlement of deserts from an Australian perspective. In: Veth P, Smith M, Hiscock P (eds) Desert peoples: archaeological perspectives. Blackwell, Oxford, pp 34–57

Holguín LR (2019) The changing Neolithic landscape of Ulaan Nuur: Modelling hydro-social dynamics in the Mongolian Gobi Desert. Ph.D. dissertation, University of Southampton

Holguín LR, Sternberg T (2018) A GIS based approach to Holocene hydrology and social connectivity in the Gobi Desert, Mongolia. Archaeol Res Asia 15:137–145

Honeychurch W (2017) The development of cultural and social complexity in Mongolia. In: Habu J, Lape PV, Olsen JW (eds) Handbook of East and Southeast Asian archaeology. Springer, New York, pp 513–532

Honeychurch W, Makarewicz CA (2016) The archaeology of pastoral nomadism. Ann Rev Anthropol 45(1):341–359

Janetski JC, Madsen DB (eds) (1990) Wetland adaptations in the Great Basin. Brigham Young University Museum of Peoples and Cultures Occasional Papers 1. University of Utah Press, Salt Lake City

Janz L (2012) Chronology of Post-glacial hunter-gatherers in the Gobi Desert and the neolithization of arid Mongolia and China. Unpublished Ph.D. dissertation, University of Arizona

Janz L (2016) Fragmented Landscapes and economies of abundance: the broad spectrum revolution in Arid East Asia. Curr Anthropol 57(5):537–564

Janz L, Odsuren D, Bukhchuluun D (2017) Transitions in palaeoecology and technology: hunter-gatherers and early herders in the Gobi Desert. J World Prehist 30(1):1–80

Janz L, Rosen AM, Bukhchuluun D, Odsuren D (2021) Zaraa Uul: an archaeological record of Pleistocene-Holocene palaeoecology in the Gobi Desert. PLoS One 16(4):e0249848

Kelly RL (1990) Marshes and mobility in the western Great Basin. In: Janetski JC, Madsen DB (eds) Wetland adaptations in the Great Basin. University of Utah Press, Salt Lake City, pp 259–276

Lourandos H (2010) Change or stability?: hydraulics, hunter-gatherers and population in temperate Australia. World Archaeol 11(3):245–264

Maher LA (2019) Persistent place-making in prehistory: the creation, maintenance, and transformation of an Epipalaeolithic landscape. J Archaeol Method Theory 26(3):998–1083

Maher LA, Macdonald DA, Allentuck A, Martin L, Spyrou A, Jones MD (2016) Occupying wide open spaces? Late Pleistocene hunter–gatherer activities in the Eastern Levant. Quatern Int 396:79–94

Mason SLR (1995) Acornutopia? Determining the role of acrons in past human subsistence. In: Wilkins J, Harvey D, Dobson M (eds) Food in antiquity. Exeter University Press, Exeter, pp 12–14

Mithen S (2010) The domestication of water: water management in the ancient world and its prehistoric origins in the Jordan Valley. Philos Trans A Math Phys Eng Sci 368(1931):5249–5274

Morgan C (2012) Modeling modes of hunter-gatherer food storage. Am Antiq 77(4):714–736

Munro ND, Atici L (2009) Human subsistence change in the Late Pleistocene Mediterranean Basin: the status of research on faunal intensification, diversification and specialisation. Before Farming 1:1–6

Nadel D, Tsatskin A, Belmaker M, Boaretto E, Kislev ME, Mienis H, Rabinovich R, Simchoni O, Simmons T, Weiss E, Zohar I (2004) On the shore of a fluctuating lake: environmental evidence from Ohalo II (19,500 B.P.). Israel J Earth Sci 53:207–223

Nicholas GP (1998) Wetlands and hunter-gatherers: a global perspective. Curr Anthropol 39(5):720–731

Olszewski DI, Al-Nahar M (2016) Persistent and ephemeral places in the Early Epipaleolithic in the Wadi al-Hasa region of the western highlands of Jordan. Quatern Int 396:20–30

Pokrandt J (2014) Water management of a Late Chalcolithic pastoral culture in Jordan's southeastern desert: case study of Qulban Beni Murra. Levant 46(2):268–284

Potts R (2012) Evolution and environmental change in early human prehistory. Ann Rev Anthropol 41(1):151–167

Ramsey MN, Jones M, Richter T, Rosen AM (2015) Modifying the marsh: evaluating early epipaleolithic hunter-gatherer impacts in the Azraq wetland, Jordan. Holocene 25(10):1553–1564

Ramsey MN, Maher LA, Macdonald DA, Rosen A (2016) Risk, reliability and resilience: phytolith evidence for alternative 'Neolithization' pathways at Kharaneh IV in the Azraq Basin, Jordan. PLoS One 11(10):e0164081

Ramsey MN, Rosen AM (2016) Wedded to wetlands: exploring Late Pleistocene plant-use in the Eastern Levant. Quatern Int 396(7):5–19

Ramsey MN, Rosen AM, Nadel D (2017) Centered on the Wetlands: integrating new Phytolith evidence of plant-use from the 23,000-year-old site of Ohalo Ii, Israel. Am Antiquity 1–21

Rosen AM (2007) Civilizing climate: social responses to climate change in the ancient near East. Altamira, Lanham, MD

Rosen AM (2010) Natufian plant exploitation: managing risk and stability in an environment of change. Eurasian Prehist 7(1):117–131

Rosen AM, Rivera-Collazo I (2012) Climate change, adaptive cycles, and the persistence of foraging economies during the late Pleistocene/Holocene transition in the Levant. PNAS 109(10):3640–3645

Rosen AM, Hart TC, Farquhar J, Schneider JS, Yadmaa T (2019) Holocene vegetation cycles, land-use, and human adaptations to desertification in the Gobi Desert of Mongolia. Veg Hist Archaeobotany 28(3):295–309

Rosen AM, Odsuren D, Bukhchuluun D, Janz L (n.d.) Holocene desertification and human resilience in the Eastern Gobi Desert. Manuscript in Preparation

Rowley-Conwy P, Zvelebil M (1989) Saving it for later: storage by prehistoric hunter–gatherers in Europe. In: O'Shea J, Halstead P (eds) Bad Year economics: cultural responses to risk and uncertainty. Cambridge University Press, Cambridge, pp 40–56

Schlanger SH (1992) Recognizing persistent places in Anasazi settlement systems. In: Rossignol J, Wandsnider L (eds) Space, time, and archaeological landscapes. Springer US, Boston, MA, pp 91–112

Schneider JS, Tserendagva Y, Hart TC, Rosen AM, Spiro A (2016) Mongolian 'Neolithic' and early bronze age ground stone tools from the northern edge of the Gobi Desert. J Lithic Stud 3(3):479–497

Smith M, Veth P, Hiscock P, Wallis LA (2005) Global deserts in perspective. In: Veth P, Smith M, Hiscock P (eds) Desert peoples: archaeological perspectives. Blackwell, Oxford, pp 1–13

Stiner MC (2004) Small game use and expanding diet breadth in the Eastern Mediterranean basin during the Palaeolithic. In: Brugal JP, Desse J (eds) Petits Animaux et Sociétés Humains: du Complément Alimentaire Aux Ressources Utlitaires: XXIVe Recontres Internationales d'Archéologie et d'histoire d'Antibes. Antibes, APDCA, pp 499–513

Stiner MC, Munro ND (2002) Approaches to prehistoric diet breadth, demography, and prey ranking systems in time and space. J Archaeol Method Theory 9(2):181–214

Stutz AJ, Munro ND, Bar-Oz G (2009) Increasing the resolution of the broad spectrum revolution in the Southern Levantine Epipaleolithic. J Hum Evol 56(3):294–306

Tindale NB (1977) Adaptive significance of the Panara grass seed culture of Australia. Stone tools as cultural markers. R. V. S. wright. Australian Institute of Aboriginal Studies, Canberra, Australia, pp 345–350

Tushingham S, Bettinger RL (2013) Why foragers choose acorns before salmon: storage, mobility, and risk in aboriginal California. J Anthropol Archaeol 32(4):527–537

Veth P (1989) Islands in the Interior: a model for the colonization of Australia's arid zone. Archaeol Ocean 24(3):81–92

Weiss E, Kislev ME, Simchoni O, Nadel D (2004a) Small-grained wild grasses as staple food at the 23,000-year-old site of Ohalo II, Israel. Econ Botany 58(Supplement):SI25–S134

Weiss E, Wetterstrom W, Nadel D, Bar-Yosef O (2004b) The broad spectrum revisited: evidence from plant remains. Proc Natl Acad Sci USA 101(26):9551–9555

Wittfogel KA (1957) Oriental despotism: a comparative study of total power. Yale University Press, New Haven, CT

Wollstonecroft MM, Ellis PR, Hillman GC, Fuller DQ (2008) Advances in plant food processing in the Near Eastern Epipalaeolithic and implications for improved edibility and nutrient bioaccessibility: an experimental assessment of *Bolboschoenus maritimus* (L.) Palla (sea club-rush). Veg Hist Archaeobotany 17:19–27

Wright J, Ganbaatar G, Honeychurch W, Byambatseren B, Rosen A (2019) The earliest Bronze Age culture of the South-Eastern Gobi Desert, Mongolia. Antiquity 1–19

Beyond Boom and Bust: Climate in the History of Medieval Steppe Empires (C. 550-1350 CE)

Nicola Di Cosmo

Abstract The use of paleoclimate data in historical work has become a new and dynamic endeavor in several areas of historical research. This chapter is concerned with the empires created by pastoral nomads in the steppe regions of Eastern and Central Eurasia over approximately three millennia, from the early appearance of complex Scythian and Siberian polities in the early first millennium BCE to the Dzungar empire of the seventeenth and eighteenth centuries. This essay aims to show how paleoclimate data may be used to illuminate connections, dynamics, and causal nexuses in the important, and yet often overlooked, historical experience of pastoral peoples and the empires they created. Paleoclimate data are especially important to supplement the scarce documentary sources left behind by pastoral nomads.

Keywords Medieval Inner Asia · Steppe nomads · Pastoralism · Mongol empire · Uyghur empire · Mongolia

The use of paleoclimate data in historical work has become a new and dynamic endeavor in several areas of historical research. This paper is concerned with one of these areas: the empires created by pastoral nomads in the steppe regions of Eastern and Central Eurasia for about three millennia, from the early appearance of complex Scythian and Siberian polities in the early first millennium BCE to the Manchus who conquered China in the seventeenth century and the Dzungar khanate of the eighteenth century. This essay aims to show how paleoclimate data may be used to illuminate connections, dynamics, and possible causal nexuses otherwise not visible in documentary or material sources. To assess and frame the relevance of climate within the broader discourse of steppe history, in the first part of this essay I will address specific issues that continue to define the historiography of steppe nomads, and in particular the role of climate within it. In the second part, I will describe three cases that illustrate salient aspects of the use of paleoclimate data in relation to nomadic empires, each of which addresses a separate climate event.

N. Di Cosmo (✉)
Institute for Advanced Study, Princeton, USA
e-mail: ndc@ias.edu

© The Author(s) 2022   177
A. Izdebski et al. (eds.), *Perspectives on Public Policy in Societal-Environmental Crises,*
Risk, Systems and Decisions, https://doi.org/10.1007/978-3-030-94137-6_12

Ethnographic literature on the environment of pastoral economies and nomadic societies have long recognized the vulnerability of semiarid rangelands to climatic downturns, a discourse often couched in terms of "disequilibrium" (Vetter 2005; von Wehrden 2012). Disequilibrium, in this context, refers to the asymmetric relationship between natural resources and economic productivity, whereby the grassland's above-average net primary production often generates an excess of animals that in turn deplete its resources through overgrazing, while below-average land productivity, as a result of climatic stresses, causes the animals' starvation, hypothermia (in winter), or illnesses, which may reduce the size of herds below the level required for human sustenance. Such conditions, among others, have militated against the establishment of a steady state between natural and human systems. A balanced exploitation of resources through rangeland management, usually achieved by the seasonal rotation of available pastures, has been essential to the life of nomads since the beginning of this particular mode of production, but their societies have remained exposed to high levels of vulnerability (Fernandez-Gimenez et al. 2015a).

Taking into consideration just the recent history of Mongolia, both winter disasters and droughts have caused catastrophic losses of animals, human starvation, and displacement of a magnitude that compelled the government to invoke international aid (Batima et al. 2005; Fernandez-Gimenez et al. 2015b; Begzsuren et al. 2004). Given the recurrent nature of such disasters, a reasonable and appropriate historical question is to ask how such events may have affected the vulnerability or resilience of the empires that nomads created, or, to put it in a different way, in what way climate, so essential to preserving life on the grassland, may be related to the nomads' social and political history. On the other hand, processes of accelerated political centralization may be favored by increased grassland productivity. Above-average precipitation and warmer temperatures, by lowering the aridity and prolonging the growing season, trigger a positive economic cycle that directly affects the well-being of the livestock, decreasing the rate of mortality and expanding the number of animals. Hypothetically, and under certain historical circumstances, augmented resources may lead to political changes. An expanded economy can keep a portion of the population permanently employed in other activities, which in premodern times were chiefly to do with military ventures. The abundance of healthy horses allowed more warriors to take the field, and such expanded military capability could translate into firmer and more durable political centralization.

Since Mongolia was the birthplace of most Inner Asian empires, its paleoclimate has become an essential source to inquire about the linkages between climatic fluctuations, environmental transformations, and political events. While most people may be familiar with the most macroscopic manifestations of nomadic history, such as the Mongol conquests, the "nomadic" or "steppe" empire as an historical category requires a few words of explanation, especially in light of various theories that have been advanced to explain their formation (Rogers 2012). In particular, the category of "steppe empire" is in itself an historical product that tends to obscure the diversity and variety that can be detected upon closer examination, and therefore it is useful to discuss briefly some of the notions that inform this category, which are also relevant to how climate has been brought into the discussion of the formation or demise of these empires.

Nomadic empires emerged in central and eastern Eurasia probably already in the early first millennium BCE, and their appearance is closely related to the domestication and riding of the horse, and to the development of stratified societies, in which political and economic power was wielded by a warrior aristocracy. Even though these empires have occupied a place of central importance in Eurasian and world history, the reasons for their recurring emergence and disappearance are still, generally speaking, as much unknown to us as they were to ancient observers on both ends of Eurasia. Climatic theories have been hovering over the interpretation of the inner dynamics of nomadic politics but without ever rising beyond the stage of conjectures based on a generic understanding of the effects of climate variability in contracting or expanding the nomadic *lebensraum*, and thus supposedly leading to migration, conflict, and conquest (Kradin 2015). Such generic notions, however, have not generated specific knowledge, nor have they produced evidence that could contribute to explaining historical processes. Nonetheless, the environmental determinism that is at the root of these theories has found a new and rather powerful voice in scientific publications, which play an important role by promoting methods and concepts in the study of the role of climate in the history of steppe nomads.

Climatologists studying Mongolia, central Asia, and the northern regions of China, have been especially involved in the phenomenon of nomadic societies' response to climate-induced stresses. The premise, sometimes explicit and sometimes unstated, is that various types of disruption of environmental conditions acted on nomads as "push factors" that elicited a social or political response. The object of these studies is to find regularities that may explain the history of interaction of nomads with China, and presumably other settled civilizations. The assumption that conflicts between nomads and China are inherently reducible to climatic variability has produced (and sometimes justified a posteriori) a research model based on correlations between certain climatic events and the presumed behavior of nomads, concentrating, naturally, on precipitation and temperature (Bai and Kung 2011; Fang and Liu 1992; Su et al. 2016; Pei and Zhang 2014; Lee et al. 2017; Pei et al. 2019). Seeking to isolate the climate anomalies that are most likely to cause political and economic disruptions, leading to migrations and conflicts with settled peoples, these studies rely on statistical analysis as evidence, while the sources for their single-point data remain opaque at best. In any case, different studies have reached different conclusions, arguing at times that the cause of nomadic invasions and other historical events should be attributed to variations in moisture, such as droughts, and at times to drops in temperature, causing frosts or frigid winters.

The wide gulf separating statistical methods (and especially statistical "proof" of causation) from historical analysis has proved to be a formidable obstacle to collaborative ventures. While most climatologists, generally speaking, would be disinclined to endorse such extreme climatic reductionism, it is also true that the largest part of climatological literature available on nomads' history reflects or subscribes to this orientation, and therefore plays a rather important role in defining climatic

approaches to the history of nomads and attending methods and tools. What is necessary, however, is an historical approach to climate that seeks to comprehend actual
historical events, which are (emphatically) not limited to questions of migrations or
conflict, but rather to the analysis of critical junctures in nomadic history, and especially the interplay between climatic fluctuations and the structures and institutions
that constituted the internal scaffolding of nomadic societies, and were essential to
the formation of empires, their sudden disintegration, or slow demise. Such questions may include governance, fiscality, military expansion, trade, diplomacy, and
other issues that need to be considered in relation to any transformational event, to
the extent that such an event may be connected to changes in the pastoral nomadic
economy due to climate-related environmental changes.

In this regard, and last in our list of preliminary observations, we need to address
more directly the very object of our inquiry, namely, what we mean when we speak
of "nomadic empire" and how we should frame questions that involve climate. As we
mentioned above, a common assumption in both historical and climatological literature is that pastoral nomads "react" to climatic changes pretty much always in the
same way, regardless of their degree of political organization, social complexity, or
economic diversification. Occasionally, such "push factors" were responsible for the
pastoral nomads' success in constituting large military forces and engaging in invasion and conquest. The formative process of these entities, however, lay beyond the
horizon of the historical visibility of literate societies, and went largely unrecorded
since it typically occurred in remote regions "beyond the pale" of civilization. It
is exactly that extreme dearth of documentation that makes climatic data potentially valuable, based on the reasonable assumption, derived from anthropological,
ethnographic, and environmental research, that severe, protracted or otherwise transformational climate events affect the pastoral economy and thus lead to deep modifications of social hierarchies and political systems. That does not mean that we
should automatically associate such variability with specific societal responses, but
that climate data (as we shall see below) can supplement our analysis in important
ways, by directly disproving unfounded assumptions or by suggesting new avenues
of inquiry, as well as otherwise unsuspected causal linkages and connections.

In particular, climatic data may shed light on processes of adaptation, resilience,
or recalibration of economic and social policies that would be otherwise difficult to
explain. In fact, the surprising diversity that can be observed within the "galaxy" of
nomadic political formations has not been properly explained other than by attributing
it to external influences and pressures. While there is certainly a degree of permeability between nomads and non-nomads, such osmotic properties neither determined
nor qualified all aspects of nomadic political and economic life. Most of the major
political formations that steppe nomads created in eastern Inner Asia were born out
of periods of chaos that spun economic crises, internal warfare, and displacement
before a new political order emerged. Even though some of the "building blocks" are
comparable, such as political features, kinship structures, cultural underpinnings (e.g.
religious authority), and especially economic bases, each case produced a different
outcome, and a different empire. In the face of such diversity, assuming that all

nomads behaved in the same manner would introduce an irremediable bias into the analysis, and that applies also to the study of climate and environmental issues.

In order not to reduce this question to another case of "splitters versus lumpers," it is important to recognize that the pastoral economy has specific requirements that are based on the ecology of rangelands and the biological lifecycle of the livestock, but the social and political contexts in which pastoral production is organized may vary a great deal. One common characteristic of steppe nomadism is the mobility inherent to seasonal shifts between pasture grounds, which also controls significant aspects of their technology and social life. The dependency of nomads on their herds, and therefore on the environment that allows the animals to survive, is a consideration that is clearly connected to climate variability, as we shall see below. Far less obvious is the relationship between such climatic variability and the political life of nomads, especially in cases of "imperialization," that is, the rise of powerful and expansive polities. While historical analysis has steered away from investigating such connections, and one may speculate whether this was due to inaccessibility of data, methodological limitations, a distrust of climatic hypotheses, or a combination of these, it would seem extremely unlikely that climate events can be eliminated from consideration when natural environment and human existence are so closely intertwined.

In sum, even though new research has greatly advanced our knowledge of the role of steppe empires in world history (Biran 2019), we are still a long way from understanding the mechanisms and the dynamics that are behind the formation of such complex and powerful polities, a process that was a necessary condition for any projection and propagation of nomadic power outside its natural habitat. Likewise, as historical scholarship moves away from theories that attribute the presumed collapse of past societies and empires to natural disasters (Butzer 2012; Haldon et al. 2020), the notion of climatic reasons for the disappearance of nomadic empires must also be questioned. What may be represented as a climatic "turn" in the history of nomadic empires seeks to accomplish, at a minimum, three objectives. First, it aims to take stock of and introduce climate data into historical analysis, without allowing climate data to dominate it. Second, climate analysis has to be strictly related to the time and location of the events under discussion, specifically and explicitly indicated. Third, environmental and climatic data need to be properly contextualized and integrated with all sources available, especially textual and material, not just to achieve a better understanding of the events, but especially because such data can generate new questions that would not be raised by other sources. The cases discussed below are examples of a climate-informed approach to the history of Inner Asian empires.

Case Studies

The three case studies presented here are meant to illustrate different types of climatic variability in their historical context. While these studies cannot be assumed to be in any way paradigmatic of the relationship between climate and nomadic empires, they may nonetheless be representative of a wider range of related issues and therefore are to be understood as exemplary, rather than normative, cases. What is especially relevant here, rather, is the implicit method of inquiry, which places a high premium on the use of climatic data for its inherent potential to allow new questions to arise, and for its uses in reconfiguring or reinterpreting other pieces of evidence associated with the historical events in question.

1. The fall of the Eastern Türk Empire (630 CE)

A widely discussed case in both the historical and climatological literature concerns the demise of the Eastern Türk empire (Zhang 2002; Fei et al. 2007; Fei 2008; Erkoç 2017b; Di Cosmo et al. 2017; Ganiev and Kukarskih 2018). This was a steppe nomadic polity, and a successor state of the Türk empire, established in Mongolia in the mid-sixth century CE. The Türk rulers expanded westward, but their state subsequently split into a western and an eastern branch after a civil war in the 580s. The Eastern Türk empire was ruled by a *qaghan* whose political and economic base was located in today's Inner Mongolia, although its sovereignty extended into the northern Mongolian region. Chinese historical sources indicate that a series of natural disasters might have been behind a famine that, beginning in 627 CE, ravaged the Türk empire, and in particular the region that was the base of the *qaghan*'s economic power and its political center, in the piedmontal region to the north of the Yellow River.

The harsh winters of 627–8 and 628-29 were described as extremely cold, causing a massive loss of herds and subsequent famine among the Türks. While the Türk ruler, Illig qaghan, had often raided in previous years the Chinese frontier and imposed a heavy tribute on the Tang dynasty, ruling in China, such raids were no longer possible due to military weakness. Increasing the fiscal pressure on other subordinate tribes led to rebellions and a weakening of the central authority of the *qaghan*, now vulnerable to internal and external challenges, such as the rise of the Xueyantuo confederacy in Mongolia. The Chinese army under general Li Jing (Erkoç 2017a), after having waited until the Türks had been severely weakened, launched their offensive in 630, bringing the Türk empire to an end. The *qaghan* himself was captured but died a few years later (Graff 2002).

This sequence of events occurred against the backdrop of an economic crisis caused by at least two extremely severe winters. Climatological analysis has indicated that the likely cause of the very low temperatures was the volcanic forcing due to an eruption dated to the year 626, which occurred during an already cold period in the Northern Hemisphere, lasting through much of the sixth and seventh centuries CE (Büntgen et al. 2016). The low temperatures caused early frosts and droughts, decimating the herds. The famine that ravaged the empire was a direct consequence of the loss of animals, causing widespread mortality among the people and reducing precipitously the economic and military power of the ruler. The rapid demise of the Eastern Türk empire under the shock of prolonged climatic adversity, and the overall unfolding of the crisis suggests a series of considerations that may clarify key aspects of Türk rulership as well as some salient causal linkages that may explain its collapse.

First, the Türk empire's fairly rapid breakdown underscores the fragility of its economy, which was ultimately based solely on pastoral production. This provided not only food and other necessities, but, most importantly, horses for military use. The other revenue streams were taxes from subordinate tribes and tributes from China, but such exactions could not be collected without the military power to enforce them. Therefore, as soon as both the Tang leadership in the south and the subordinate peoples in Mongolia realized the state of weakness of the Türk leadership, they began to take military action against the *qaghan*. Unable to support economically its people, and with an army increasingly feebler and less numerous, there were no options left to the *qaghan*, whose attempt to lead his people to areas with better pastures also failed due to internal political struggles, and probably because of the extent of the territory affected by the disaster. Long-distance migration was prevented by the deployment of Tang armies on the Chinese frontier, and by hostile nomads in the north.

In conclusion, the evident linkages between the natural disasters that ravaged the Eastern Türk empire from 627 to 629 and its subsequent demise allow us to contextualize the events in terms that connect economic, military and political aspects, and identify more clearly the vulnerability of the empire and its ability to withstand deep and sudden crises. The identification of clearly dated and spatially explicit climate data leave no doubt about the origin of the crisis, which was clearly identified also in the written sources. Unlike what statistically based studies on nomadic interaction with settled societies might suggest, the crisis did not lead to migrations, or to the nomads becoming more aggressive. On the contrary, they suffered a crushing defeat after nearly three years of a deeply debilitating crisis.

2. The Uyghur drought (783-747 CE)

Climate data are especially useful when they raise questions about which written sources are silent. This is the case for the discovery of the longest and most severe drought in the climate history of Mongolia, documented through two tree ring-based chronologies from central Mongolia. The period of the drought, between 783 and 850 CE, coincides for the most part with the rule of the Uyghur empire (744-840), whose capital, Qara Balgasun, was based in the Orkhon valley (Di Cosmo et al. 2018; Hessl et al. 2012). Contrary to the climate events reported during the Eastern Türk empire, this drought is not mentioned in the sources, and thus was never discussed in the history of the Uyghurs as a possible contributing factor to the demise of the empire. In sum, the evidence from tree-ring data points to environmental conditions that alter the historical narrative of the rise and fall of the Uyghur empire. The severe and protracted drought did not trigger any of the effects often assumed to be associated with droughts in a grassland environment, such as famine and migration, or the invasion, pillaging, and conquest of neighboring communities. On the contrary, the most significant events in Uyghur history were the diplomatic efforts to secure trade relations and peace.

The Uyghur empire was significantly different from the previous one, the Türk empire, in at least two ways, both of which are still poorly documented, namely, urbanization and economic complexity. The Uyghur elite included central Asian families of Sogdian origin who were involved in long-distance trade between China and Central Asia. The state-regulated trade between the Tang dynasty in China and the Uyghur empire, through which Uyghur horses were exchanged for Chinese silk, constituted a major state revenue for the Uyghur *qaghan* and doubtless nourished an extensive international trade. Moreover, the Uyghur capital is the largest urban center ever built by a nomadic polity, and other urban centers and palace complexes have been documented as well. Finally, documentary sources attest to the existence of agriculture. Nonetheless, the state gradually lost military power, and eventually fell, partially because of internal dissent, but especially because it could not resist against the invading armies of the Kirgiz, another nomadic people. Historical sources indicate that a winter disaster similar to that suffered by the Türks compounded the crisis, but we have no independent supporting climate evidence, since tree-ring data only apply to the summer season.

Factoring the drought into this summary analysis of the evolution of the Uyghur empire and its salient traits can potentially transform our understanding in several ways. First, it might provide an explanation, albeit hypothetical, for the diversity and greater complexity of the Uyghur economy. If the pastoral economy suffered from the drought, it stands to reason that this situation would have a doubly negative impact, by hurting not just society at large but also the military, since horses (the essential component of all nomadic armies) would be fewer and less healthy. The poor quality of Uyghur horses has indeed been documented in Chinese sources. Lacking an efficient army, the Uyghur state resorted to diplomacy to secure commercial agreements. Moreover, agriculture was developed, which in droughty conditions would have to be done by artificial irrigation. The apparent urban expansion of Uyghurs could also point to market towns and production centers. However, since the archaeology of the Uyghur period is still in its early stages, such matters are as yet poorly known. Second, the progressive weakening of the Uyghur economy might have had negative effects on the state's political stability and its military effectiveness, precipitating the collapse of its leadership in the face of the attack by the Kirgiz, located to the north of the Uyghurs. Finally, the pastoral economy, weakened by decades of drought, would have been hit even harder than under normal circumstances even by moderate winter disasters of the type mentioned above. Eventually, the Uyghur empire was wiped out by the Kirgiz, and the survivors fled to northern China and to the Tarim basin (Drompp 2002).

In conclusion, the climate data provide critical evidence that allows us to "read" apparent anomalies as resulting from environmental stresses or as the product of reactions and adaptations to the same. The empire's diminished reliance on military activity in the latter part of its existence, and coincidental with the duration of the drought, could be placed in direct relation with the assumed downturn in the pastoral economy, while the diversification of its economic base could be interpreted as a response to the same, by way of compensating for the contraction of the pastoral economy with other forms of income and food production. In sum, on the basis of paleoclimatic data, we need to question whether, notwithstanding the silence of the textual sources, we might couch the history of the Uyghur empire as a history of resilience, defined by the attempt to overcome difficulties by building a different economic structure and concurrently modifying their civil and military institutions. Such resilience may also stem from the formative process of Uyghur elites, which included commercial families, and their predisposition towards a more diversified economy. Framing the history of the Uyghur empire as a resilience narrative, in any case, advances our understanding of nomadic empires by focusing on economic questions that would otherwise not come to the surface.

3.   The rise of the Mongol empire (1206-1226).

The assumption that environmental reversals are a necessary concurrent cause, if not the primary cause, of the formation and expansion of nomadic empires has no empirical basis. As we have seen above, understanding climatic stresses is indeed essential to framing their historical significance. A case in point has been evidenced by recent research on the climatic history of Mongolia. Tree-ring data from the lava field in the Orkhon Valley (Central Mongolia) show the occurrence of a "pluvial" period spanning fifteen years that coincided with the zenith of the military activity of Chinggis Khan and his nascent empire. After Chinggis Khan assumed supreme command of all Mongols he engaged in a series of campaigns that involved not only Mongolia but also northern China, against the Jin dynasty, the Xixia kingdom (in today's northwest China) and the kingdom of Khwarezm (in today's Uzbekistan and neighboring regions). Moreover, a Mongol army was sent on an exploratory mission into Caucasia and Russia, traveling around the Caspian Sea before returning to Mongolia. It is this intense military activity that lay the foundations for the later expansion of the Mongol empire into China, Iran, and Russia. The concurrence of the pluvial phase, lasting from 1211 to 1225, and the "burst" of Mongol armies outside the Mongol steppes during Chinggis Khan's lifetime, which occurred between 1209 and 1226, raises the question of the possible relationship between the two (Pederson et al. 2014; Putnam et al. 2016).

The pluvial period, which also took place during a relatively warm phase, increased the productivity of the steppes by mitigating the natural aridity of the region, increasing the water supply, and therefore supporting above-average growth of plants and nutrients, and a concurrent increase of biomass. The augmented energy in the local ecology, in turn, favored the reproductive cycle of animals, reduced their mortality rate, and produced a net increase in livestock and horses. As a secondary effect, the pluvial allowed for greater carrying capacity, that is, for a higher ratio between animals and land area, so that more animals could be supported by a relatively smaller territory for a longer period of time. This is an especially important aspect when we consider the requirements of the Mongol army, whereby each warrior had to be supported by several horses and sheep. If mounting an expedition required the concentration of a certain number of troops in a given region, higher carrying capacity would have eased logistic considerations.

The Mongol empire is anomalous with respect to other steppe empires because of its unprecedented territorial extension, and while most conquests occurred after Chinggis Khan's lifetime, no persuasive explanation has been given so far about the ability of the Mongols to sustain a long period of challenging military expeditions and campaigns outside of their ecological zone, with a relatively limited population, and scant economic resources. However, based on the climate evidence and inferred environmental effects, it is possible to argue that the increased grassland productivity acted upon on-going political processes by extending and expanding the traditional potentialities of a nomadic empire. This hypothesis, which we may call "high-yield grassland productivity hypothesis," would not be, in itself, a "cause" for the Mongols' campaigns but would function as the material basis by which the nomads' potential for military expansion could be expressed and fulfilled for a longer time and over greater distances. This hypothesis would include, at the very least, the following four points.

First, the Mongol empire was unified under Chinggis Khan in 1206 after several decades of internecine war, which caused widespread economic and social distress. The new political order of Mongolia, based on centralized leadership, could thereafter rely on a long unbroken period of positive environmental conditions, and, through its military successes, acquire the revenues necessary to sustain the economic recovery and the new political establishment. Arguably, a negative climatic impact on the environment would have made it impossible for the Mongol society and economy to recover speedily, and thus threatened the existence of its newly-forged centralized state.

Second, a long strung of campaigns, in a traditional nomadic army like the early Mongols, required the constant supply of horses, who become mature for military use between three and four years of age. A lush grassland and low mortality would translate into a steady increase in the horse population, which meant that armies could be larger and horses lost in war could be quickly replaced. As we have seen above, the sudden loss of horses, in the case of the Eastern Türk empire, or the gradual dwindling of horse production, in the case of the Uyghur empire, reduced military capabilities, causing collapse in the first case and the loss of military preparedness as well as propensity for war in the second. In the case of the early Mongol expansion the opposite phenomenon took place, unleashing the expansion of military potential and its sustainability over a long period of time.

Third, on the political level, the increased carrying capacity of the land allowed Chinggis Khan himself to concentrate in his own estate a large number of soldiers, inclusive of his own large bodyguard (*keshik*) and other troops. The concentration of political and military power was an essential prerequisite for political centralization and the prevention of challenges from other leaders. The Orkhon valley was also used, according to some sources, as a gathering place for the army from which to organize and launch military campaigns. Political centralization and military logistics would thus directly benefit from the enhanced productivity of the land.

Fourth, there may have been an increase in agricultural production (Rösch et al. 2005). While at this stage such a hypothesis is purely conjectural, the concurrence of more abundant rainfall for irrigation and the availability of captives from raids into agricultural areas to be used as forced labor in Mongolia might have spurred the development of agriculture. Such a hypothesis is consistent with significant evidence of agricultural production in the area of Karakorum, which became the capital of the Mongol empire under Chinggis Khan's son and successor Ögödei.

In sum, the evidence from tree-ring data not only allows us to discard previous hypotheses that linked Mongol invasions to droughts (Jenkins 1974), but it also and most significantly provides insights into the connections between military activity, political power, and economic resources. Studying the underlying nexuses between such matters as horse supply, land productivity and campaign logistics, for instance, is not just central to any comprehensive historical analysis of Chinggis Khan's military success, but is especially important to achieve a higher degree of understanding of nomadic empires in general. The "discovery" of the pluvial period in the paleoclimate data provides unique evidence to reassess the extraordinary success of the Mongol conquest in its early phase.

Concluding Remarks

The close connection between the pastoral economy practiced by steppe nomads and environmental variability has always had profound implications for the fortunes of Inner Asian empires, but their degree of vulnerability and resilience in the face of climatic adversities, or their ability to leverage environmental advantages, have seldom been gauged. Extensive research on the ecology of rangelands, as well as ethnographic and anthropological studies on traditional uses of pastoral resources, show the degree and characteristics of the dependency of the nomads on the delicate balance between climate and ecology that controls their animals' life cycles. Historical cases, however, show how nomads who reached a higher level of political organization and greater social complexity strove to overcome the limitations imposed by that dependency. Nomadic empires availed themselves of an increasingly wide range of fiscal and other instruments to support their political structures (Di Cosmo 1999). The development of some of those instruments may find its roots in the search for security and survival that makes steppe nomadism a specific form of human adaptation to ecological conditions and to climate variability. The diversity of the cases presented here shows, on the one hand, the measures implemented to counter unfavorable conditions, and, on the other, the ability of nomadic leadership to use surplus productivity for political ends. Historical studies have concentrated especially on the consequences of nomadic empires' expansiveness and the connectivity they promoted in a "Eurasian exchange" in which armies, traders, and germs crossed the continent (Allsen 1997; Green 2020). Such a "world-history" approach has been fruitful, but does not answer the question of the formation of nomadic polities, as well as their relative diversity and complexity. The connection between environmental conditions, economic structures, and political configurations is key to a new approach to steppe history, integrating natural, material, and documentary data.

References

Allsen T (1997) Commodity and exchange in the Mongol empire: a cultural history of Islamic Textiles. Cambridge University Press, Cambridge

Bai Y, Kung JKS (2011) Climate shocks and Sino-nomadic conflict. Rev Econ Stat 93(3):970–981

Batima P, Natsagdorj L, Gombluudev P, Erdenetsetseg B (2005) Observed climate change in Mongolia. AIACC Work Pap 12:1–26

Begzsuren S, Ellis JE, Ojima DS, Coughenour MB, Chuluun T (2004) Livestock responses to droughts and severe winter weather in the Gobi Three Beauty National Park, Mongolia. J Arid Environ 59(4):785–796

Biran M (2019) Introduction: mobility transformations and cultural exchange in Mongol Eurasia. J Econ Soc Hist Orient 62(2–3):257–268

Büntgen U, Myglan VS, Ljungqvist FC, McCormick M, Di Cosmo N, Sigl M, ... & Kirdyanov AV (2016) Cooling and societal change during the Late Antique Little Ice Age from 536 to around 660 AD. Nat Geosci 9(3):231–236

Butzer KW (2012) Collapse, environment, and society. Proc Natl Acad Sci 109(10):3632–3639

Di Cosmo N (1999) State formation and periodization in inner Asian history. J World Hist 10(1):1–40

Di Cosmo N, Oppenheimer C, Büntgen U (2017) Interplay of environmental and socio-political factors in the downfall of the Eastern Türk Empire in 630 CE. Clim Change 145(3–4):383–395

Di Cosmo N, Hessl AE, Leland C, Byambasuren O, Tian H, Nachin B, ... & Cook ER (2018) Environmental stress and steppe nomads: rethinking the history of the Uyghur Empire (744–840) with paleoclimate data. J Interdisc Hist 48(4):439–463

Drompp M (2002) The Uighur-Chinese conflict of 840–848. In: Di Cosmo N (ed) Warfare in inner Asian history (500-1500). Brill, Leiden, pp 73–103

Erkoç Hİ (2017a) The impact of general Li Jing's military thought on the fall of the Eastern Türk Qaghanate. Çanakkale Araştırmaları Türk Yıllığı 15(22):41–52

Erkoç Hİ (2017b) Ecologic and economic factors in the fall of the Eastern Türk Qaghanate (627–630). Archivum Eurasiae Medii Aevi 23:69–80

Fang JQ, Liu G (1992) Relationship between climatic change and the nomadic southward migrations in eastern Asia during historical times. Clim Change 22(2):151–168

Fei J, Zhou J, Hou Y (2007) Circa AD 626 volcanic eruption, climatic cooling, and the collapse of the Eastern Turkic Empire. Clim Change 81(3):469–475

Fei J (2008) Circa AD 627 climatic cooling and the sudden collapse of the Eastern Turkic Khanate. Ganhan qu ziyuan yu huanjing (J Arid Land Resourc Environ 22(9):37–42 (In Chinese)

Fernandez-Gimenez ME, Baival B, Fassnacht S. R, Wilson D (2015a) Building resilience of Mongolian rangelands: a trans-disciplinary research conference—preface. In: Building resilience of Mongolian Rangelands: a trans-disciplinary research conference, June 9–10, Ulaanbaatar, pp 9–15

Fernandez-Gimenez ME, Batkhishig B, Batbuyan B, Ulambayar T (2015b) Lessons from the Dzud: community-based rangeland management increases the adaptive capacity of Mongolian herders to winter disasters. World Dev 68:48–65

Ganiev R, Kukarskih V (2018) Climate extremes and the Eastern Turkic Empire in Central Asia. Clim Change 149(3):385–397

Graff D (2002) Strategy and contingency the tang defeat of the Eastern Turks, 629-630. In: Di Cosmo N (ed) Warfare in inner Asian history: 500-1800. Brill, Leiden, pp 33–71

Green MH (2020) The four black deaths. Am Hist Rev 125(5):1601–1631

Hessl AE et al (2012) Khorgo Lava Pine Tree Ring Chronology. https://www.ncdc.noaa.gov/paleo/study/16773. Accessed 10 Nov 2016

Haldon J, Chase AF, Eastwood W, Medina-Elizalde M, Izdebski A, Ludlow F, ... & Turner BL (2020) Demystifying collapse: climate, environment, and social agency in pre-modern societies. Millennium 17(1):1–33

Jenkins G (1974) A note on climatic cycles and the rise of Chinggis Khan. Central Asiatic J 18(4):217–226

Kradin NN (2015) Nomadic empires in inner Asia. In: Bemmann J, Schmauder M (eds) Complexity of interaction along the Eurasian steppe zone in the first millennium CE, Bonn, pp 11–48

Lee HF, Zhang DD, Pei Q, Jia X, Yue RP (2017) Quantitative analysis of the impact of droughts and floods on internal wars in China over the last 500 years. Sci China Earth Sci 60(12):2078–2088

Pederson N, Hessl AE, Baatarbileg N, Anchukaitis KJ, Di Cosmo N (2014) Pluvials, droughts, the Mongol Empire, and modern Mongolia. Proc Natl Acad Sci 111(12):4375–4379

Pei Q, Zhang D (2014) Long-term relationship between climate change and nomadic migration in historical China. Ecol Soc 19(2):68

Pei Q, Lee HF, Zhang DD, Fei J (2019) Climate change, state capacity and nomad–agriculturalist conflicts in Chinese history. Quatern Int 508:36–42

Putnam AE, Putnam DE, Andreu-Hayles L, Cook ER, Palmer JG, Clark EH, … & Broecker WS (2016) Little ice age wetting of interior Asian deserts and the rise of the Mongol empire. Quatern Sci Rev 131:33–50

Rogers JD (2012) Inner Asian states and empires: theories and synthesis. J Archaeol Res 20(3):205–256

Rösch M, Fischer E, Märkle T (2005) Human diet and land use in the time of the Khans—Archaeobotanical research in the capital of the Mongolian Empire, Qara Qorum Mongolia. Veget Hist Archaeobotany 14(4):485–492

Su Y, Liu L, Fang XQ, Ma YN (2016) The relationship between climate change and wars waged between nomadic and farming groups from the Western Han Dynasty to the Tang Dynasty period. Clim Past 12(1):137–150

Vetter S (2005) Rangelands at equilibrium and non-equilibrium: recent developments in the debate. J Arid Environ 62(2):321–341

von Wehrden H, Hanspach J, Kaczensky P, Fischer J, Wesche K (2012) Global assessment of the non-equilibrium concept in rangelands. Ecol Appl 22(2):393–399

Zhang C (2002) Natural disasters and the decline and fall of the Eastern Turk in the Early of Tang dynasty. Natl Res Qinghai 13(4):53–56 (In Chinese)

Lessons for Modern Environmental and Climate Policy from Iron Age South Central Africa

Kathryn de Luna, Matthew Pawlowicz, and Jeffery Fleisher

Abstract How do we develop effective environmental and climate policy for regions of the world with few—if any—relevant paleoclimate, vegetation, and hydrological reconstructions and, therefore, impoverished models of the environmental and human impacts of future climate change? What if such regions are in countries with limited financial, institutional, or instrumental infrastructure to generate those records? Research in historical disciplines offer direct and indirect evidence of the relationships between societal change and past environmental and climate change, without resorting to bald instrumentalism, but, as this study shows, we need to broaden our historical toolkit if we are to develop such work in regions of the world where oral cultures and less monumental, less permanent material cultural traditions prevailed.

Keywords Southern Africa · Paleoclimate · Environmental History · Policy · Climate Change · Zambia · Archaeology · Historical Linguistics

Introduction

How do we develop effective environmental and climate policy for regions of the world with few—if any—relevant paleoclimate, vegetation, and hydrological reconstructions and, therefore, impoverished models of the environmental and human impacts of future climate change? How is the task made more complex when such poorly documented regions are also in countries with limited financial, institutional, or instrumental infrastructure to generate those records?

K. de Luna (✉)
Department of History, Georgetown University, Washington, USA
e-mail: deluna@georgetown.edu

M. Pawlowicz
School of World Studies, Virginia Commonwealth University, Richmond, USA

J. Fleisher
Department of Anthropology, Rice University, Houston, USA

© The Author(s) 2022
A. Izdebski et al. (eds.), *Perspectives on Public Policy in Societal-Environmental Crises*, Risk, Systems and Decisions, https://doi.org/10.1007/978-3-030-94137-6_13

In many areas that have seen little or no field research by scholars in the natural sciences, research in historical disciplines—notably historical linguistics and archaeology (including subspecialties like geoarchaeology and archaeobotany)—offer direct and indirect evidence of the relationships between societal change and past environmental and climate change (on the emerging preinstrumental climate of Africa, see Hachigonta and Reason 2006; Giannini et al. 2008; Tierney et al. 2010, 2013; Anchukaitis and Tierney 2013; Sletten et al. 2013; Nash et al. 2016; Voarintsoa et al. 2017; Lüning et al. 2018). There are distinct advantages to these historical archives compared to the paper, vellum, and papyrus archives of human–environment and human-climate interactions in the past used by most historians. Significantly, historical linguistics and archaeology generate historical data on geographical and temporal scales commensurate with (1) the larger geographical and temporal scales of most natural scientists' evidence of past climate and environmental change; (2) the vast scales on which most models of future climate change work; and (3) the national and regional scales on which policy must function for effective interventions in the future. The large chronological spans inherent to historical linguistic and archaeological data allow for long-term analysis of human-environmental interaction, but, leveraged against one another, they also connect the hyperlocal focus of archaeological excavation up through the regional and even continental geographies of language change, much as a speleothem or ice core sampled from a specific site can be connected to the expansive scales of shifts in global atmospheric circulations. Thus, historical linguistics and archaeology serve as key direct and indirect archives of past human–environment interactions, strategies, and outcomes at scales relevant to those governing policy-making. They also offer a way to recover past environmental constraints that may be beyond the ken of living memory or documented conditions. Finally, these archives offer indirect proxies of past periods of environmental change, with implications for identifying periods and places most in need of natural science research.

This article explores a case study grounded in historical linguistic and archaeological data in order to explore past societies' strategies of adaptation to changing hydroclimatic conditions in south central Africa over the last 1500 years. It opens with a description of the region and an introduction to the methods of historical linguistics, which are likely to be less familiar to readers than the methods of archaeology. The chapter then moves into the case study, exploring changing practices of mobility, subsistence, and trade across fifteen centuries and several climate fluctuations. Finally, the chapter returns to the implications of both this specific history for policy in south central Africa, but also these methods for the body of evidence mustered in service of producing effective policy, particularly around climate change.

Background

This study focuses on south central Africa—more specifically the region inhabited today by speakers of a broad language family called 'Botatwe.' Today, speakers of

Botatwe languages reside primarily in central, southern, and western provinces of Zambia, with some speakers also residing in lands to the south of the Zambezi River, in northern Zimbabwe, northern Botswana, and the northeastern Caprivi Strip of Namibia (Fig. 1). Generally, the vegetation of this region is characterized as wooded savanna. In the northern regions of Botatwe settlement, the dominant vegetation is miombo, while it is mopane in the south, closer to the Zambezi River. Within these two broad forms of wooded savanna, anthills, ledges and rock outcroppings, perennial rivers, annual streams, seasonally flooded marshes (dambos), and larger floodplains create a variety of microenvironments. The region is characterized by a rainy season (November–April), a cool, dry winter (May–August), and hot, dry summer (September–November).

Today, most rural inhabitants are subsistence farmers, often mixing maize and other cereal crops with fowl, small stock (particularly goats), and some cattle. Currently, the isohyetal lines marking the limits of rainfed agriculture fall south of the region, but in the past, the movement of these limits changed with the extent of the migration of the Intertropical Convergence Zone, leading to variability in the characteristics of seasonality over time (Vansina 2004; de Luna 2016). The region sits at the intersection of the low-level moisture circuits of both the Atlantic and Indian Oceans, which are affected by the 'Angolan Low' low-pressure system and the 'Benguela Niña/Niño' events (Reason et al. 2006). This seasonal weather unfolds in the context of wider climate shifts relating to ENSO phases and the Indian Ocean Dipole, particularly in southern Zambia (Rouault 2003; Hachigonta and Reason 2006; Hurrell et al. 2006; Hardesty 2007; Plisnier et al. 2008) In this way, the region

Fig. 1 Location of extant Botatwe Languages c. 1900

sits at a cross-roads of several systems affecting the quality of the rainy season and, thereby, cereal harvests. While farmers were usually able to harvest cereals, the size and quality of harvests shifted with changing climate conditions, particularly in the marginal, drier southern areas (those now characterized by mopane wooded savanna vegetation). The limited organic materials in soils require a swidden agricultural regime that is increasingly difficult to sustain in the face of modern land tenure regimes and changing opportunities for and ideas about mobility (e.g. Cliggett 2005), thereby compounding the vulnerability of rural farmers to hydroclimatic shifts (Jain 2006; Riché 2007).

Current models predict decreasing rainfall and increased storminess with more erratic rainfall when precipitation does occur (Hulme et al. 2001; Giannini et al. 2008; Kay and Washington 2008; Shongwe et al. 2009; McSweeney et al. 2010; Christensen et al. 2013). Models of the future impact of global climate change are usually rooted in paleoclimate reconstructions based on paleoclimate proxies, few of which exist in Zambia or the wider region. This dearth of paleoclimate proxies confounds our ability to understand the vulnerability of different populations in the region, revealing the need for indirect evidence of past human–environment interaction in the context of hydroclimate variability in this region. The case study in this chapter and the wider ongoing project from which it stems seeks to address this problem.

An Introduction to Historical Linguistics

Languages change over time as the lives of their communities of speakers change (Crowley and Bowern 2010; Dimmendaal 2011; Ehret 2011). Speakers invent their own new words or adopt new words from other languages to name novel (usually borrowed) technologies, practices, or ideas. Speakers' pronunciations shift, taking on regional forms through 'natural' processes like lenition or under the influence of speakers of other near-by languages. Indeed, changes occur across all language features—grammar, syntax, semantics, and so on—accumulating slowly over time in a process that eventually renders a form of speaking a 'language' in one place unintelligible from the way the same language is eventually spoken in another region. In the premodern past, such changes were often the result of the movement of speakers of a language into new regions or the adoption of the language into new regions by the people living in that area. In other words, over long periods of time— centuries or millennia—languages diverge or dissolve into new languages, much like Latin's parentage of French, Romanian, Italian, Spanish, and Portuguese (as well as many further extant and lost dialects). Dissolution is always related to human dynamics, whether the small scale incremental movements of swidden farmers or large-scale imperial expansion (such as the rapid expansion of Quechuan languages at the expense of Aymaran languages in the context of the expansion Inca rule in the Andes; Heggarty and Beresford-Jones 2010).

The fact that languages (or ancestral languages, protolanguages) diverge over time in part through changes in vocabulary offers an historical framework for reconstructing the past worlds and experiences of societies who left no written record and for regions where there has been little to no archaeology. This method assumes that words are historical artifacts attesting to the existence of the idea or object to which they refer. Briefly, a word's phonological shape and distribution in extant languages determines its place in particular "branches" of a language family tree (Fig. 1). Languages undergo predictable sound changes—a point not lost on anyone who speaks more than one language in a family of languages—which are identified through the comparative study of sound correspondences in cognate words across languages (and historical documents). Consider, for example, Grimm's law. Simplified, Proto-Indo-European initial consonant *p was conserved in Romance languages but shifted in a process called lenition to an /f/ in Germanic languages in the position at the beginning of a word stem (for example, consider French and German cognates to father, foot, fish). Thus, a word's phonological shape and distribution within a branch of a language family can also tell us which of three historical processes is responsible for its presence in one or more "generations" (protolanguages) of that the language family: inheritance, internal innovation, or borrowing from other languages. A historical linguist determines when within a language family's history a word was produced and by what process.

Reconstructions of past climate and environment from the archives of the natural sciences are absent from the Botatwe-language speaking areas of south central Africa on which this chapter focuses (Fig. 1). A recent climate reconstruction for the region's Iron Age seeks to reconcile the disparate—and dispersed—records (lake and speleothem records) available beyond the Botatwe zone (Degroot et al. 2021, esp. Fig. 6) in order to track patterns of wetter/warmer and drier/cooler conditions. This synthesis identifies a period of wetter, warmer conditions at the turn of the millennium, with two peaks—the first in the eighth or ninth century and a possible second spike in the thirteen or fourteenth century. These peaks of wetter, warmer weather in the region correspond in stunning fashion to key periods of linguistic diversification independently dated through both glottochronology and robust direct associations with the archaeological record (de Luna 2012a, 2016; Pawlowicz et al. 2018; de Luna and Fleisher 2019). More specifically, the synchronicity of regional climate shifts and language change in the form of the divergences of Central Eastern Botatwe around the mid-eighth century and, later, Kafue around the mid-thirteenth century (Fig. 2) raise questions about the relationships between these two historical processes and the possibility that language data—particularly in conjunction with archaeology—might yield insights into past human–environment relationships in the context of climatic change.

Proto-Botatwe (57–71% [100–900]; 64% median [500]).

I. **Greater Eastern Botatwe** (63–74% [500–1000]; 68.5% median [750])

    a. **Central Eastern Botatwe** (70–77% [800–1100]; 73/5% median [950]

        i. **Kafue** (78–81% [1200–1300]; 79.5% median [1250])

Outline Classification of Botatwe Languages (de Luna 2010, 2016)
(cognation rates, cognation medians, and Common Era dates of divergence in parentheses; protolanguages in **bold**, extant languages <u>underlined</u>)

Proto-Botatwe (57-71% [100-900]; 64% median [500])

I. **Greater Eastern Botatwe** (63-74% [500-1000]; 68.5% median [750])
    a. **Central Eastern Botatwe** (70-77% [800-1100]; 73/5% median [950]
        i. **Kafue** (78-81% [1200-1300]; 79.5% median [1250])
            1. <u>Ila</u>
            2. <u>Tonga</u>
            3. <u>Sala</u>
            4. <u>Lenje</u>
        ii. **Falls** (91% [1700])
            1. <u>Toka</u>
            2. <u>Leya</u>
        iii. <u>Lundwe</u>
    b. <u>Soli</u>
II. **Western Botatwe** (76-81% [1100-1300]; 78.5% median [1200])
    a. **Zambezi Hook** (83% [1400])
        i. <u>Shanjo</u>
        ii. <u>Fwe</u>
    b. **Machili** (84-58% [1400-1450]; 84.5% median [1425])
        i. <u>Mbalangwe</u>
        ii. <u>Subiya</u>
        iii. <u>Totela</u>

Fig. 2 **Outline Classification of Botatwe Languages** (de Luna 2010, 2016) (cognation rates, cognation medians, and Common Era dates of divergence in parentheses; proto-languages in **bold**, extant languages <u>underlined</u>)

        1.   *Ila*
        2.   *Tonga*
        3.   *Sala*
        4.   *Lenje*
     ii.   **Falls** (91% [1700])
        1.   *Toka*
        2.   *Leya*
    iii.   **Lundwe**
  b.   **Soli**

II.   **Western Botatwe** (76–81% [1100–1300]; 78.5% median [1200])

    a.   **Zambezi Hook** (83% [1400]).
     i.   *Shanjo*
     ii.   *Fwe*
    b.   **Machili** (84–58% [1400–1450]; 84.5% median [1425])
     i.   *Mbalangwe*
     ii.   *Subiya*
     iii.   *Totela*

Both linguistic and archaeological data attest to the significance of population movement in the linguistic divergences of Botatwe protolangauges during this period, highlighting the value of these forms of historical evidence for not only firming up 'cautious inferences' from wider regional records in the reconstruction of past climate but also the foundations for societal resilience. Evidence of human mobility related to climate change is implied in the synchronicity of the chronologies of climate and language change; population movement was a key driver of linguistic differentiation and divergence in the premodern world. Linguistic evidence teaches us that from the last quarter of the first millennium through the first quarter of the second millennium, speakers of eastern Botatwe languages spread from the predictably moist floodplains of the Kafue River to regions to the north and south of that floodplain as landscapes that had been marginal to settlement grew more attractive with increased rainfall extending the limits of rain-fed agriculture. This process was mirrored in western Botatwe language groups in the same period—and into more marginal lands (de Luna 2016; consider similar processes in Vansina 2004). Each of the trajectories of expansion involved movement into territories with substantial rivers and wetlands: the Lukanga and Busanga swamps to the east and west of the floodplain, respectively, and the Kalomo, Machili, Lunsemfwa, and Zambezi Rivers, among many smaller annual and perennial streams. These expansions, and the cultural similarities across space they engendered, are reflected in the similarities between ceramics recovered from sites across the area during this time period. The ceramics from the Kafue floodplain, most frequently closed-mouth, spherical vessels with bands of comb-stamping or incised decorations just below the rim (Fagan 1967, Fagan et al. 1969; Derricourt 1985; Pawlowicz et al. 2018), bear similarities to those found on the Batoka Plateau (Fagan 1967, Fagan et al. 1969; Huffman 1989), as well as the upper Zambezi valley (Vogel 1971). Both linguistic and archaeological evidence demonstrate that the expansion of Botatwe speaking communities occurred at the expense of preexisting populations, which were displaced or absorbed. This way, the Botatwe case study serves as an example of resilience and successful, strategic engagements with changing environmental and hydroclimatological conditions.

Thus, the linguistic, archaeological, and climatic chronologies are synchronous even as the linguistic and archaeological evidence attest to the significance of mobility in societies' abilities to engage changing environmental conditions. The centrality of mobility to societal resilience has implications for the sorts of sweeping policies that will be needed to mitigate the threats of future climate change in the region—whether those changes yield wetter or drier conditions or simply increasing unpredictability, as was the case in the west African Sahel (McIntosh 1993, 2005; McIntosh et al. 2000; Maley and Vernet 2015) and is expected to be the case in south central Africa. The specific reasons mobility featured as such an important strategy are essential to our understanding of how the environment changed in the past in the face of shifting hydroclimatic conditions. The words invented by speakers of past languages and the materials they crafted to live successfully in their world offer insights into the causes and consequences of mobility as a strategy of resilience.

The historical development of vocabulary for features of the environment or climate serves as an indirect proxy of changing conditions. The historical development of vocabulary naming changes in a society's actions and practices taken in and on the environment are even less direct as a body of evidence of changing conditions, but they are significant as evidence of strategies for engaging environmental change in the context of hydroclimatic shifts. The centuries around the turn of the first millennium coincided with a transformation of subsistence technologies as reflected in the invention of lexicons naming new subsistence techniques and technologies and the faunal remains recovered from regional sites (de Luna 2016; Pawlowicz et al. 2018; de Luna and Fleisher 2019). Counter-intuitively, an earlier first millennium shift to cereal agriculture sustained great innovation in farmers' hunting and fishing practices during the centuries before and after the turn of the millennium. Importantly, new vocabulary named forms of fishing undertaken in receding floodwaters and with spears thrown from canoes in rapid currents of more swiftly moving rivers. These linguistic innovations were matched by abundant catfish remains from settlements along the Middle Kafue, for example (Fagan 1978; Pawlowicz et al. 2018). Within the domain of hunting, new lexicons named a novel social status of 'celebrated hunter' and forms of hunting based around spearcraft (demonstrating overlap in the technologies of hunting and fishing; de Luna 2012b, 2016). The social scale of hunting scaled up as larger communities worked together using fire and human beaters in large-scale game drives, particularly targeting waterbuck (de Luna 2016). This shift in the Botatwe speaking region was contemporaneous with the emergence of differentiated food consumption habits—some focused on waterbuck—in regions to the south, including the Kalahari and eastern Botswana savannas (Denbow et al. 2008).

The social aspirations fulfilled by those with access to game and fish were gendered in ways that shaped the experiences of plenty, precarity, and dependency in local communities. Yet, the overall political structure of communities remained persistently decentralized, even under pressure from communities experimenting with political centralization and social stratification on the fringes of Botatwe-speaking communities (de Luna 2016). Indeed, the ephemeral—even seasonal—nature of the influence of high-status hunters and fishers—of a politics of reputation and fame that emerged with new hunting and fishing technologies—likely contributed to the persistence of the long-standing durable, flexible political culture structuring the way local communities responded to hydroclimatic anomalies. By diversifying the realms of work and wealth privileged as socially and economically valuable, Botatwe speaking communities ensured that control of agriculture did not, in fact, follow in lock-step with power and extreme social differentiation, as they did in other parts of the medieval world. As a result, vulnerability was more dispersed at the household level, if not at the level of the individual due to the gendered access to food, fame, and wealth.

There was a significant change in the nature of Botatwe speakers' participation in central African trade networks around the thirteenth century, overlapping with the possible second spike of warmer/wetter conditions in the climate records. Prior to that point, the beads acquired by communities in the middle Kafue were made of ostrich and other shell, and some were produced locally (Pawlowicz et al. 2018). By the end of the thirteenth century, these have been replaced by copper bangles and glass beads, the latter likely part of the Mapungubwe oblate series known from

South African materials (Wood 2011). The beginnings of this shift might be marked by the presence of a single, exceptional glass bead of the K2 Garden Roller series at Miyoba (Pawlowicz et al. 2018). That bead, the only such bead found in a stratified context north of the Zambezi, was likely made at K2 in South Africa or a related site in the twelfth century. This shift also captures the broader expansion of trade networks at the time, stretching between the Upemba Depression and the Zimbabwe Plateau and increasingly linked with Indian Ocean commerce (Denbow et al. 2015; Stephens et al. 2020). These shifts in trade and subsistence were undoubtedly related (as individuated by the loanwords describing such activities; de Luna 2012b, 2016) and both contributed to the maintenance of decentralized political power, as was also the case in the unpredictable climate and articulated subsistence mosaic that emerged in the Inland Niger Delta region of west Africa several centuries earlier (McIntosh 1993, 2005).

Mobility was the lynchpin in Botatwe strategies of resiliency in the warmer, wetter conditions of the centuries around the turn of the millennium—particularly to weathering the spikes of moisture that would have transformed seasonal flood patterns as well as the relationship between the characteristics of rain and the varietals of millet and sorghum planted. Mobility was essential to the work of hunters and fishers as they worked during seasonal lulls in the agricultural calendar, often traveling to distant hunting and fishing grounds, perhaps to reunite with clan members and extended family (de Luna 2016). It was a determining factor in the location of settlements in the Middle Kafue. Both settlement mounds and more ephemeral sites were preferentially located at a transition zone between grasslands of the floodplain and higher elevation woodlands. This pattern also existed in the twentieth century, presumably for the same reason: to ensure access to subsistence resources in each environmental zone (Derricourt 1985). Striking a balance between avoiding the flood during the wet and accessing waterways during the dry was surely another consideration. But mobility featured in more sustained ways than these daily or weekly traverses. The archaeological and linguistic data also demonstrate that packing up house and moving to new territories was a key strategy during this period.

Comparison of medieval strategies to later strategies implicated in responses to the cooler and drier conditions centuries later remind us that the nature of climate change—warmer/wetter or cooler/drier—did not determine societal response. Some of the paleoclimate reconstructions discussed for the wider region of central and southern Africa (Degroot et al. 2021) register a shift to drier, cooler conditions in the second half of the seventeenth century. Once again, this corresponds to the last period of linguistic divergence in the Botatwe language family as the Falls protolanguage diverged into Toka and Leya, now spoken around Livingstone in southern Zambia. This divergence during drier, cooler conditions supports the idea that mobility underlay strategies to accommodate all forms of hydroclimatic anomalies—not just periods of increased moisture and temperature. These movements south, toward the Zambezi River may also have related to the extension and intensification of trade networks south the river—networks that ultimately linked to the Indian Ocean and to Europe. Recent work on the Little Ice Age in Europe draws out trade—particularly long-distance trade—as one resilient adaptation to the effects of climate anomalies within the Dutch context (Degroot 2018). It may be that such strategies were repeatedly used by distinct societies in different historical contexts. After all, speakers of the Central Eastern Botatwe and Kafue protolanguages similarly reconfigured their trade networks, facilitating the population movement and linguistic divergence corresponding to the pulses of warmer, wetter climate, discussed above. The borrowing of non-Botatwe vocabulary for fishing and hunting techniques during this period supplies the proof that such activities connected Botatwe speaking and non-Botatwe speaking communities, often across great distances, revealing the intersection of subsistence and interregional trade.

The amplification of trade as a strategy to mitigate the threats of climate change to men and women's access to status, wealth, or even simply their definition of well-being connected very distant communities. Falls speakers' strategies were indirectly connected through Indian Ocean networks as certainly as they were directly connected to developments in the copper mines south of the Zambezi River or the trading hub downriver at Ingombe Ilede (summarized in de Luna 2016). These historical examples offer stark reminders of the extraordinary, rippled geographies of interlocking, articulated responses to environmental and climate change that we should anticipate as we plan our policies to address human-induced climate change and its unequal impacts. They should also remind us that many of the most resilient strategies—resilience here demonstrated by the fact that Botatwe speaking communities absorbed or pushed out neighbors with other languages, material cultures, and lifeways—are coupled together, as we see with developments in mobility, trade, and subsistence with the Botatwe story. Indeed, it may be instructive to remember that the conceptual distinction between 'subsistence' and 'trade' was an invention of European moral philosopher's ruminations on political economy during the seventeenth and eighteenth centuries, inviting our reflection on the many ways societies' ideas about wealth, political influence, and well-being intersected with reconfigurations of political economies during periods of climate change (de Luna 2016).

Forgotten Strategies for Future Policy

Established histories and ongoing research tracking adaptation and change in the face of climate and environmental change over the last millennia and a half in south central Africa opens a number of pressing questions for policy makers working on issues of resilience and equity on the continent.

Mobility. This is not the first study to point to the significance of mobility and the capacity to negotiate access to land and other resources (like game, fish, and salt) as a mechanism for addressing climate and environmental change. Anthropologists and archaeologists in West Africa have shown in great detail how the climatic variation characterized foremost by instability can create the conditions for complex systems with subsistence specializations (farming, herding, fishing, hunting) that are integrated through mobility and exchange (Park 1992; McIntosh 2005). As places that connect diverse and seasonally active environments, floodplains offer unique contexts for such complex systems. Case studies in the Inland Niger Delta, the Middle Senegal Valley (McIntosh 1999), and the Kafue floodplain and wetlands of south central Africa demonstrate the way that such diverse and integrated subsistence systems can be forces for resilience in the face of climate instability. One of the challenges of pressing such historical models into service for contemporary policy, however, is that these systems also rely on highly decentralized decision-making structures, with key connections distributed *across* the system rather than concentrated in nodes of authority and power. Such studies remind us that the invention of private property and the nation state model of governance has come at a price of increasing environmental vulnerability for many societies; the history of land reform at the national level teaches us that large-scale structural change in land access in Africa is not impossible, though, to date, such policies have largely benefitted those in power.

New Currencies and Social Values Diffusing Political Influence. Historical case studies from 'premodern' contexts also bring into stark relief the importance of wider conversations about capitalism, wealth, and well-being. Stories like the Botatwe case in which new climate regimes also created new opportunities for wealth accumulation are hardly surprising reminders that there will be a redistribution of wealth with future climate change. But, the Botatwe story also highlights the way that the development of new, non-material 'currencies' reflecting changing societal values (hunter's fame, for example) are essential for the success of any form of resilience in the face of climate-driven environmental change, particularly if the policy goals include a more equitable, ephemeral/seasonal distribution of political influence and material resources. The decentralization of ephemeral social influence to include novel practices with the potential to generate material wealth seasonally helped to sustain a decentralized political organization across centuries and millennia among Botatwe-speaking societies, even as neighboring societies experimented with statecraft and the concentration of wealth and power (de Luna 2016).

Identifying Unknown Vulnerabilities. In the face of climate change, south central Africa is vulnerable to environmental threats like wildfires that will shift in frequency and character as vegetation and storminess changes across the region. This is particularly telling given indications of a long history of both wild and anthropogenic fires in the geoarchaeological record from the region (Pawlowicz et al. 2020). Although current policy largely seeks to take fire management out of the hands of local settlements, such work has been highly localized in the past and carefully articulated with forms of animal, game, and land husbandry rooted in immediate conditions. In Zambia, knowledge of human-fire-non-human interactions has been lost in some areas, resulting from the establishment of national parks, hunting and fishing laws and licenses, and new hybrid crops and forms of veterinary care (replacing the use of fire to manage insect born livestock diseases, for example). We are in need of research on past fire management practices and their effects on local vegetation, disease, and animal populations in this region of the world. We are also in need of policies that facilitate the circulation of such knowledge at the local level in countries facing the fire threats of increased storminess with global climate change.

Identifying Sites and Foci of Policy-Relevant Research across Disciplines. Precisely because historical case studies have the potential to identify unimagined future environmental threats and past strategies for adaptation that are (for the time periods considered here) low-tech and often, therefore, low-cost strategies, practitioners of the historical disciplines that generate such studies should be at the table in policy discussions. Historians, archaeologists, historical linguists, and other specialists in historical sciences—particularly those who work on far earlier periods and the expansive scales of chronology and geography that match climate change—are also in a unique position to help determine and prioritize the geographical and chronological foci of natural science research on paleoclimate and environmental reconstruction. Such contributions to the natural science research agenda will improve our modeling of past and therefore future climate and environmental change in contexts for which we might also be able to develop low-tech, low-cost mitigation policies. This is of vital importance in contexts like rural Zambia.

References

Anchukaitis KJ, Tierney JE (2013) Identifying coherent spatiotemporal modes in time uncertain proxy paleoclimate records. Clim Dyn 41:1291–1306

Christensen JH et al (2013) Climate phenomena and their relevance for future regional climate change supplementary material. In: Stocker TF et al (eds) Climate change 2013: the physical science basis. Contribution of working group I to the fifth assessment report of the intergovernmental panel on climate change. Cambridge University Press, Cambridge

Cligget L (2005) Grains from grass: aging, gender, and famine in rural Africa. Cornell University Press, Ithaca, New York

Crowley T, Bowern C (2010) An introduction to historical linguistics, 4th edn. Oxford University Press, Oxford

de Luna KM (2010) Classifying Botatwe: M.60 and K.40 languages and the settlement chronology of south central Africa. Africana Linguistica 10:65–96

de Luna KM (2012a) Surveying the boundaries of history and archaeology: Early Botatwe settlement in south central Africa and the "sibling disciplines" debate. Afr Archaeol Rev 29:209–251

de Luna K (2012b) Hunting reputations: talent, individualism, and community-building in precolonial south central Africa. J Afric Hist 53:279–299

de Luna KM (2016) Collecting food, cultivating people: subsistence and society in central Africa. Yale University Press, New Haven

de Luna KM, Fleisher JB (2019) Speaking with substance: methods of language and materials in African history. Springer, Berlin

Degroot D (2018) The frigid golden age: climate change, the little ice age, and the Dutch republic, 1560–1720. Cambridge University Press, Cambridge

Degroot D et al (2021) Toward a rigorous understanding of societal responses to climate change. Nature 591:539–550

Denbow J et al (2008) Archaeological excavations at Bosutswe, Botswana: cultural chronology, paleo-ecology and economy. J Archaeol Sci 35:459–480

Denbow J et al (2015) The glass beads of Kaitshaa and early Indian Ocean trade into the far interior of southern Africa. Antiquity 89:361–377

Derricourt RM (1985) Man on the Kafue: the archaeology and history of the Itezhitezhi area of Zambia. L. Barber Press, New York

Dimmendaal GJ (2011) Historical linguistics and the comparative study of African languages. John Benjamins, Amsterdam

Ehret C (2011) History and the testimony of language. University of California Press, Berkeley

Fagan BM (1967) Iron age cultures in Zambia: Kalomo and Kangila. Chatto & Windus, London

Fagan BM (1978) Gundu and Ndonde Basanga and Mwanamaimpa. Azania 13:127–134

Fagan BM et al (1969) Iron age cultures in Zambia: Dambwa, Ingombe Ilede, and the Tonga, vol 2. Humanities Press, New York

Giannini A et al (2008) A global perspective on African climate. Clim Change 90:359–383

Hachigonta S, Reason CJC (2006) Interannual variability in dry and wet spell characteristics over Zambia. Climate Res 32:49–62

Hardesty DL (2007) Perspectives on global-change archaeology. Am Anthropol 109:1–7

Heggarty P, Beresford-Jones D (2010) Agriculture and language dispersals: limitations, refinements, and an Andean exception? Curr Anthropol 51:163–191

Huffman TN (1989) Ceramics, settlements and late Iron Age migrations. Afric Archaeol Rev 7:155–182

Hulme M et al (2001) African climate change 1900-2100. Clim Res 17:145–168

Hurrell JW et al (2006) Atlantic climate variability and predictability: a CLIVAR perspective. J Clim 19:5100–5121

Jain S (2006) An empirical economic assessment of impacts of climate change on agriculture in Zambia, Pretoria, South Africa. University of Pretoria, Centre for Environmental Economics and Policy in Africa (CEEPA), CEEPA Discussion Paper, No. 27. http://www.ceepa.co.za/docs/CDP No27.pdf

Kay G, Washington R (2008) Future southern African summer rainfall variability related to a southwest Indian Ocean dipole in HadCM3. Geophys Res Lett 35:L12701

Lüning S et al (2018) Hydroclimate in Africa during the medieval climate anomaly. Palaeogeogr Palaeoclimatol Palaeoecol 395:309–322

Maley J, Vernet R (2015) Populations and climate evolution in north tropical Africa from the end of the neolithic to the dawn of the modern era. Afric Archaeol Rev 32:179–232

McIntosh R (1993) The pulse model: genesis and accommodation of specialization in the Middle Niger. J Afric Hist 34:181–220

McIntosh SK (1999) Floodplains and the development of complex society: comparative perspectives from the west African semi-arid tropics. Archeol Pap Am Anthropol Assoc 9:151–165

McIntosh RJ (2005) Ancient middle Niger: urbanism and the self-organizing landscape. Cambridge University Press, Cambridge

McIntosh RJ et al (eds) (2000) The way the wind blows: climate, history, and human action. Columbia University Press, New York

McSweeney C et al (2010) The UNDP climate change country profiles improving the accessibility of observed and projected climate information for studies of climate change in developing countries. Bull Am Meteor Soc 91:157–166

Nash DJ et al (2016) African hydroclimatic variability during the last 2000 years. Quatern Sci Rev 154:1–22

Park TK (1992) Early trends toward class stratification: chaos, common property, and flood recession agriculture. Am Anthropol 94:90–117

Pawlowicz M et al (2018) Contingent mobility in south central Africa: Preliminary report of the 2018 field season of the Bantu mobility project at Basanga, Zambia. Report Submitted to Zambian National Heritage Conservation Commission

Pawlowicz M et al (2020) Capturing people on the move: spatial analysis and remote sensing in the Bantu Mobility Project Basanga, Zambia. Afric Archaeol Rev 37:69–93

Plisnier PD et al (2008) Impact of ENSO on east African ecosystems: a multivariate analysis based on climate and remote sensing data. Glob Ecol Biogeogr 9:481–497

Reason CJC et al (2006) Seasonal to decadal prediction of southern African climate and its links with variability of the Atlantic Ocean. Bull Am Meteor Soc 87:941–955

Riché B (2007) Climate change vulnerability assessment in Zambia. The World Conservation Union (IUCN), Climate Change and Development Project, Pilot Phase Geneva

Rouault M (2003) South east tropical Atlantic warm events and southern African rainfall. Geophys Res Lett 30:8009

Shongwe ME et al (2009) Projected changes in mean and extreme precipitation in Africa under global warming. Part I: Southern Africa. J Clim 22:3819–3837

Sletten HR et al (2013) A petrographic and geochemical record of climate change over the last 4600 years from a northern Namibia stalagmite, with evidence of abruptly wetter climate at the beginning of southern Africa's Iron Age. Palaeogeogr Palaeoclimatol Palaeoecol 376:149–162

Stephens J et al (2020) Lead isotopes link copper artefacts from northwestern Botswana to the Copperbelt of Katanga Province, Congo. J Archaeol Sci 117:105124

Tierney JE et al (2013) Multidecadal variability in east African hydroclimate controlled by the Indian Ocean. Nature 493:389–392

Tierney JE et al (2010) Late-twentieth-century warming in Lake Tanganyika unprecedented since AD 500. Nat Geosci 3:422–425

Vansina J (2004) How societies are born: governance in west central Africa before 1600. University of Virginia, Charlottesville

Voarintsoa NRG et al (2017) Stalagmite multi-proxy evidence of wet and dry intervals in north-eastern Namibia: linkage to latitudinal shifts of the inter-tropical convergence zone and changing solar activity from AD 1400 to 1950. Holocene 27:384–396

Vogel JO (1971) Kumadzulo, an early iron age village site in southern Zambia. Oxford University Press for the National Museums of Zambia, London

Wood M (2011) Interconnections: glass beads and trade in southern and eastern Africa and the Indian Ocean. In: 7th to 16th centuries AD. SAU, Uppsala

Crisis and Recovery

Systemic Risk and Resilience: The Bronze Age Collapse and Recovery

Luke Kemp and Eric H. Cline

Abstract In this chapter we apply the concepts of resilience theory and systemic risk to the Bronze Age Collapse. We contend that this was a case of synchronous failures driven by both long-term trends in interconnectedness and inequality, as well as external shocks such as climate change, warfare (including from hostile migration), rebellion, and earthquakes. This set off a chain reaction as the loss of key cities destabilised the trade-network and undermined state revenue, leading to further rebellion, migration, and warfare. Eventually, enough cities were destroyed to undermine the economic, cultural, and political fabric that held the Bronze Age together. Many states recovered and displayed resilience through the Bronze Age systems collapse. No two states were alike in their resilience. The Neo-Assyrians persisted by moving from a strategy of trade to conquest. The surviving Hittites in northern Syria, in contrast, relied on the modularity of their semi-feudal structure. Systemic risk and resilience are helpful lens for viewing the Bronze Age collapse and recovery, as well as taking lessons for the modern globalised world. It at least provides historical grounds for believing that synchronous failures can happen and can be lethal to states.

Keywords Resilience theory · Systemic risk · Synchronous failure · Bronze Age · Collapse

Introduction

Systemic risk may be a ubiquitous danger for complex systems, whether they be in an age of bronze or silicon. We define systemic risk as the ability for a single disruption, or a series of individual disruptions, to cascade into a systems wide failure. This often occurs as a 'critical transition' in which there is a relatively rapid change

L. Kemp
Cambridge University, Cambridge, UK

E. H. Cline (✉)
George Washington University, Washington, DC, USA
e-mail: ehcline@email.gwu.edu

© The Author(s) 2022   207
A. Izdebski et al. (eds.), *Perspectives on Public Policy in Societal-Environmental Crises*,
Risk, Systems and Decisions, https://doi.org/10.1007/978-3-030-94137-6_14

from one state to another (Scheffer et al. 2012). Critical transitions are well studied for ecosystems (the typical term here being 'regime shifts'), including pollinator communities and coral reefs (Bellwood, Hughes, and Folke 2004; Holbrook et al. 2016), and financial systems (Haldane and May 2011; Acemoglu, Ozdaglar, and Tahbaz-Salehi 2015; Beale et al. 2011). Far less research exists for the modern world system (Helbing 2013; Keys et al. 2019), or historical state systems. Yet there are significant fears that the current globalised world is self-organising towards a state susceptible to systemic crises (Centeno et al. 2015; Nyström et al. 2019).

The Bronze Age State System is one of the most well-known and studied historical cases of a gradual systems collapse. The Bronze Age collapse involved the fragmentation and loss of multiple states that were deeply culturally, politically, and economically interconnected. The collapse coincided with multiple shocks hitting cities throughout the system within the space of approximately a century (Cline 2014, 2021).

In this article we apply key concepts in the systemic risk and socio-ecological resilience literature to the case of state collapse and recovery in the Bronze Age system. Was the decline of the Bronze Age a case of contagion, a domino effect due to the dropping of a key state or city (critical node loss), or the terrible misfortune of multiple shocks that reinforced each other? Or was it all three? Did recovering states exhibit any key attributes of resilience?

Systemic Resilience and Risk in Societies

Resilience

Resilience generally refers to the ability for a system to withstand and recover from different perturbations without losing its fundamental identity (Gunderson and Holling 2002; Walker, Salt, and Reid 2012; Folke et al. 2016; Cumming and Peterson 2017). It is the ability for a system to be robust to different shocks, recover from damage, and resist fundamental change to a new system state (the three Rs) (Grafton et al. 2019). Antifragility is a similar, but subtly different concept. It is not simply the ability to withstand random shocks, but to actively benefit from them (Taleb 2014). The idea is already used in several fields, including in biology through hormesis: the ability for biological systems to gain from small doses of toxins or stress, but damaged by sufficiently large quantities. Antifragility can be roughly thought of as resilience (persisting through stress) in combination with learning and adaptation (changing system structure based on learning).

In applying resilience, we need to ask "resilience for what and to what"? In this chapter we are focusing on the resilience of political and economic systems (primarily states) in the Bronze Age from both internal and external stressors. Resilience is agnostic. It says nothing about whether the system is moral and worth sustaining. Our analysis is agnostic too, but there are reasons to question whether many of the

Table 1 Principles of socio-political resilience

Principles	Description
Adaptive capacity	The ability for a society to change existing behaviour based on observation and (social and organizational) learning to better fit the environment
Diversity	Diversity refers to the variety of elements in a system, the balance between different elements and their disparity (how much elements differ from one and other)
Interconnectedness	The number and intensity of linkages between different elements of a system. This can range from information (social connections, sharing of administrative data) through to material (trade) flows
Modularity	Modularity measures how densely connected 'hubs' within a network can be decoupled. A network is usually more robust if a key node can be destroyed with little impact on the remaining network
Redundancies	Repeated and replicated components within a system. Like kidneys, if one fails, another can take over

states of the Bronze Age were worth preserving. It did, after all, contain empires of domination that were dependent, at least in part, on stratification and slavery.

There is no universally accepted set of resilience principles. Nor is there any agreement on how and in what contexts different principles should be applied (Grafton et al. 2019). This is especially true for states, as most of the resilience literature has been applied to either engineering or socio-ecological systems more broadly. There have been a few attempts to provide a set of key principles or considerations for resilience (Walker, Salt, and Reid 2012; Grafton et al. 2019; Biggs et al. 2012; Yu et al. 2020; Ungar 2018). These tend to both overlap and diverge. There remains neither full consensus nor agreement on which principles can be fruitfully applied to states. Importantly, the evidence for some commonly cited principles for resilience, such as leadership or experimentation, are mixed (Biggs et al. 2012). In Table 1 we have summarised several of the key principles of resilience. These include diversity, interconnectedness, modularity, redundancies, and adaptive capacity (response abilities). These are based on the existing proposals for resilience principles, but with an eye to those which are both applicable to states and can be potentially observed or measured for the Bronze Age.

Systemic Risk

Systemic risk first became a subject of interest due to stock market flash crashes and contagions in the financial system. It was originally intended to explore the possibility that the loss of a single company could cause the destabilization of an entire market or even economy. There is now increasing acknowledgement that systemic risk is relevant across a wide range of areas, ranging from ecology to politics. Systemic risk challenges traditional notions of risk by placing greater emphasis on the vulnerability of a system over the threats it faces. Ultimately, systemic risk is as much a product of structure as shocks.

Many structural features operate as 'tipping points' when it comes to systemic risk. That is, they can operate in a threshold manner in which a certain amount is beneficial for resilience, but too much creates fragility. There is a pernicious and uncertain threshold beyond which features like interconnectedness and diversity can become harmful rather than helpful to a system. This is similar to, but subtly different from, the notion of 'hypercoherence': that as a society becomes more complex and interdependent it also becomes more liable to collapse and disaggregation (Dark 2016).

Any such threshold is unlikely to be a simple matter, but also hinge on the challenges that a system is facing. Interdependence, for instance, may be a blessing during times of stability, but an Achilles heel if there is a large enough shock. In one study of financial networks, Acemoglu et al. (2015) put forward the case that densely connected financial networks tend to buffer against small disturbances, but perpetuate sufficiently large ones. There is broader evidence for this trade-off, in which greater interconnectedness and scale tends to mean resilience against local disruption and more vulnerability towards systemic failure (Scheffer et al. 2012). The system can shift resources to cover small, local losses, yet a large enough shock is more likely to be amplified. There is an unknown tipping point past which a shock goes from being locally bearable to systemically spread. In short, the conditions for resilience can, under different circumstances, become the seeds of systemic risk.

Systems which are dependent on a few key nodes tend to be more vulnerable than those with a more distributed structure. Systems can typically be thought of as 'fat-tailed' or 'random'. Random networks are characterized by a roughly equal distribution of connections between nodes. A typical example is the US road system. The opposite to random networks are 'scale-free' arrangements in which connections have a power-law distribution. Yet such systems with a complete power-law distribution are rare in both human and natural systems (Broido and Clauset 2019; Stumpf and Porter 2012). We opt for the more telling and practical designation of 'fat-tailed': whether linkages tend to be dominated by a few nodes. The network of flights in the US is an archetypal example, with a few hubs such as the John F. Kennedy airport in New York having an order of magnitude more activity than the average airport (Homer-Dixon 2008; Ferguson 2018). Many human networks exhibit such a skew. Industries characterized by network externalities, such as social media platforms, are particularly prone to these 'winner takes all dynamics' (Schilling 2002). These are rarely natural occurrences, but are reinforced by economic and political positive feedback loops. For example, banks that have been labelled as 'too big to fail' have not shrunk since the 2008 Global Financial Crisis, due to consolidation, lobbying, and (unchanged) favourable regulations (Ioannou, Wójcik, and Dymski 2019). Indeed, wealth in general appears to follow a pattern of becoming increasingly fat-tailed and unequal, due to high returns on capital and the ability for the rich to politically reinforce their position. Natural systems can also have such patterns. Disease mortality, for instance, appears to follow a fat-tail distribution throughout history (Cirillo and Taleb 2020). Such systems, whether biological or social, tend to become more susceptible to systemic risk.

Systemic risk appears to have multiple failure modes: contagion, synchronous failures, and critical node loss. Contagion involves an initial shock and its effects being transmitted to the rest of the system. Financial contagion is one of the clearest examples of this. It is most likely to occur in highly interconnected, homogenous and tightly-coupled systems. Critical node loss also occurs in more interdependent, fat-tailed systems. It entails the loss of a key node(s), radically disrupting system functioning. This can happen for electricity networks in which the loss of a key transmitter can cause a region-wide blackout. Synchronous failure is an overarching framework in which multiple, reinforcing stressors cause different, interacting systemic crises which combine to create larger intersystem crises (Homer-Dixon et al. 2015). These tend to be composed of long-term growing vulnerabilities leading to disasters ('long-fuse, big-bang') coinciding with simultaneous stresses, and then culminating in self-reinforcing crises. The 2008 Global Financial Crisis, which was driven by long-term trends in market deregulation and inequality leading to the US sub-prime mortgage crisis and its contagious spread into other markets, as well as an oil price surge, is one example. These different failure modes are not mutually exclusive. Synchronous failures are indeed likely to occur through both critical losses leading to a process of contagion (or, vice versa).

There is some emerging evidence of scale impacting the speed in which a critical transition can occur. One analysis of 42 ecosystems found that regime shifts occurred disproportionately more quickly in larger systems (Cooper, Willcock, and Dearing 2020). Why this is the case is unknown, but logically could be linked to mechanisms of systemic risk. Further study is needed and whether this can be applied to state systems is unclear. When combined with ideas of systemic risk and the presence of a threshold effect for connectivity, there are grounds for believing that the larger and more interconnected a system is, the more susceptible it is to a catastrophe and the quicker and more complete will be its eventual disaggregation.

The Bronze Age

In this section, we will briefly map both the interconnected nature of the Bronze Age state system and the different shocks involved in its slow demise. It should be noted that we are focused here on what can be considered as the 'core' of the Bronze Age, meaning the urban consumers as well as economic and state elites. Our analysis rarely touches upon the more populous 'periphery', namely the largely rural communities that provided raw materials including grain and ores. This is a necessity due to the limited nature of the historical data, which may be common across most historical collapses. We have a high-resolution image for elites and urban areas, but far less for the rural periphery and non-elites at the end of the Bronze Age in the Aegean and Eastern Mediterranean as well.

System Interconnectivity: Economic, Political and Cultural

It is difficult to describe in just a few words the extent of the economic, political, and cultural interconnectivity among the various societies and civilizations in the Bronze Age Aegean and Eastern Mediterranean, among whom we can count the Mycenaeans, Minoans, Hittites, Cypriots, Canaanites, Egyptians, Mitanni, Assyrians, and Babylonians (Cline 2014, 2021). Suffice it to say that we are looking here at a Small World Network, in Social Network Analysis (SNA) terms, wherein even if one of the civilizations was not in direct contact with one of the others, it was in indirect contact though one or more of the others. If there are less than three such "hops" required to connect all members of the network, then it qualifies as a Small World Network, which this does. This is simply a mathematical way of saying that this was an international and globalized world system, for its time and area.

The interconnectedness of the Bronze Age to facilitate trade in metals, particularly copper and tin, was a defining characteristic of the time. It was facilitated by both the demand for such metals, as well as new long-distance mobility technologies such as sail and the chariot, and shared traditions of traders (Kristiansen 2016).

That such economic, political, and cultural interconnectivity was all intertwined is clear from both textual evidence and the archaeological record. For instance, we can see that the various societies were not completely self-sufficient, but rather were dependent to a certain extent upon each other for items ranging from raw materials such as gold, silver, copper, and tin to finished goods and even basic supplies such as grain and dried fish.

We read this, for instance, in letters dating to the eighteenth century BCE found at the site of Mari in ancient Mesopotamia. These mention the importation of finished goods such as gold daggers inlaid with lapis lazuli and leather shoes sent from Minoan Crete (referred to as "Caphtor" in the texts) as well as the arrival of raw tin from Afghanistan and its subsequent exportation onward to the site of Ugarit in what is now coastal northern Syria, where it was traded to Minoan merchants and sent on to Crete. Many of these traded goods were critical. Tin, an essential component of bronze, after which the age was named, was as essential to their world as oil (petroleum) is to ours today (Bell 2012).

The archive of royal letters dating to the time of Pharaohs Amenhotep III and his son Akhenaten in the fourteenth century BCE, found at the site of Amarna in Egypt in 1887, similarly document substantial economic activity at the highest levels. These mention the shipments of tons of raw copper and large quantities of gold in addition to tremendous numbers of finished goods, all couched in the guise of "gift giving" between the Great Kings of Egypt, Assyria, Babylonia, and Mitanni, as well as extensive (and expensive) dowries for the weddings of royal princesses sent by those kings to marry the Egyptian pharaohs (Cline and Cline 2015). We also find similar discussions and lists of exchanged goods between royal elites in the archives of the Hittites at their capital city of Hattusa, as well as in the archives maintained by private merchants working on behalf of the palace in the city of Ugarit, mentioned above, dating to the fourteenth through twelfth centuries BCE.

These exchanges at the highest levels likely masked the vast majority of economic transactions which were doubtless being conducted at a much lower level, perhaps under the cover of the elite exchanges. This is much like the Trobriand Islanders who were participating in the so-called Kula Ring in the South Pacific, whom Malinowski studied in the 1920s. There, while the chiefs of each island ceremonially exchanged valued armbands and necklaces made of shells, the crewmembers of the canoes who had transported the chiefs were busy trading with the locals on the beach for food, water, and other necessary staples of life. Such commercial transactions were the real economic motives underlying the ceremonial gift exchanges of the chiefs, just as the merchants and diplomatic messengers of the Egyptian and Near Eastern kings traveled with caravans containing more mundane but necessary staples of life alongside the royal gifts (Cline 2014, 2021).

Shocks and Interconnections

This interconnected world came to a surprisingly sudden end just after 1200 BCE. Within a matter of a few decades, and one to two generations at most, the populations of these areas were devastated and the economic, political, and cultural ties between the various societies and civilizations destroyed. The shocks suffered by the various societies and civilizations were many and varied. Climate change was especially impactful, manifested in the form of a severe mega-drought lasting between 150 and 300 years in an area stretching from what is now northern Italy across modern Greece and Turkey to Cyprus, the Levant, Egypt, and even as far to the east as Iran and Iraq. The evidence is now undeniable, coming from all of the above-named countries as a series of proxy data derived from studies of lake sediments, stalagmites in caves, and coring from lakes and lagoons. While significant, the local impacts of this climatic shift varied substantially across time and space (Riehl et al. 2014; Finné et al. 2017). Still, it appears to have played a key role, with agent-based modelling exercises already suggesting that it was a primary driver towards the Greek Dark Ages (Vidal-Cordasco and Nuevo-López 2021).

There are also indications of wide-spread famine, as recorded specifically and plaintively in texts found in the archives of the merchants in Ugarit as well as the royal Hittite archives at Hattusa. Societal inequality may have come into play here, as one text from an Ugaritic official says to the king: "grain staples from you are not to be had! (The people of) the household of your servant will die of hunger! Give grain staples to your servant!" (Cohen 2021; Cline 2021). Possible internal uprisings, hypothesized at sites ranging from Hazor in Canaan to Mycenae on mainland Greece, may also have been an additional direct result of both the famine and social inequality at the time.

The mega-drought is most likely the driving factor underlying the migrations of specific groups of people across the region at this time, apparently moving from the western Mediterranean across to the Aegean and thence to the eastern Mediterranean. Some were hostile, such as the so-called Sea Peoples who are featured in the records of the Egyptian pharaohs Merneptah and Ramses III in 1207 and 1177 BCE respectively, as well as unnamed invading forces mentioned in newly published texts from Ugarit. Others were perhaps more peaceful and assimilated with the local population, as indicated archaeologically at sites like Ashkelon, located in modern-day Israel.

There were also natural disasters, such as a series of earthquakes over a fifty-year period (known as an "earthquake storm"), which factored into the overall collapse, with evidence for such at sites ranging from Mycenae to Troy and elsewhere (Nur and Cline 2000). These undoubtedly contributed to the overall Collapse, but also may have contributed to events such as internal rebellions at some of the cities.

Not to be left out, but without conclusive evidence for it yet, is the fourth Horseman of the Apocalypse: pestilence. It has long been known that an epidemic wiped out most of the members of the royal Hittite family ca. 1350 BCE, including the ruling king Suppiluliuma I, and that a later Egyptian pharaoh, Ramses V, apparently died from smallpox ca. 1140 BCE. There is currently no indication yet of any sort of epidemic contributing to the actual Collapse. There are, for instance, no mass graves that have been found and there are no mentions in the contemporary texts. Yet such an additional catastrophe cannot be discarded out of hand. Epidemics do not often leave a clear archaeological trace and disease has frequently accompanied other calamities throughout history. Given that trade and war are two of the key conduits of disease (Scott 2017), there is a non-trivial chance that this played a role during the collapse.

Mapping the Systems Collapse

It is difficult to actually map the systems collapse in real time, for it is nearly impossible to accurately date the demise and fall of each city and/or civilization. While it seems likely that Ugarit fell by ca. 1185 BCE, and Troy by approximately 1180 BCE, even these are just estimates based on a variety of different factors and sources. Even the destruction of Megiddo at the end of the Bronze Age, which now has dozens of radiocarbon dates from the relevant levels, is still the subject of active debate, for possible suggestions range from ca. 1177–1130 BCE or even later for the termination of Stratum VII at the site.

Complicating matters even more is that life did not necessarily come to a complete halt on a particular day even in cities which fell, unlike Pompeii which was buried in a matter of hours or days by Mt. Vesuvius in 79 CE and never occupied again. Hattusa in Anatolia, for instance, is now thought to have been abandoned by the royal family well before the date of its final destruction. Mycenae on the Greek mainland may have been occupied by squatters, and there may have even been a substantial number of surviving inhabitants in the case of Tiryns, perhaps for decades after the cities' elite and most of the citizens were killed or fled for their lives.

Nevertheless, we can certainly say that life as they knew it in the interconnected and globalized Small World Network of the Late Bronze Age Aegean and Eastern Mediterranean essentially came to end shortly after 1200 BCE. By 1100 BCE, their world was quite different, and by 1000 BCE it was so different as to be almost unrecognizable to a former occupant of Mycenae, Tiryns, Troy, Hattusa, or Ugarit. Now many of those cities lay empty and abandoned; some, such as Ugarit, would not recover for centuries.

Is there any evidence that what happened in the Bronze Age was a case of systemic risk? We cannot with any strong confidence suggest that this was a critical node loss. Nor can we provide a quantitative analysis of whether there was a threshold in interconnectivity that undid the Bronze Age. The lack of a clear time series means that for now we cannot model or know if the loss of a critical city destabilized the entire network. Yet, there are signs of other systemic risk archetypes being at play. We can reasonably conclude that the different hazards were not independent. In fact, it may even be somewhat misleading to simply call the Bronze Age collapse a 'perfect storm,' since to many that implies a chance convergence of independent different shocks. Instead, there are good reasons to view the Bronze Age Collapse as a systems disaster in which certain shocks reinforced others, rather than being unrelated. It was, in the language of systemic risk, more likely to be, and perhaps more accurately described as, a case of 'synchronous failures'.

Figure 1 depicts one way of viewing this interconnected crisis in the Bronze Age. To put it briefly, as cities—the key economic and political nodes in the Bronze Age Network—were lost, the remaining urban centers became more vulnerable and new threats worsened. Climatic change appears to have not only decreased crop yields in key areas, but it also likely drove both benign and hostile migration, which in turn threatened trade-routes and further stressed the food system. As different cities failed, they would have likely triggered greater scarcity in the remaining network,

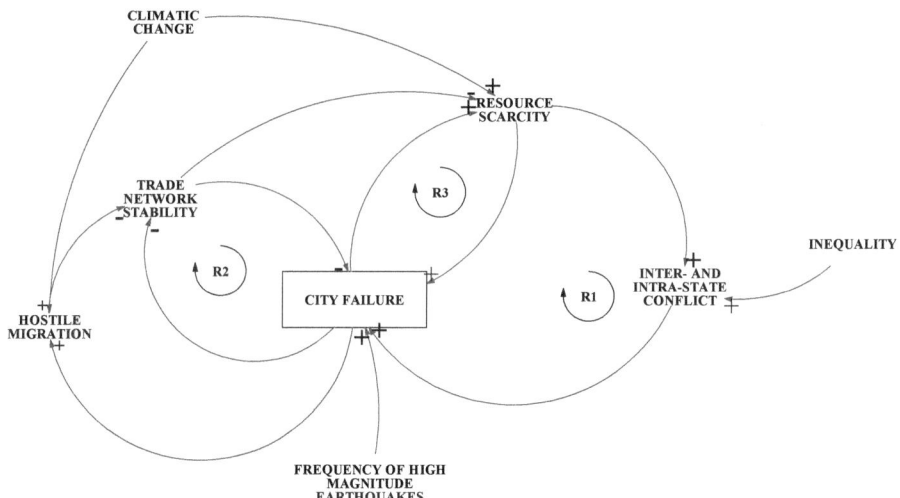

Fig. 1 A causal-loop diagram of the bronze age collapse

creating an impetus for competition (warfare) and worsening underlying vulnerabilities (debt crisis and inequality triggering rebellion), including to other shocks such as earthquakes.

We identify three key reinforcing synchronous failures. The first is resource scarcity, driven by decreasing crop yields, driving conflict. This will not have been unique to the Bronze Age, of course; in the modern world, decreasing food security and resource scarcity has been linked to socio-political violence stretching from Egypt to Syria (Natalini, Bravo, and Jones 2019), to give just one example. Modern case studies and literature have provided evidence that shocks to food security can influence emigration, conflict, and natural mortality rates (Richards, Lupton, and Allwood 2021). Moreover, the decrease in grain yields likely led to higher rates of slavery and debt peonage as peasants failed to repay loans measured in wheat or barley (Graeber 2011). Factors such as debt (Graeber 2011), declining economic opportunity (Lawson 2019), and inequality (Wilkinson and Pickett 2010) have all been linked to social friction and even rebellion. Ancient regimes were also notorious for demanding too much of farmers and peasants during difficult times, causing flight or fights, as was the case for the Third Dynasty of Ur (Scott 2017). We also know that in the period leading up to the systems collapse, inequality was worsening in the Bronze Age world (Scheidel 2017). Thus, cities and states faced a potent mix of factors that made them ripe for either violent upheaval or for turning towards conflict to secure scarce resources in an increasingly chaotic situation. In addition, the transport of grains, metal (particularly tin), and other goods would have been disrupted as cities fell, worsening resource scarcity.

The second synchronous failure is hostile migration and piracy driven by city failure. The initial burst of migration likely came from movements by those whom we have come to call the Sea Peoples. Piracy within the Bronze Age network likely drew upon these migrants, as well as peasants from the Mediterranean, mercenaries, disillusioned warriors, and workers (Hitchcock and Maeir 2014, 2016). The multicultural marauders settled among existing peoples, but also preyed upon the already fraying trade network. As more cities fell, the number of peasants, workers, mercenaries, and others looking for alternative work grew. Thus, piracy was likely driven to a certain extent by the loss of cities. And, as such sea-borne skirmishers increased, the trade network was further destabilised. If disease was also an active part of the Bronze Age crisis, then the emigration of citizens whether it be through piracy or more traditional mobility could have also acted as a disease vector. Microbes often spread alongside chaos.

Third, the loss of trade links and increasing food scarcity would have undermined the revenue sources of both cities and their overarching governments. As grain supplies shrunk, so too did government coffers. This would have likely led to further restrictions in trade, as well as higher unemployment, and hence driven both the adoption of piracy and further migration.

Each of these synchronous failures relies on the interconnectedness of the Bronze Age acting as a conduit for chaos. The reliance of cities on the trade network for strategic resources such as tin, as well as grain, provided a buffer against most perturbations, but likely became a liability during the combination of megadrought

Table 2 Synchronous failures in the bronze age

Causal loop	Explanation
R1	Conflict (both inter-state war and intra-state rebellion) due to resource scarcity from drought further undermines grain stockpiles
R2	Piracy and hostile migration undermine cities as well as the overall trade network. As cities fall, further migration and piracy spreads
R3	The failure of trade networks due to the loss of key cities and the presence of piracy and invaders. This leads to economic decline in surviving hubs

and conflict. We have, for example, cities and states requesting grain and aid from others during the Bronze Age collapse, as demonstrated by a recently translated text from Ugarit which quotes the king of that city beseeching the Egyptian pharaoh: "May my lord save [the land of Ugarit], and may the king give grain to save my life…and to save the citizens of the land of Ugarit" (Cohen 2021; Cline 2021). Similarly, the proximity and easy maritime connections provided a highway for diplomacy and trade, but also eventually for invaders, migrants and pirates.

These different shocks had varying timeframes and severities. Inequality, increasing interconnectivity, and reliance on trade can be thought of as 'long-fuse, big-bang' phenomena. They took decades or centuries to build-up, but eventually the pressure erupted. These coincided with 'simultaneous stresses' including climate change, migration, disease, and earthquakes, which unfolded over the space of decades, years, or even days. The 'reinforcing' crises were then the warfare, piracy, and the other maladaptive and misbegotten responses to disaster. Vulnerability festered in the Bronze Age system, before it was revealed by a few ill-timed disasters (Table 2).

Bronze Age Recovery

It is easier, in some ways, to detail how the different societies and civilizations did or didn't recover from the collapse, and how the world system changed. But it is a complicated scenario to document which states collapsed, which ones recovered or transformed, and which ones were superseded by new entities that arose in the aftermath of the Collapse.

For instance, some, like the Neo-Assyrians and the Egyptians, were able to hold on longer than the others, maintaining the position of the king/pharaoh and his retainers, the trappings of the elites surrounding them, the administrative structures of the state, and even such basics as the ability to write and continue to record the transactions and activities of their society. Eventually, however, even the Neo-Assyrians succumbed for almost two hundred years to the ravages of continued drought, famines, and the encroachment of new enemies such as the Arameans. It would not be until the ninth century BCE that they were able to reassert and re-establish themselves, replacing the trade of the Late Bronze Age with military campaigns and warfare to get the basic supplies that they needed to survive. The Egyptians, meanwhile, also beset with drought and problems with the Nile, devolved into rival factions and, occasionally, rival Pharaohs, especially during the Twenty-First and Twenty-Second Dynasties. Yet, the strong embedded ideological structure of Egypt, combined with its lesser exposure to drought, ensured that there was some degree of continuity despite internal struggle.

Others, like the Mycenaeans and Minoans in the Aegean, the Hittites in central Anatolia, and most of the small Canaanite kingdoms in the central and southern Levant, collapsed quickly and almost entirely, never to rise in its original form again. In each case, however, the key word is "almost," for in every instance, some segment of the society managed to survive, by transforming and adapting to the new reality.

The Phoenicians, for instance, apparently were able to take advantage of the chaos and thrive in the sudden absence of Ugarit, such that they now began venturing out into the waters of the Mediterranean virtually unopposed and eventually re-establishing the links to the Aegean and western Mediterranean. Similarly, the rump state of Carchemish, located in northern Syria and ruled by a member of the Hittite royal family, survived alongside of other small city-states and kingdoms, to collectively take part in the new societal order as the Neo-Hittites, known from the Hebrew Bible as well as from Neo-Assyrian records. In Greece, the surviving population slowly rebuilt their society from the ground up, although they had lost almost everything, including writing, and took longer than any of the other societies to make their way back out of the Dark Age in which they had suddenly found themselves.

Although we cannot be sure, it looks like the previous interconnectivity of the Bronze Age Small World Network did not necessarily aid in the recovery of the key societies and states, for not all survived the centuries after the Collapse. While trade networks did persist, albeit by breaking down into smaller networks, many of the recovering states did not rely on trade for their rejuvenation. However, overall, we may still probably utilize Holling's 'Adaptive Cycle' for at least portions of the network, such as the transformation of the surviving Canaanites into the Phoenicians on the coastal regions of what is now Lebanon. Some, especially those who argue that the Collapse is actually more of a transformation, might even argue that the 'Adaptive Cycle' as a whole can be seen in the eventual reconnection of this entire region, even though the culmination of the cycle took at least four long centuries, until the end of the eighth century BCE, to come to fruition (see also Newhard and Cline, this volume).

Notably, there is no evidence that these states or peoples collectively realized that they were in a collapse. There are, however, indications that some of the individual entities were aware that they were facing a disaster. For instance, the Egyptian administration in Canaan bred heartier cattle and increased their grain production during the beginning of the arid period. They extended dry farming in the Southern and Eastern fringes of the Levant to supply food to the harder hit area of the northern Near East (Finkelstein et al. 2017). Clearly the Egyptians were not blind to their predicament; they were aware of the drought and were taking adaptive steps. They, however, could not have been aware of the severity or duration of the drought, nor the wider context of the unfurling transformation.

As noted, none of the collapses seem to have been total. While there was undoubtedly multiple site destruction, the loss of particular forms of writing, and the severing of many trade networks, both settlements and governance lived on in some more decentralized form. Collapse in the Bronze Age was frequently a process of political disaggregation (Scott 2017). This can also be seen as a triumph of modularity: in lieu of the previous larger, overarching networks of trade and governance, in the aftermath of the Collapse, smaller entities now took on a prominent role in carrying forward culture and populations. However, this should not detract from the genuine human and cultural loss that apparently accompanied the collapse of the Bronze Age.

Each state appears to have relied on a different principle or attribute of resilience to aid in its recovery from the Bronze Age collapse. The Hittites had an almost semifeudal structure[48] which provided some basis for modularity. The rump states that continued on in the absence of the Hittite empire had enough of a cultural and political connection to carry on into the Neo-Hittites, but were modular enough to ensure that they did not fall apart alongside the Empire. The Neo-Assyrians, in contrast, appear to have relied on a more sinister approach of adaptation from trade to conquest. As trade proved to be an insufficient and unpredictable source of capital for the state, it instead adopted a more reliable and lucrative strategy of warfare. This may have been spurred by the adoption of iron weaponry and the annexing of crucial iron ore mines, though this remains to be further investigated (Mann 1986).

The Neo-Assyrians proved to be not only resilient, but sinisterly anti-fragile. On the other hand, as noted, Egypt suffered a prolonged period of political fragmentation and conflict after the Collapse. However, it likely avoided a far worse fate by having sufficient diversity in ecology such that areas less exposed to drought were able to provide coverage for those that suffered greater losses. Overall, both the Canaanites/Phoenicians and Neo-Assyrians exhibited anti-fragile properties as they gained significantly in the aftermath of the collapse. The Neo-Assyrians grew beyond the borders of the previous Assyrian Empire, while the Phoenician city-states found tremendous commercial success and left a long legacy, including the alphabet. In ecology there is the notion of a 'niche': the position of a species in its ecosystem and the conditions to which it is adapted. The Phoenicians essentially expanded their niche in the absence of Ugarit, as did the Neo-Assyrians due to the vulnerability of their neighbors.

These different approaches to resilience can be broadly thought of as evolution or involution. Evolution (adaptive change through competition) was clearly key to the militaristic rise of the Neo-Assyrians. 'Involution,' i.e., adaptive change through cooperation, communication and symbiosis (Hustak and Myers 2012), was far more apparent in the trade-based growth of the Phoenicians. Hence not all forms of resilience for one state are equally beneficial for their neighbors or the wider system (Table 3).

Table 3 Examples of different strategies of resilience in the iron age recovery

State	Process	Primary resilience strategy
Neo-Hittites	Continuity	Modularity through semi-feudal structure
Neo-Assyrians	Expansion	Adaptation through conquest (niche construction)
Egyptians	Recurring fragmentation	Diversity and differing exposure
Canaanites/Phoenicians	Transformation	Adaptation through expansion overseas in the absence of Ugarit

Conclusions: Parallels to the Modern Globalised World System

The modern world has an intriguing number of echoes with the Bronze Age State System. First, like the Bronze Age, it is deeply politically and economically interconnected despite the presence of multiple competing and cooperating states. Second, in both systems there was one particularly essential item of trade which drove the world system. For the Bronze Age, it was tin imported primarily from Afghanistan; for the Modern World, it is oil coming from a handful of key exporters. Third, many of the hazards faced by the two world systems are surprisingly similar. The world of the Bronze Age had drought brought on by climatic variation, while we have modern anthropogenic global warming. The Bronze Age was an unequal age marked with rebellion at its end. Similarly, intra-country inequality has surged persistently over the past decades (Piketty and Goldhammer 2017).

We note, however, that there are a number of obvious differences, which may be either good or bad. For instance, on the plus side, oil doesn't face the same bottlenecks as tin likely did in the Bronze Age. On the other hand, however, there are deep concerns over the declining quality of oil and its energy return on energy invested (EROI).

Furthermore, the modern world appears to have the same long fuses as the Bronze Age in terms of its interconnectedness and inequality. It also has many of the same impending shocks by way of climate change and migration stemming from environmental disruptions. The key difference lies, hopefully, in adaptive capacity. The modern world has far greater potential for adaptation and learning since it can foresee oncoming crises, rely on better technical capabilities, and learn from the past. Hopefully it can adapt without replicating the bloody path of the Neo-Assyrians.

It is also reasonable to point out that the modern world system is likely both more densely interconnected and far faster operating than the Bronze Age World System. Such modern systems, when paired with a world of random shocks, may be simply prone to failure. This idea of 'normal accidents' has already been put forward for complex, tightly-coupled systems such as nuclear power plants. In such systems, disasters are both inevitable and unforeseeable (Perrow 1999). There may similarly be a phenomenon of 'normal collapse' at play for densely connected networks of states. Thus, if the wrong events transpire and conspire in another 'perfect storm' or series of 'synchronous failures', we could quite easily end up replicating the Bronze Age Collapse—modernized and updated, but with the same outcome for all intents and purposes.

Overall, however, the Bronze Age Collapse will likely always be mired in some fog of societal failure. We may never have precise enough data to confidently assert what kind of network failure occurred, what were the most impactful threats, or why some states recovered while others did not. Nonetheless, the existing data and theories concerning systemic risk and resilience can give us the ground for informed estimates. For example, the Bronze Age does appear to be an example of synchronous failures, with the loss of cities driving an economic downturn, trade destabilization, war, privacy, and rebellion, stimulating the further loss of urban centers. This was spurred by climatic change as well as long-term trends in increasing inequality and interconnectedness. It was an example of 'long fuses, big bangs', simultaneous stresses and reinforcing failures (Homer-Dixon et al. 2015), combining together to form the crash.

While the fall of the Bronze Age societies had commonalities in risk, the stories of recovery all vary. Each land was unique in how it did, or didn't, navigate the challenges of the collapse. Some, such as the surviving remnants of the Hittites, relied on existing patronage and semi-feudal networks to ensure that some political and cultural unity was carried forward. Others, such as the Neo-Assyrians, took a more brutal approach of abandoning trade with its decreasing yields, and taking up arms to prey on its weakened neighbors.

Worryingly, many of the vulnerabilities and threats present in the Bronze Age echo through to today. The difference is that their scale and intensity is heightened, with a vastly more interconnected world armed with nuclear weapons, dependent on oil and facing extreme climate change. There are other differences, most importantly in our technology and knowledge of both oncoming crises and past systems collapse. Whether these prove to be enough for resilience remains to be seen. Importantly, the kind of resilience matters. A modern state with nuclear weapons taking a Neo-Assyrian approach to recovery is far from desirable.

References

Acemoglu D, Ozdaglar A, Tahbaz-Salehi A (2015) Systemic risk and stability in financial networks. Am Econ Rev 105:564–608

Beale N et al (2011) Individual versus systemic risk and the Regulator's Dilemma. Proc Natl Acad Sci 108:12647–12652

Bell C (2012) The merchants of ugarit: oligarchs of the late bronze age trade in metals? In Kassianidou V, Papasavvas G (eds) Eastern mediterranean metallurgy and metalwork in the second Millennium BC: a conference in Honour of James D. Muhly; Nicosia, 10th–11th October 2009. Oxford, Oxbow Books: 180–87

Bellwood DR, Hughes TP, Folke C (2004) Confronting the coral reef crisis. Nature 429:827–833

Biggs R et al (2012) Toward principles for enhancing the resilience of ecosystem services. Annu Rev Environ Resour 37:421–448

Broido AD, Clauset A (2019) Scale-Free networks are rare. Nat. Commun. 10:1017

Centeno MA, Nag M, Patterson TS, Shaver A, Windawi AJ (2015) The emergence of global systemic risk. Annu Rev Sociol 41:65–85

Cirillo P, Taleb NN (2020) Tail risk of contagious diseases. Nat Phys 16:606–613

Cline EH (2014) 1177: The year civilization collapsed. Princeton, Princeton University Press

Cline EH (2021) 1177: The year civilization collapsed. Princeton, Princeton University Press, Revised and updated edition

Cline EH, Cline DH (2015) Text messages, tablets, and social networks in the late Bronze Age Eastern Mediterranean: the small world of the Amarna letters. In: Myrnarova J (ed) Egypt and the near east: crossroads II. Proceedings of an international conference on the relations of Egypt and the near east in the Bronze Age, Prague, September 2014. Prague: Charles University, 17–44

Cohen Y (2021) The "Hunger Years" and the "Sea Peoples": preliminary observations on the recently published letters from the "house of Urtenu" Archive at Ugarit. In Machinist P, Harris RA, German JA, Samet N, and Ayali-Dashan N (ed) Ve-'Ed Ya'aleh (Gen 2:6): Essays in Biblical and Ancient Near Eastern Studies Presented to Edward L. Greenstein. Atlanta: SBL Press, 47–61

Cooper GS, Willcock S, Dearing JA (2020) Regime shifts occur disproportionately faster in larger ecosystems. Nat Commun 11:1175

Cumming GS, Peterson GD (2017) Unifying research on social-ecological resilience and collapse. Trends Ecol Evol 32:695–713

Dark K (2016) The waves of time: long-term change and international relations. Bloomsbury Academic, London

Ferguson N (2018) The square and the tower: networks and power, from the freemasons to facebook. Penguin Press, New York

Finkelstein I, Langgut D, Meiri M, Sapir-Hen L (2017) Egyptian imperial economy in Canaan: reaction to the climate crisis at the end of the late bronze age. Ägypt Levante Egypt Levant 27:249–260

Finné M et al (2017) Late bronze age climate change and the destruction of the Mycenaean palace of Nestor at Pylos. PLOS ONE 12:e0189447

Folke C, Biggs R, Norström A, Reyers B, Rockström J (2016) Social-ecological resilience and biosphere-based sustainability science. Ecol Soc 21(3):41. https://doi.org/10.5751/ES-08748-210341

Graeber D (2011) Debt: the first 5,000 years. Brooklyn, Melville House

Grafton RQ et al (2019) Realizing resilience for decision-making. Nat Sustain 2:907–913

Gunderson LH, Holling CS (eds) (2002) Panarchy: understanding transformations in human and natural systems. Washington DC, Island Press

Haldane AG, May RM (2011) Systemic risk in banking ecosystems. Nature 469:351–355

Helbing D (2013) Globally networked risks and how to respond. Nature 497:51–59

Hitchcock LA, Maeir AM (2014) Yo-ho, yo-ho, a seren's life for me! World Archaeol. 46:624–40

Hitchcock LA, Maeir AM (2016) A Pirate's life for me: the maritime culture of the sea peoples. Palest Explor Q 148:245–264

Holbrook SJ, Schmitt RJ, Adam TC, Brooks AJ (2016) Coral reef resilience, tipping points and the strength of herbivory. Sci Rep 6:35817

Homer-Dixon T (2008) The upside of down: catastrophe, creativity, and the renewal of civilization. Island Press, Washington DC

Homer-Dixon T et al (2015) Synchronous failure: the emerging causal architecture of global crisis. Ecol. Soc. 20: art6

Hustak C, Myers N (2012) Involutionary momentum: affective ecologies and the sciences of plant/insect encounters. differences 23:74–118

Ioannou S, Wójcik D, Dymski G (2019) Too-Big-To-Fail: why megabanks have not become smaller since the global financial crisis? Rev Polit Econ 31:356–381

Keys PW et al (2019) Anthropocène risk. Nat Sustain 2:667–673

Kristiansen K (2016) Interpreting Bronze Age Trade and Migration. In: Knappett C, Kiriatzi E (eds) Human mobility and technological transfer in the prehistoric Mediterranean. Cambridge University Press, Cambridge, pp 154–180

Lawson G (2019) Anatomies of revolution. Cambridge University Press, Cambridge

Mann M, The sources of social power I: a history of power from the Beginning to AD 1760. Cambridge, Cambridge University Press

Natalini D, Bravo G, Jones AW (2019) Global food security and food riots—an agent-based modelling approach. Food Secur 11:1153–1173

Nur A, Cline EH (2000) Poseidon's horses: plate tectonics and earthquake storms in the late Bronze Age Aegean and Eastern Mediterranean. J Archaeol Sci 27:43–63

Nyström M et al (2019) Anatomy and resilience of the global production ecosystem. Nature 575:98–108

Perrow C (1999) Normal accidents: living with high-risk technologies. Princeton, Princeton University Press

Piketty T, Goldhammer A (2017) Capital in the twenty-first century. MA, Harvard University Press, Cambridge

Richards CE, Lupton RC, Allwood JM (2021) Re-framing the threat of global warming: an empirical causal loop diagram of climate change, food insecurity and societal collapse. Clim Change 164:49

Riehl S, Pustovoytov KE, Weippert H, Klett S, Hole F (2014) Drought stress variability in ancient near eastern agricultural systems evidenced by 13C in barley grain. Proc Natl Acad Sci 111:12348–12353

Scheffer M et al (2012) Anticipating Critical Transitions. Science 338:344–348

Scheidel W (2017) The great leveler: violence and the history of inequality from the stone age to the twenty-first century. Princeton, Princeton University Press

Schilling MA (2002) Technology success and failure in winner-take-all markets: the impact of learning orientation, timing, and network externalities. Acad Manage J 45:387–398

Scott JC (2017) Against the grain: a deep history of the earliest states. Yale University Press, New Haven

Stumpf MPH, Porter MA (2012) Critical truths about power laws. Science 335:665

Taleb NN (2014) Antifragile: things that gain from disorder. Random House Trade Paperbacks, New York

Ungar M (2018) Systemic resilience: principles and processes for a science of change in contexts of adversity. Ecol Soc 23(4):34

Vidal-Cordasco M, Nuevo-López A (2021) Resilience and vulnerability to climate change in the Greek Dark Ages. J Anthropol Archaeol 61:101239

Walker B, Salt D, Reid W (2012) Resilience thinking: sustaining ecosystems and people in a changing world. Island Press, Washington DC

Wilkinson RG, Pickett K (2010) The spirit level: why equality is better for everyone. Penguin Books, New York

Yu DJ et al (2020) Toward general principles for resilience engineering. Risk Anal 40:1509–1537

Panarchy and the Adaptive Cycle: A Case Study from Mycenaean Greece

James M. L. Newhard and Eric H. Cline

Abstract In this brief paper, we consider and apply the concept of Panarchy and the Adaptive Cycle to a case study from the ancient world, specifically the Mycenaeans at the end of the second millennium BCE. We suggest that the collapse of elite Mycenaean society can be conceptualized as a result of its over-reliance on a hyper-networked international system, whose disintegration brought about a cascading event upon the Aegean World. It may be useful to view the events in this area in terms of regional adaptive cycles and their engagement within and upon broader interconnected systems (Panarchy).

Keywords Panarchy · Adaptive cycle · Bronze age · Aegean · Mycenaeans

Introduction

Resilience theory has frequently been applied to the study of how systems (both human and environmental) are structured and how those structures respond to their surrounding conditions.[1] Within these constructs, some scholars have argued that there is a cycle of rise and fall, collapse, restructuring, rebirth, etc., that every system goes through. Known as the "Adaptive Cycle," this is usually depicted as an infinity loop, with "reorganization, growth, conservation, and release" as its four quadrants (see Fig. 1 and, e.g., Holling 2001; Holling and Gunderson 2002; Redman and Kinzig

---

[1] The authors would like to thank John Haldon for the opportunity to participate in the online seminar during Fall 2020 and for the invitation to contribute to this volume. Some of the ideas and thoughts contained in this contribution are based on, and anticipate, material in the forthcoming volume by Cline entitled *After 1177: The Rebirth of Civilization* (Princeton University Press), which is tentatively scheduled for publication in 2023.

J. M. L. Newhard
College of Charleston, Charleston, SC, USA

E. H. Cline (✉)
George Washington University, Washington, DC, USA
e-mail: ehcline@email.gwu.edu

A. Izdebski et al. (eds.), *Perspectives on Public Policy in Societal-Environmental Crises*, Risk, Systems and Decisions, https://doi.org/10.1007/978-3-030-94137-6_15

Fig. 1 Visualization of the "Adaptive Cycle" (following Holling and Gunderson 2002, Fig. 1) Reproduced by permission of Island Press, Washington, DC

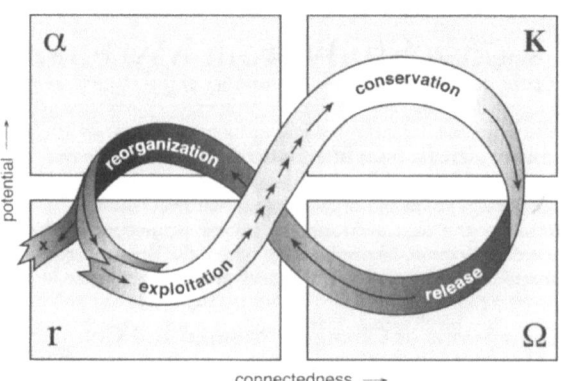

2003). The extent to which the system maintains the ability or flexibility to avoid a dramatic release phase is considered a measurement of its resilience and adaptability.

Of particular relevance to this paper is the 'Release', or Ω phase, which occurs when a system, as a part of its 'Conservation' (K) phase, has produced efficiencies and structures to such an extent that it becomes inflexible in its response to external pressures and changes. The result is a process of devolution or disintegration, resolving into a period of restructuring (α phase). As part of this, it has been further suggested that resources might be reorganized into a new system, in order to take advantage of opportunities created by the disaster (see, e.g., Redman and Kinzig 2003; Redman 2005, 72–74; among others).

Within the eastern Mediterranean at the end of the twelfth century BCE, a series of disruptions are observable in the archaeological and historical record. These disruptions appear at different times, are of different levels of intensity, and impact the socio-political and economic systems to varying degrees. Proposed causes of these disruptions also vary, with no one single culprit identified. The variability in time, scale, and impact have left a disparate array of theories as to what, exactly, happened. In this paper we suggest that the Adaptive Cycle and construct of Panarchy can assist in framing the impacts upon the eastern Mediterranean at this time. In particular, the construct of Panarchy is a useful heuristic to contextualize the differing effects found in the eastern Mediterranean writ large, as well as to discuss impacts at regional and local scales.

Brief Overview of the Bronze Age Collapse

Just after the beginning of the twelfth century BCE, at the end of what archaeologists call the Late Bronze Age in the Aegean and Eastern Mediterranean and across an area which stretched from what is now modern Italy to Afghanistan and from modern Turkey to Egypt, a thriving and interconnected group of societies or civilizations suddenly experienced a catastrophic collapse. By 1100 BCE, at the absolute latest, the interconnected world which they had known for several centuries by that point had broken down. For some, the existing socio-economic and political system was removed and supplanted; others proved more resilient and were able to adjust and otherwise survive, but almost always in a new permutation. The groups in question included the Mycenaeans on mainland Greece and the Minoans on Crete; the Cypriots and the Egyptians; the Hittites in Anatolia (modern Turkey); the Canaanites in the Levant (modern Israel, Lebanon, Syria, and Jordan); and the Mitanni, Assyrians, and Babylonians in Mesopotamia (modern Iraq).

However, the question of what actually caused the Collapse (as it is known) has not ever been answered to everyone's satisfaction, despite numerous attempts at an explanation. Hypotheses that have been suggested over the years include drought, famine, earthquakes, and invaders. Especially favored was the notion of sweeping destructions by the so-called Sea Peoples, whom the Egyptian records claim overran everyone and everything that was in their way in both 1207 BCE and 1177 BCE, during the reigns of the Pharaohs Merneptah and Ramses III. "No land could stand before their arms," said Ramses III on his mortuary temple at Medinet Habu, near the Valley of the Kings in Egypt, "from Khatte, Qode, Carchemish, Arzawa, and Alashiya on. … They were coming forward toward Egypt, while the flame was prepared before them. Their confederation was the Peleset, Tjekker, Shekelesh, Danuna, and Weshesh, lands united" (translation following Wilson 1969, 262–63). Of all the countries, areas, and groups mentioned in these inscriptions, only the Egyptians themselves were able to hold off the invaders, at least according to their own account.

Most recently, however, one of us has suggested that the Sea Peoples were simply one manifestation of the larger problem(s) during that period, and that it is more likely to have been a combination of all of the above—a perfect storm of calamities, as it were—that led to the collapse and breakdown of the globalized society that had flourished for centuries by that point (Cline 2014, 2021). Moreover, it took several decades, and perhaps as much as a century, for the globalized network as a whole to completely collapse, for the various civilizations went down at slightly different times and in slightly different ways, depending upon how resilient they were.

The subsequent period of transition, which saw various processes of rebirth, reorganization, or replacement in different parts of the affected region as a whole, proceeded at varying speeds and by diverse means in each area. While some, like the Assyrians, held out longer than others and recovered faster than most, it was to be nearly four hundred years, not until ca. 800 BCE, before we can truly speak of the globalized network to have been reestablished. There was much transformation involved during these centuries, including new players in several areas, such as the kingdoms of Israel and Judah in what had been southern Canaan. Among those who adapted to the "new normal" were pockets of surviving populations, such as the Phoenicians in what is now primarily coastal Lebanon, who may have taken advantage of the discord and chaos in the years following 1177 BCE and the destruction of cities such as Ugarit in northern Syria (see also Kemp and Cline, this volume).

Resilience/Adaptive Cycle/Panarchy

The Bronze Age Collapse is an event that requires teasing out the processes which regulate continuity, change, and collapse within a system.

Complex systems operate at a series of scales, whether this be a leaf on a tree within a forest, or a household in a village of a larger state system. Each of these units holds the capacity for operating within an Adaptive Cycle, yet the processes within each will be affected by actions taken by other systemic cycles with which they are associated—either at larger or smaller scales (hierarchical, Fig. 2) or at similar scales (heterarchical). This interdependent, networked system of adaptation is termed 'Panarchy' (Holling 2001; Holling, Gunderson, and Peterson 2002) and serves as a means by which one can conceptualize complex adaptive systems and their impact upon other associated systems of organization—both vertically in terms of hierarchical systems of organization, and horizontally in relation to other similarly-scaled systems that are operating at the same time in close proximity or association.

Furthermore, as Holling, Gunderson, Redman, and others have pointed out, it is possible—and potentially very relevant in this case—to discuss "collapse" within the construct of "Panarchy." As Redman and Kinzig (2003) have stated: "many historical studies see the collapse of a civilization as the end of the discussion, because the period, dynasty, or even society appears to be at an end. Happily, resilience theory recognizes that collapse, release, and reorganization are just as integral as the exploitation/growth phase. ... Sometimes the shell of the old system, e.g., the government or social organization, survives the collapse and is reused. At other

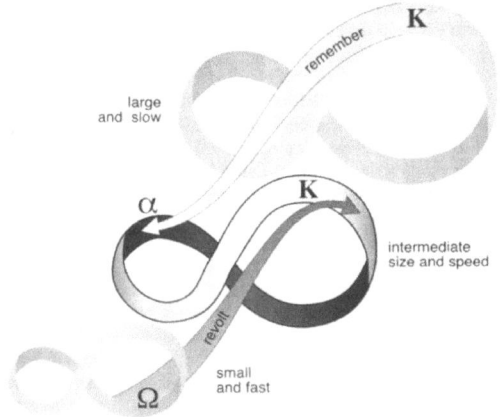

Fig. 2 In this example, a smaller system, as part of its Ω phase, foments revolution, which impacts a larger system undergoing a hyper-extended phase (K), thereby contributing to an ensuing Ω phase in the larger system. Alternatively, a larger system in a K phase contributes previously forgotten constructs of organization to a smaller system, facilitating restructuring (α phase), Fig. 2) (following Holling and Gunderson 2002. Reproduced by permission of Island Press, Washington, DC.)

times after a collapse, the system devolves to a lower level with potential subsequent reintegration. During periods of reorganization of human systems, the focus is usually on social, political, or economic organization."

It is at this critical Ω phase where the hierarchical interdependence (or lack thereof) of systems comes into critical play. During and after the process of release, lower and higher systems of organization serve as structures that can either buffer against complete systemic failure or as guides towards reorganization. Figure 2 presents a hypothetical example, where an intermediate system undergoing collapse brought on by disturbances at a lower level of organization is assisted by 'memory' (expected patterns/practices) from a higher-level system.

Drawing from the historical/archaeological record, an Ω phase observed widely in Anatolia in the eighth century CE (Roberts et al. 2018) was mitigated in a specific region in central Anatolia, given the imposition of the Byzantine state for strategic purposes (Newhard et al., 2021). Despite the environmental and archaeological records showing decreases in agriculture and human activity throughout Anatolia, the region around the city of Avkat (north-central Anatolia) maintained if not increased in the amount of human activity as a result of its importance in the defensive network of the Byzantine Empire.

In this instance, engagement with the higher-order Byzantine system buffered the Avkat region against stresses which would have otherwise been felt. Contrarily, it as has been pointed out by various scholars that frequently it is small, and often very rapid, changes in just one part of the structure that will end up bringing down the entire system (see, e.g., Redman 2005, 73–74). This occurs in systems where the hierarchical relationships are inherently weak, have otherwise become weakened via systemic stress, or conversely, are overdependent upon that component of the structure (see Kemp and Cline, this volume).

We would suggest that the twelfth to tenth centuries BCE in the Eastern Mediterranean can be seen within the construct of the Adaptive Cycle and of Panarchic collapse and reconfiguration (i.e., the K–Ω–α phases at various levels). Viewing the eastern Mediterranean as a series of heterarchical systems interconnected by a series of hyper-networked elite relationships, the multiple disruptions occurring would impact each of these systems differently, based upon their own idiosyncratic configurations. If so, then we might begin by restating the facts as follows: after centuries of sustained interaction and international trade and diplomatic contacts between the Great Powers of the Late Bronze Age (e.g., Hittites, Egyptians, Assyrians, Babylonians, Cypriots), as well as the smaller kingdoms of the Canaanites, during what can be described as the K phase of the Adaptive Cycle, the system came crashing down within a few decades—two to three generations at most—in what seems to be rather clearly the "release" or Ω phase of the cycle.

What we see next is the recovery, or reorganization, part of the cycle (α phase). During this period, which varied in length depending upon the area and society involved, the various civilizations either successfully transformed/adapted or, having been unsuccessful, continue to collapse and reconfigure into new systems. Thus, for example, the Neo-Assyrians—who had successfully sustained themselves for longer than most of the rest of the Great Powers—finally suffered through two centuries of instability, drought, famine, and attacks from groups such as the Arameans, before remerging as the dominant power in the Near East beginning in the ninth century BCE.

An alternative scenario was followed by the Hittites in Anatolia, whose system collapsed entirely, as part of the "release" or Ω phase of the cycle. While the main part of the empire did not recover or reorganize, one small part carried on—namely the rump states located in northern Syria, such as the small kingdom of Carchemish, which was ruled by a descendant of the Hittite king.

Similarly, most of the small Canaanite kingdoms and city-states simply vanished and were eventually assimilated or replaced by new entities such as the kingdoms of Israel, Judah, Edom, and Moab. The small city-states on the coast of what is now Lebanon—including Byblos, Sidon, Tyre, and Arwad—not only survived but flourished amid the chaos that followed the Collapse. They now became "anti-fragile," to invoke a term coined by Taleb (2014), and took over the role that had previously been filled by entities such as Ugarit, expanding out into the Mediterranean and eventually sending their ships as far as the Aegean and beyond, to north Africa and even Spain.

While Taleb's term is useful, we note that some scholars have suggested modifying the α phase within the adaptive cycle, so that it describes a period during which "resources are reorganized into a new system to take advantage of opportunities" (Redman and Kinzig 2003). Such a description would aptly fit the reorganization on the part of the newly reborn Phoenicians. (We note that a similar discussion may also be relevant for Cyprus, and its transition from the Bronze Age to the Iron Age.)

The Aegean Region as an Example of Panarchic Collapse

The aftermath of the Collapse was problematic for more political entities than just the Hittites. For instance, Mycenaean society on Mainland Greece, as they had known it for centuries by that point, essentially came to an end in the decades after 1200 BCE. The Collapse dealt a blow to the Mycenaean civilization which, until then, had been flourishing since the seventeenth century BCE. While the full nature of the political and socio-economic system is still under scrutiny and likely varied from region to region (Parkinson 2007), if evidence from Pylos can stand as a rough proxy, it appears that elite elements of society, established at regional centers in a heterarchical system, engaged in a process of 'prestige finance' by which resources (local and exotic) were mobilized and redistributed as symbolic capital between themselves and others within their individual, regional (likely) hierarchical systems (Nakassis 2011). During the Collapse, the ability to mobilize both would have been severely curtailed, leading to the loss of political power. Most of the main centers suffered destruction, and the elite systems that had most likely administered the bulk of long-distance exchange—along with the central administrative centers, the need for specialized labor, and administrative tools such as writing—disappeared.

Cumming and Peterson (2017) lay out four criteria for defining collapse: (1) elements of the system must be lost, (2) the loss is quickly executed (within a generation), (3) the loss includes significant amounts of social capital (broadly defined), and 4) the consequences must be long-lasting. The Mycenaean world of the twelfth century BCE meets all four of these criteria. Key actors (*wanax*), system components (writing/administration), and interactions (long-distance ties with key trade partners) were lost. This loss happened relatively quickly—likely within a generation—and included elements required for maintaining the prestige economy (social capital) of the Mycenaean elites. The removal of the Mycenaean palatial system limited long-distance engagement and flattened social hierarchies.

Although partial habitation continued at some of the cities, such as at Mycenae, and some lucky or skilled few may have even successfully continued some sort of life as they had known it previously for a few decades, e.g., in the Lower City at Tiryns, the Mycenaean palatial system, for all intents and purposes, failed to adapt and transform to the new normal. What remained were those systems that operated independently (or could operate independently) of the elite system (Arena 2015). It would be centuries before the same level of socio-political complexity was regained. In sum, the vast majority of the elite centers, such as Mycenae, Pylos, and Thebes, seem to have succumbed rather quickly, along with all of their societal norms, ranging from palatial administration and feasting to their writing system. While the collapse of this system may have been welcomed by some, as has been suggested by Weiberg and Finné (2018), who see the large-scale Mycenaean building projects in particular as straining the economy to its breaking point, overall, it can only be viewed as a catastrophe for the majority of the Mycenaean elites and certainly for their society as a whole, which ended rather abruptly.

However, we would also note that the extent of the "Collapse" is surely one of perspective, at least in terms of socio-political organization. As has been noted, the construct of collapse is often a construct of identity (Lantzas 2016; Strunz, Marselle, and Schröter 2019; Cumming and Peterson 2017). For instance, if we were able to ask the *wanax* (king) of Pu-ro (Bronze Age Pylos) whether the world as he knew it had fallen apart, he would definitely reply in the affirmative. The system in which he primarily operated was gone. However, if we were to ask the same of a farmer living at the same time in the periphery, they would probably say that nothing had really changed, unless of course they had been personally affected by the drought.

Further, it has long been apparent to most scholars that the Collapse was neither total nor complete in terms of broader, international connections. It is now apparent that long-distance trade did not completely disappear during the transition from the Bronze Age to the Iron Age on the Greek mainland, particularly at sites like Lefkandi in Euboea; it may also be the case that it was these links to the Eastern Mediterranean that helped contribute to the eventual economic, political, and societal reconfigurations that resulted in the Greek societies of the Archaic and Classical periods.

In short, while catastrophic to the upper echelons and elite of the Mycenaeans, if we view the events in the Aegean at the end of the Bronze Age in terms of the longue durée, the disappearance of Mycenaean society and the end of the Bronze

Age civilizations in the Aegean did not spell the end of inhabitation on the Greek mainland, nor on Crete and the Cycladic islands for that matter. Instead, it may be useful, once again, to suggest that the events in this region should be viewed in terms of the extent to which the idiosyncratic structures of regional systems adapted and changed—in other words, regional adaptive cycles and their engagement within and upon broader interconnected systems (Panarchy).

While viewing the impacts within the eastern Mediterranean within these constructs is useful to conceptualize the pattern and identify ramifications of hierarchical and heterarchical relationships, the Adaptive Cycle and Panarchy do not render information as to *how* those patterns and relationships work. Further work into the individual systems from a construct of heterarchy, which assesses both hierarchical (vertical) and networked (horizontal) relationships, will enable us to better review and understand the impacts of events such as those experienced in the eastern Mediterranean at the end of the Bronze Age (Cumming 2016). Combined with a stated framework for describing collapse (c.f., Cumming and Peterson 2017), the processes afoot in the eastern Mediterranean would have the capacity for deeper cross-cultural comparison with other collapse events—ancient and modern.

Discussion/Conclusions

By way of brief conclusions, we may summarize the situation as follows. First and foremost, it is clear that the Late Bronze Age in the Aegean and Eastern Mediterranean ended in either disruptions to or the elimination of the hyperconnective adaptive system that had been in place for centuries by that point, leading to cascading disruptions to, and/or the elimination of, lower levels of organization. These impacts, and the subsequent α phases, vary by area and specific society, depending upon a variety of factors. Among others, these include their own internal systemic organization (e.g., Aegean vs. Egypt), their own panarchic organizational structure (i.e., the connections between state/regional/local systems, including political, economic, religious, and social), and their dependence upon the hyperconnective adaptive system that had just collapsed. In the case of the Aegean, once the wider eastern Mediterranean networked system was removed, those regions that were highly mobilized towards this engagement were deeply impacted.

In addition, there are some broad takeaways and observations that we might be able to generalize from what we see in the Aegean and Eastern Mediterranean during and immediately after the Collapse. For example, the panarchic collapse that we suggest has taken place came about precisely because of the high levels of interconnectivity and multiple 'stresses.' Furthermore, given the differences in lower-level systemic structures, the impacts to those systems will vary. That is to say, a climatic 'dry phase' may more severely impact one system than another, while the disruption in long distance trade may hit one system hard but may have less effect on another. It does seem clear that the numerous and varied pressures present at the end of the Late Bronze Age brought everyone into a process of reorganization and that it may have been the 'failure' of one or more of the interconnected societies that began the unraveling which ultimately resulted in the Collapse.

Overall, from the particular vantage point of the collapse of the Small World network that was in place within the Aegean and Eastern Mediterranean areas at the end of the Late Bronze Age, we would suggest that 'collapse' per se is not uniform in scale. Furthermore, (a) it is not simultaneous in occurrence; (b) it is not instantaneous (although it is also not a long drawn-out affair); (c) it is not necessarily bad for absolutely everyone concerned (some may in fact celebrate the downfall of the central institutions and economy); and d) it was not necessarily recognizable from within the maelstrom at the time that it was taking place.

It seems clear that although some may have been aware at a local level that things were going very wrong (e.g., textual references to drought, famine, and invaders), the vast majority, if not all, of the affected societies were unaware that their larger interconnected network as a whole was about to fail. There was also no indication at the time that it would take centuries for the system to once again achieve the level of complexity that had been in place ca. 1200 BCE. It was only in retrospect that the Collapse was identifiable; and it is only now, millennia later, that we recognize that it was also most likely inevitable, as part of the larger, and ongoing, panarchic cycle in these regions.

In conclusion, it is clear that further study of the Collapse and its aftermath in the Aegean and Eastern Mediterranean lends itself well to the key points addressed in this volume. We should be able to assess the level of resilience, or lack thereof, in the various societies that were affected in this region. We can also, from studying the texts that have been left to us, as well as from the occasional proxy indicator, determine the extent to which the people in those societies were able to perceive or understand the major challenges and the risks that they faced, including whether their reactions were systematic and organized or random and inchoate.

Finally, it seems apparent, to us at least, that there are lessons to be learned from what happened to these societies more than three thousand years ago and that a better historical understanding of their responses, both positive and negative, to the threats that they faced could potentially help us grapple with the comparable risks and challenges that we currently face today, most particularly but not limited to climate change and the other stresses that arise from such a global threat to the environment. Now it is not just a specific region in the Aegean and Eastern Mediterranean that will be affected, but potentially our entire globalized society that is at risk if we do not listen to and learn from what happened to our predecessors at the end of the Late Bronze Age.

References

Arena E (2015) Mycenaean peripheries during the palatial age: the case of Achaia. Hesperia 84(1):1–46. https://doi.org/10.2972/hesperia.84.1.0001

Cline EH (2014) 1177 B.C.: the year civilization collapsed. Princeton: Princeton University Press

Cline EH (2021) 1177 B.C.: the year civilization collapsed. Revised and Updated. Princeton: Princeton University Press

Cumming GS (2016) Heterarchies: reconciling networks and hierarchies. Trends Ecol Evol 31(8):622–632. https://doi.org/10.1016/j.tree.2016.04.009

Cumming GS, Peterson GD (2017) Unifying research on social-ecological resilience and collapse. Trends Ecol Evol 32(9):695–713. https://doi.org/10.1016/j.tree.2017.06.014

Holling CS (2001) Understanding the complexity of economic, ecological, and social systems. Ecosystems 4(5):390–405. https://doi.org/10.1007/s10021-001-0101-5

Holling CS, Gunderson LH (2002) Resilience and adaptive cycles. In: Gunderson LH, Holling CS (eds) Panarchy: understanding transformations in human and natural systems, pp 25–62. London: Island Press

Holling CS, Gunderson LH, Peterson GD (2002) Sustainability and Panarchies. In: Gunderson LH, Holling CS (eds) Panarchy: understanding transformations in human and natural systems, pp 63–102. London: Island Press

Lantzas K (2016)"Reconsidering collapse: identity, ideology, and Postcollapse settlements in the Argolid. In: Faulseit RK (ed) Beyond collapse: archaeological perspectives on resilience, revitalization, and transformation in complex societies, pp 459–85. Center for Archaeological Investigations Occasional Paper 42. Carbondale: Southern Illinois University Press

Nakassis D (2011) Reevaluating staple and wealth finance at Mycenaean Pylos. In: Pullen DJ (ed) Political economies of the Aegean bronze age. papers from the Langford conference, Florida State University, Tallahassee, 22024 February 2007, pp 127–48. Oxford: Oxbow

Newhard J, Elton H, Haldon J (2021) Assessing continuity and change in the sixth to ninth century landscape of north-central Anatolia. In: Roosevelt C, Haldon J (eds) Winds of change: environment and society in Anatolia. In: Proceedings of the 15th annual international ANAMED symposium, 29–30:141–157, April, 2021. Istanbul: Koç University

Parkinson WA (2007) Chipping away at the mycenaean economy: obsidian exchange, linear B, and palatial control in late bronze age Messenia. In: Galaty ML, Parkinson WA (eds) Rethinking Mycenaean Palaces II: revised and expanded Second Edition, 87–101. Cotsen Institute of Archaeology at UCLA Monograph 60. Los Angeles: University of California

Redman CL (2005) Resilience theory in archaeology. Am Anthropol 107(1):70–77

Redman CL, Kinzig AP (2003) Resilience of past landscapes: resilience theory, society, and the long Durée. Conserv Ecol7(1). http://www.ecologyandsociety.org/vol7/iss1/art14/

Roberts N, Cassis M, Doonan O, Eastwood W, Elton H, Haldon J, Izdebski A, Newhard J (2018) Not the end of the world? post-classical decline and recovery in rural Anatolia. Hum Ecol 46(3):305–322. https://doi.org/10.1007/s10745-018-9973-2

Strunz S, Marselle M, Schröter M (2019) Leaving the 'sustainability or collapse' narrative behind. Sustain Sci 14(6):1717–1728. https://doi.org/10.1007/s11625-019-00673-0

Taleb NN (2014) Antifragile: things that gain from disorder. Paperback. New York: Random House

Weiberg E, Finné M (2018) Resilience and persistence of ancient societies in the face of climate change. World Archaeol 50(4):584–602. https://doi.org/10.1080/00438243.2018.1515035

Wilson JA (1969) The war against the peoples of the sea. In: Pritchard JB (ed) Ancient near eastern texts relating to the old testament, 3rd Edition with Supplement, pp 262–63. Princeton: Princeton University Press

Managing the Roman Empire for the Long Term: Risk Assessment and Management Policy in the Fifth to Seventh Centuries

John Haldon, Hugh Elton, and Adam Izdebski

Abstract This chapter analyses the reasons for the survival of the eastern Roman state from three different but complementary angles: imperial administration, the environmental conditions impacting land-use for the period, and the ability of the state to leverage resources. We conclude that a major contributory factor in survival was the effective use of natural resources and a self-reinforcing social-ecological system through which the state and its elites and infrastructure facilitated the survival of landscapes, generating the resources necessary for the state's continued existence. In areas where this broke down—as in the western part of the empire—the Roman state in the long term disappeared.

Keywords Administration · System · Agency · Structure · Environment · Complex adaptive systems · Ecological continuity

At the beginning of the fifth century, as it had for centuries, the Roman Empire stretched from northern Britain to the Red Sea and from the Straits of Gibraltar to the Caucasus. Seventy-five years later, control of western Europe had been lost, and by the mid-seventh century Syria and Egypt had fallen into Arab hands. And yet the Roman state in the East survived the loss of well over half of its territory and tax income. The east Roman state was at its maximum extent in the middle of the sixth century, following Justinian's reconquests of territory in N Africa and Italy. But it was overextended, with its political center at Constantinople and reaching westward as far as the Balearic Islands, including N. Africa as far as the straits of Gibraltar, along with most of Italy (with Sardinia, Corsica and Sicily) and the Balkans up to the

J. Haldon (✉)
Princeton University, Princeton, NJ, USA
e-mail: jhaldon@princeton.edu

H. Elton
Trent University, Ontario, ON, Canada

A. Izdebski
Max Planck Institute for the Science of Human History, Jena, Germany

Jagiellonian University in Krakow, Krakow, Poland

© The Author(s) 2022
A. Izdebski et al. (eds.), *Perspectives on Public Policy in Societal-Environmental Crises*,
Risk, Systems and Decisions, https://doi.org/10.1007/978-3-030-94137-6_16

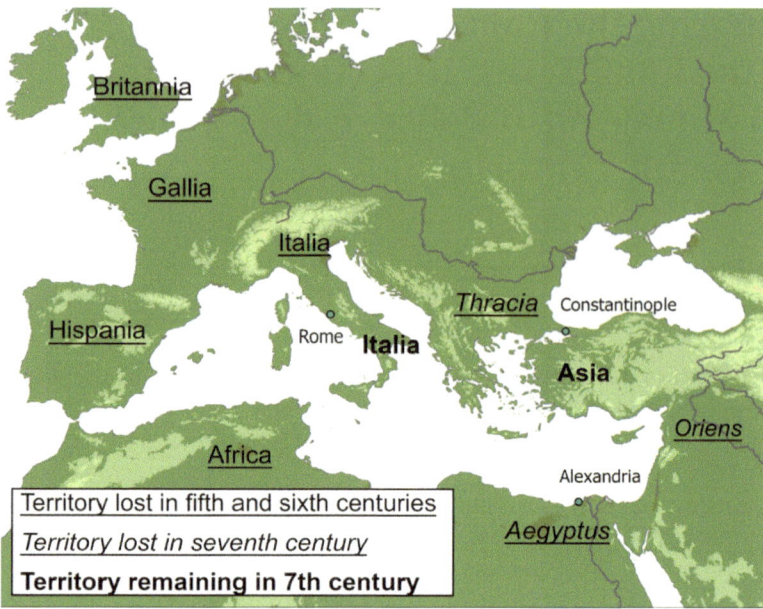

Fig. 1 Territorial changes in the Roman empire between 400 and 700 CE

Danube. In the east it included Egypt and greater Syria (modern Syria and N. Iraq, much of Jordan, all of modern Israel and Lebanon) (Johnson 2010; Mitchell 2007). But already in the 560 s this edifice began to break down (Whittow 2010; Maas 2005). Much of Italy was lost to the invading Lombards from the late 560 s; most of the central and northern Balkan provinces were lost to Slavic and then Turkic invaders from the 560 s; while between 634 and 642 the Arab-Islamic invasions resulted in the loss of the richest provinces in the east, Egypt and Greater Syria (Fig. 1).

By the mid-seventh century, imperial political and military control was confined to the southern Balkan coastal regions, northern and western Anatolia and the central plateau, and the Aegean. The wealthiest tax-generating provinces were lost. By the same token, however the empire was also reduced to a hard, defensible core, focused on Constantinople with its triple walls and sea-defenses, and protected by an Anatolian hinterland in turn covered by the Taurus and Anti-Taurus mountains in the south (Whittow 2010; Howard-Johnston 2010). How did the empire survive, having lost some two-thirds of its territory and up to three-quarters of its revenue within a period of fewer than ten years (C.E. 633–641) (Kaegi 1992; Hendy 1985: 64–167, 616–618)? Its history has often been described in terms of collapse, but on closer examination this is not an appropriate term in the context, for the empire not only survived, it recovered and became a major international power dominating the eastern Mediterranean basin by the tenth century.

Although the debate has generally moved past mono-causal explanations, much recent work has focused on the impact of climate change and of the impact of the sixth-century plague (Harper 2017; Haldon et al. 2018; Sessa 2019; Eisenberg and Mordechai, this volume). While we accept that both climate change and pandemic affected the Late Roman Empire, focusing on these sorts of challenges alone risks denying the crucial role of human agency and all too often replaces analysis of causation with description of chronological correlation.

The Late Roman Empire: An Administrative Approach

The eastern Roman Empire in the fifth century was run by the emperor, usually resident in Constantinople. Fifth and early sixth-century Constantinople was a huge city (c. 500,000), so big that it could not be supplied from its own hinterland but was dependent on the state-managed import of grain from Egypt, at least until it was lost to first the Persians and then the Arabs in the early to mid-seventh century (Teall 1959). Roman emperors of this period were usually resident in Constantinople and appear primarily concerned with warfare and religious politics. However, the majority of the Roman population (most scholars accept about 90%) were subsistence farmers and the problems of the emperor and the capital city were not their day-to-day concern. These two sub-systems, of imperial administration and of agricultural practice, were linked by the state extraction of surpluses. Any changes in agricultural productivity, regardless of cause, had the potential to affect the supply of food and taxes which the emperor and capital needed.

The Empire extracted surpluses in money, manpower, and goods from its population. The majority of taxation in the fifth century was based on the amount of land held and was not progressive, very different from modern systems based on individual productivity. This structure of taxation meant that minor variations in inter-annual productivity had little effect on imperial income, i.e. the risk was transferred from the state to the farmers, with rich farmers better able to buffer this than less prosperous individuals. Major variations as a result of war, natural disaster (flood, earthquake, drought), or plague usually had greater direct impacts on cities than on the countryside, and the larger the city, the greater the impact, so that Constantinople suffered more heavily from the Justinianic plague than most villages (Mordechai and Pickett 2018). Imperial economic mitigation measures were focused on short-term problems, with the emperor providing reduction of tax assessments, repair and rebuilding support, or grain import to major cities suffering from famines.

When making decisions, emperors were generally well-informed as a result of the information sent to the centre by provincial governors (as well as by letters from those wishing to influence policy). And the conciliar process of decision-making allowed most decisions to be discussed by informed individuals before the emperor made a decision (Elton 2009, 2018). These realities were often misunderstood or dismissed by contemporary critics asserting the control of government by favourites and of emperors kept in ignorance. Despite such accusations, the major imperial concern in the fifth and sixth centuries appears to have been the effectiveness of government rather than ideological purity. Thus in 400 the emperor Arcadius (395–408) permitted a successful general named Fravitta to continue to worship as a pagan, despite a law prohibiting pagans from holding office. And in 433, during the aftermath of the First Council of Ephesus, when the emperor Theodosius II (408–450) suggested in the imperial council that bishops in the region of Cilicia should recognize the authority of their patriarch or be exiled, this was opposed by the eastern praetorian prefect, responsible for taxation, who warned that it would cause disturbances.

In the decisions that they did make, emperors tended to focus on short-term economic and military effectiveness, a focus which sometimes had significant long-term consequences. The collapse of the western Empire is thus often attributed to the loss of Africa, the result of pragmatic short-term decisions by the eastern Empire to make peace in 434 and then to abandon an African invasion in favour of defending against a Hunnic crisis in 441. In the long-term, loss of African resources is often seen as critical. (Wickham 2009: 78; Kelly 2008: 119–129).

These imperial priorities were different from those of farmers, and though war might depress regional levels of production and of distribution, it seems less likely to have changed what farmers chose to grow. During the fifth to eighth centuries there was a general reduction in the scale of cereal-centered agriculture, either gradual or abrupt, at different points in time, particularly in Anatolia (Roberts 2018). How do we relate this data to questions of state and farmer resilience, and to climate change and the Justinianic plague? And was the reduction in the population of Constantinople from the mid-sixth century the result of plague, of difficulties in feeding its population, the result of losing territory, or part of a pattern of urban decline in the eastern Mediterranean?

Our understanding of imperial and farmers' decisions is based on the source material that we have. The majority of literary sources that focus on politics and decision-making are anecdotal and interested in short-term events over a few years or an imperial reign. With these, we have little certainty that all or even the most significant events are described at an imperial level, and they say very little about agriculture. Literary sources are very different from archaeological evidence that usually handles time in terms of centuries and is more applicable to the *longue durée* (Decker 2009). Neither of these types of evidence is well-suited for understanding decisions by farmers as to what to grow. For many crops, especially cereals and vegetables, these were annual decisions, though the planting of fruit trees, especially olives which can take up to a decade to reach fruition, might be an expression of confidence in military and economic stability. Changes in what to plant could be based on many factors, of which short-term variations in markets and security were probably more critical than long-term changes in climate which were not visible to contemporaries (Elton 2021).

Landscape and Climatic Change in the Late Roman Empire: An Environmental Approach

Given these limitations in the potential of the historical and archaeological sources to describe the dynamic process of maintaining resilience of the intertwined state and agricultural systems in the face of natural and societal pressures, a recent approach tries to combine these more traditional sources with large amounts of natural scientific data, in particular coming from the palaeoenvironmental studies. Paleoenvironmental data come from the natural archives, that is different locations in the natural environment, such as lakes, peat bogs, or caves, where sediments ("remains" of biological and physical processes) accumulate over time. By using a wide variety of laboratory techniques to study their physical and chemical composition, it is possible to reconstruct landscape changes—in terms of both land morphology and vegetation cover—as well as climatic oscillations with sub-centennial, at times even decadal, precision (Haldon et al. 2014; Izdebski et al. 2016).

Not all former Roman lands in the Central and Eastern Mediterranean boast a large number of well-studied natural archives, but there is enough of them to understand broader patterns of climatic and environmental change (Fig. 2). What emerges is a highly regionalized pattern, in which trajectories of climatic change do not necessarily overlap with transformations in the landscape. Figure 3 shows the direction

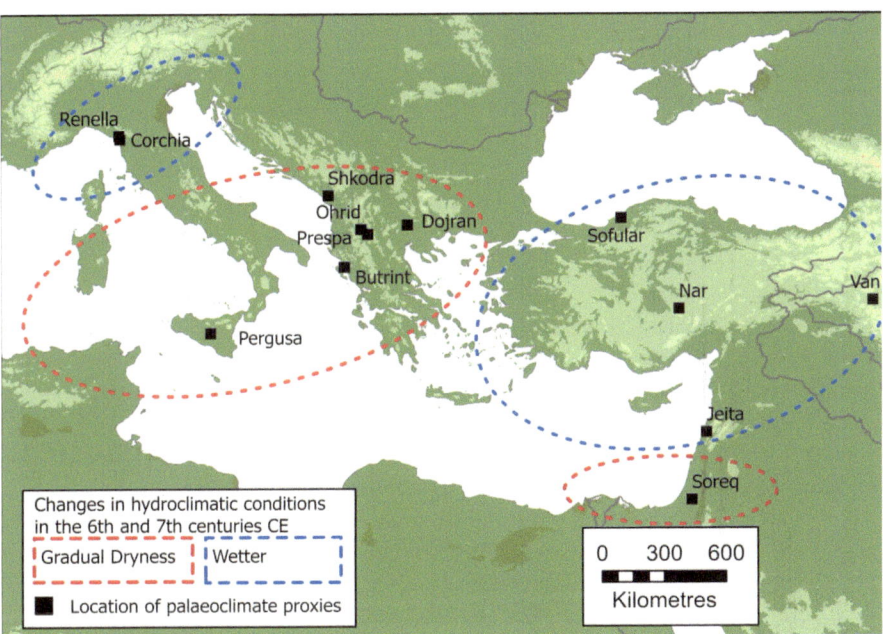

Fig. 2 Regional patterns of hydroclimatic change in the Central and Eastern Mediterranean at the end of Antiquity, based on the proxy evidence from lakes and caves (after Labuhn et al. 2018)

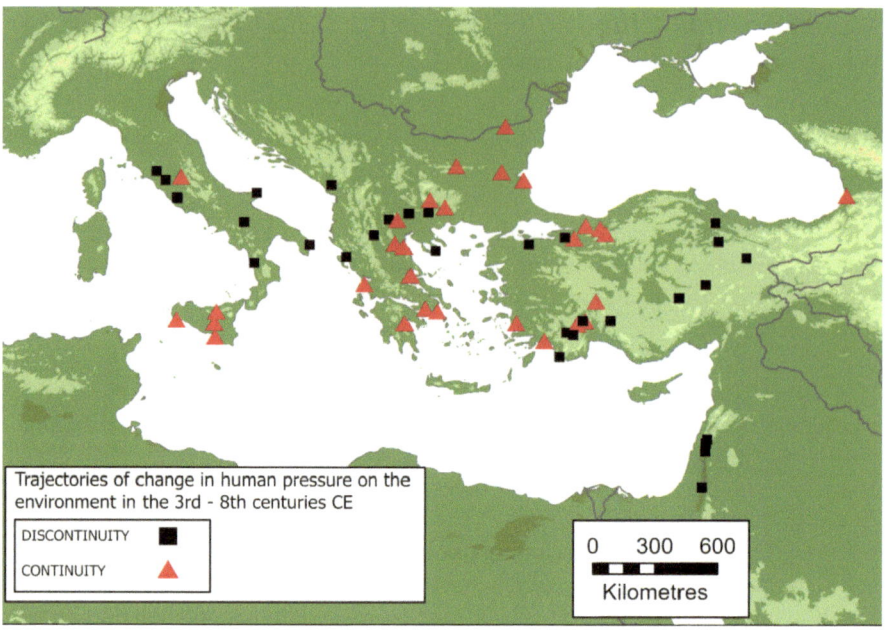

Fig. 3 Trajectories of landscape change in the Central and Eastern Mediterranean at the end of
Antiquity, based on the pollen data (after Izdebski 2020)

of change in hydroclimatic conditions—based on proxy records from more than a
dozen lakes and caves located across the Mediterranean—on the Roman lands in
the final centuries of Antiquity (6th–7th c.). There was no single trend for the entire
Mediterranean, and different regions of the Empire were experiencing contrasting
trajectories of climatic variability. For instance, in the southern Italian-Balkan area
the amount of rainfall was gradually decreasing, while in Anatolia and the Northern
Levant these centuries were characterized by wetter than usual conditions. However,
if we look at Fig. 3, which synthesizes data on vegetation cover change in all key areas
of the Eastern Roman Empire, no obvious correlation between landscape and climatic
changes occur. Within the same "climatic change" zone, for instance, Anatolia, its
different parts showed either continuity in human pressure on the landscape (levels
of agricultural activity were maintained, even if farming strategies were modified in
some areas), or discontinuity (less agricultural activity or land abandonment). Put it
another way, we have areas of continuity both in regions that experienced increasing
dryness and those where more rainfall became available. In parallel, human activity
in the landscape was discontinued both where it was becoming significantly drier,
but also where it was becoming wetter.

If climatic variability is not an explanation for significant transformation
of Mediterranean landscapes, where do these patterns come from? Strikingly,
comparing a geographical distribution of vegetation changes (Fig. 3) and political
transformations (Fig. 1) turns out to be more revealing than comparing landscape and

climate at that time. Ecological continuity can be observed primarily in core areas of the rampant Roman State, known as the Byzantine Empire. Sicily, Southern Greece or Western Anatolia where all areas which remained under relatively tight control of the Eastern Roman government well into the 8th c. The maintenance of the previous levels of human pressure on the landscape and hence agricultural productivity on one hand enabled the Roman state to survive the loss or destruction of several other provinces, while on the other hand it was the continued existence of this state that encouraged and enabled the ecological continuity (Izdebski 2021).

The Late Roman Empire in the East: A Systems Approach

How this was possible can be understood through the lens of a complex adaptive systems framework, in which five overlapping themes provide a helpful focus through which to interrogate the historical, archaeological and palaeoenvironmental sources: first, the nature and quality of the empire's natural capital (water, agrarian and pastoral resources, people); second, the nature and quality of the physical capital over which it disposed (labor, infrastructure); third, its human capital (skills, competences, attributes, including belief systems); fourth, the ways in which access to and/or control over resources was structured; and fifth, the level of redundancy built into the system as a whole—the degree to which there exists a plurality of functionally effective options for achieving key outcomes for survival (Levin et al. 2013; Scheffer 2009). Given the historical context, we add a sixth element, namely the broader international context which formed the context within which the empire existed.

The rump of the eastern Roman state possessed several natural advantages in respect of its strategic geography and the natural frontiers that an invader had to overcome, including strong seasonal weather patterns and especially extremes of temperature and environmental conditions on the central Anatolian plateau. The combination of these aspects enabled the state to organise an effective defence based on minimal central expenditure, led by and relying heavily upon local elites loyal to the centre (Whittow 1996: 15–37; Haldon 1999: 34–66). In addition, a generally unstable climatic and environmental context actually benefited the empire by fortuitously favouring grain production and livestock breeding at just the moment at which these were essential to supporting the military and supplying the capital, Constantinople (Haldon 2019). The state also maintained an effective administrative apparatus for the efficient extraction, distribution and consumption of resources to best advantage (Haldon 2016b; Brandes 2002). Fourth, the 'political theology' of the Christian Roman state was deployed consistently to maintain and reinforce imperial authority and legitimacy, thus maintaining a solid grip over provincial elites who managed and administered fiscal and other resources (Brandes 2013; Magdalino 2010; Cameron 2007). This was especially the case with the elites in Anatolia, the south Balkan coastal zones that remained under direct imperial control, but also Sicily and southern Italy. Finally, from the middle of the seventh century social/cultural status and peer-recognition became increasingly focused on the imperial court and on personal connection with the ruler; an influx of dependent elites from non-traditional origins, together with an effective logistics and communications network, facilitated maximum state control (Brubaker and Haldon 2011: 573–598). In Anatolia this permitted constant re-occupation of sites/key points, roads, and other resources captured by the invaders (Haldon 2016a: 159–282).

The high degree of infrastructural and ideological cohesion and identity thus contributed to the maintenance of system identity and systemic complexity. Only in terms of spatial extent was there a significant simplification, in terms of territorial control. But this seems in fact to have contributed to sustainability and resilience by reducing the state's operational costs and permitting a high level of central control. Adaptive capacity in this case was articulated through the geographical and geopolitical advantages the state enjoyed, the incidental benefits of (unperceived) climatic/environmental factors, substantial organizational advantages, continuing central control over the Anatolian, Balkan, and Italian/Sicilian elites, and ideological cohesion. Last, but by no means least, its major political/ideological enemy in the seventh century, the Umayyad Caliphate, had to contend in its own domain with both high levels of regionalisation and a dispersal of resources, as well as its own internal conflicts (Robinson 2010: 202–224; Kennedy 2004: 90–98).

Conclusion

The survival of the eastern Roman state can be analysed in different ways. The approaches taken here cover similar ground, the effectiveness of the state, but from slightly different angles, one focusing on imperial administration and patterns of taxation, another on the nature of the environmental conditions and observable trends in land-use for the period, the last on the ability of the state to leverage the resources that it had. The changes in the agricultural productivity of the empire are key, and provide an indication of the nature of the available resources as well as of the environmental impact of both human activity (in particular of warfare on the one hand and farmers' response to changing fiscal and market conditions on the other) and of minor shifts in climatic conditions. Survival was made possible by effective use of natural resources, grounded in the maintenance of inherited patterns and levels of land use but with modifications that were required due both to climatic variability and to the needs of the state. A virtuous circle was set up: a self-reinforcing social-ecological system whereby the surviving state and its elites and infrastructure enabled the survival of landscapes which in turn provided the resources necessary for the state's continued existence. In areas where this broke down, the Roman state in the long term disappeared. Importantly, while there was clearly no ecological thinking or awareness as understood today, this virtuous circle or feedback mechanism was not a chance outcome, but rather the result of established or institutionalized practices of governance, underpinned by a powerful ideology that reinforced existing social hierarchies and their impact on the natural environment, thus ensuring the survival of the entire socio-economic system.

The continued survival of the eastern Roman Empire provides a good example of a geo-strategic shrinkage that in effect aided stabilisation, resilience and recovery, as well as of the ways in which contemporary and near-contemporary observers and participants understood and explained how their world was changing around them.

References

Brandes W (2013) Taufe und soziale/politische Inklusion und Exklusion in Byzanz. Rechts-geschichte/Legal History 21:75–88

Brandes W (2002) Finanzverwaltung in Krisenzeiten. Untersuchungen zur byzantinischen Administration im 6.-9. Jahrhundert (Frankfurt am Main)

Cameron Av (2007) Enforcing orthodoxy in Byzantium. Studies in Church History 43 (= K. Cooper and J. Gregory, eds., Discipline and diversity. Woodbridge) 1–24

Decker M (2009) Tilling the hateful earth: agricultural production and trade in the late Antique East (Oxford)

Elton HW (2021) Agricultural decision making on the south coast of the black sea in classical antiquity. In: Coskun A (ed) Ethnic constructs, royal dynasties and historical geography around the black sea littoral (Stuttgart), pp 343–355

Elton HW (2018) The late roman empire in late antiquity: a political and military history (Cambridge)

Elton HW (2009) Imperial politics at the court of Theodosius II. In: Cain A, Lenski N (eds) The power of religion in late antiquity (Aldershot), pp 133–142

Harper K (2017) The fate of Rome (Princeton)

Haldon JF (2019) Some thoughts on climate change, local environment and grain production in Byzantine northern Anatolia. In: Izdebski A, Mulryan M (eds) Environment and society in the long Late Antiquity (Late Antique Archaeology 12. Leiden), pp 200–206

Haldon JF (2016a) The empire that would not die: the paradox of East Roman survival 640–740 (Cambridge MA)

Haldon JF (2016b) Bureaucracies, elites and clans: the case of Byzantium, ca 500–1100. In: Crooks P, Parsons T (eds) Empires and bureaucracy in world history. From late Antiquity to the twentieth century (Cambridge), pp 147–169

Haldon JF (1999) Warfare, state and society in the Byzantine world, pp550–1204 (London)

Haldon J, Huebner SR, Izdebski A, Mordechai L, Elton H, Newfield TP (2018) Plagues, climate change, and the end of an empire: a response to Kyle Harper's The Fate of Rome", (1): Climate", History Compass. 16.12 (2018); e12508. https://doi.org/10.1111/hic3.12508; (2): Plagues and a crisis of empire, History Compass. 16.12 (2018); e12506. https://doi.org/10.1111/hic3.12506; (3): Disease, agency, and collapse, History Compass 16.12 (2018); e12507. https://doi.org/10.1111/hic3.12507

Haldon J, Roberts N, Izdebski A et al (2014) The climate and environment of Byzantine Anatolia: integrating science, history and archaeology. J Interdisc Hist 45:113–161

Hendy MF (1985) Studies in the Byzantine monetary economy c.300–1450 (Cambridge)

Howard-Johnston JD (2010) The rise of Islam and Byzantium's response. In: Oddy A (ed) Coinage and history in the seventh-century Near East (London), pp 1–9

Izdebski A (2021) Ein vormoderner Staat als sozioökologisches System: Das oströmische Reich 300–1300. Sandstein, Dresden

Izdebski A, Holmgren K, Weiberg E et al (2016) Realising consilience: How better communication between archaeologists, historians and natural scientists can transform the study of past climate change in the Mediterranean. Quatern Sci Rev 136:5–22

Johnson S (ed) (2010) The oxford handbook of late antiquity. Oxford-New York

Mitchell S (2007) A history of the late Roman empire (Oxford)

Kaegi WE (1992) Byzantium and the early Islamic conquests (Cambridge)

Kelly C (2008) Attila the Hun (London)

Kennedy H (2004) The prophet and the age of the Caliphates. The Islamic Near East from the sixth to the eleventh century (London)

Levin SA et al (2013) Social-economic systems as complex adaptive systems: modelling and policy implications'. Environ Dev Econ 18(2):111–132

Magdalino P (2010) Orthodoxy and Byzantine cultural identity. In Rigo A, Ermilov P (eds) Orthodoxy and heresy in Byzantium. The definition and notion of Orthodoxy and some other studies on the heresies and the non-Christian religions (Quaderni di Nea Rome 4. Rome), pp 21–46

Mordechai L, Pickett J (2018) Earthquakes as the quintessential SCE: methodology and societal resilience. Human Ecol 46. https://doi.org/10.1007/s10745-018-9985-y

Roberts N (2018) Revisiting the Beyşehir occupation phase: land-cover change and the rural economy in the eastern Mediterranean during the first millennium AD. In: Izdebski A, Mulryan (eds) Environment and society in the long late antiquity, 53–68

Robinson CF (2010) The rise of Islam, 600–705. In Robinson CF (ed) New Cambridge History of Islam, 1: The formation of the Islamic world, sixth to eleventh centuries (Cambridge)

Scheffer M (2009) Critical transitions in nature and society (Princeton)

Sessa K (2019) The new environmental fall of rome: a methodological consideration. J Late Antiquity 12(1):211–255

Teall J (1959) The grain supply of the byzantine empire, 330–1025. Dumbarton Oaks Papers 13:87–139

Whittow M (1996) The strategic geography of the Near East. In: idem, The making of Orthodox Byzantium, 600–1025 (London)

Wickham C (2009) The inheritance of Rome (London)

Success and Failure in the Norse North Atlantic: Origins, Pathway Divergence, Extinction and Survival

Rowan Jackson, Jette Arneborg, Andrew Dugmore, Ramona Harrison, Steven Hartman, Christian Madsen, Astrid Ogilvie, Ian Simpson, Konrad Smiarowski, and Thomas H. McGovern

Abstract In this chapter, we examine the iconic disappearance of the Medieval Norse Greenlanders and use qualitative scenarios and counterfactual analysis to produce lessons for policymakers. We stress the role that archaeologists and historians have in adding context to contemporary social and environmental challenges and use human-environmental histories as 'natural experiments' with which to test scenarios. Rather than drawing direct analogies with discrete historical case studies

R. Jackson (✉) · J. Arneborg · A. Dugmore
Geography, School of GeoSciences, University of Edinburgh, Drummond Street, Edinburgh E8 9XP, Scotland, UK
e-mail: rowan.jackson@ed.ac.uk

R. Jackson
Department of Archaeology and Heritage Studies, School of Culture and Society, University of Aarhus, Moesgård Allé 20, 8270 Højbjerg, Denmark

J. Arneborg · C. Madsen
Middle Ages, Renaissance and Numismatics, National Museum of Denmark, 1220 Copenhagen, Denmark

A. Dugmore · I. Simpson · T. H. McGovern
Human Ecodynamics Research Centre and Doctoral Program in Anthropology, The Graduate Center, City University of New York, 365 Fifth Avenue, New York, NY 10016-4309, USA

R. Harrison
Department of Archaeology, History, Cultural Studies, and Religion, University of Bergen, 5020 Bergen, Norway

S. Hartman
Department of Tourism Studies and Geography, Mid Sweden University, 83125 Östersund, Sweden

National Museum of Iceland, Reykjavík, Iceland

A. Ogilvie
INSTAAR (The Institute of Arctic and Alpine Research), University of Colorado, Boulder, USA

The University of the Highlands and Islands, Orkney, Scotland

I. Simpson
Department of Archaeology, Durham University, Durham DH1 3LE, UK

A. Izdebski et al. (eds.), *Perspectives on Public Policy in Societal-Environmental Crises*, Risk, Systems and Decisions, https://doi.org/10.1007/978-3-030-94137-6_17

such as Norse Greenland, such cases form complete experiments with which to ask 'what if' questions and learn from a range of real (retrofactual) and alternative (counterfactual) scenarios. By testing a range of scenarios associated with climate impacts and adaptive strategies, evidence from the past might be used to learn from unanticipated changes and build a better understanding of theory and concepts, including adaptation and vulnerability, and their application to the present. The Norse Greenland case study illustrates an important lesson for climate change adaptation scenarios; even a highly adaptive society can, over the course of several centuries, reach limits to adaptation when exposed to unanticipated social and environmental change.

Introduction

The prospect of ecological or even societal collapse has become a common post-millennium discourse in western media (Diamond 2005; Vogelaar et al. 2018). Ecological collapse, resource conflict, existential risk, and social and ecological breakdown are common headlines in environmental columns and increasingly popular in academic discourse (Hulme 2008, 2016; Bostrom and Cirkovic 2011). But in spite of the increased attention to divergence from historical trends, such as species extinction (Ceballos et al. 2015, 2017), it is only in recent years that the disciplines of history and archaeology have gained attention in global change research (GCR). As Ortman (2019:1) explains, this is unsurprising seeing as it is only in recent years that "archaeologists have conducted research that is explicitly designed to address contemporary issues" (see also Jackson et al. 2018a). However, this does not mean archaeological and historical research traditions have been solely focused on the past. These research traditions have always been influenced by contemporary issues—influencing interpretation and knowledge production—and have sought to contextualise relevance through analogy and the identification of historical antecedent (cf. Lucas 2019; Cronon 1992). Instead, since the turn of the millennium we have seen archaeological research highlight the concept that, while the past may provide far from optimal analogues to current and future trends, these new social-ecological developments and attendant crises can be influenced by processes similar to those experienced by societies in the past (Dawdy 2009; Riede 2017; Hartman et al. 2017). Metaphors such as 'laboratory of the past' have highlighted the utility of historical and archaeological data as 'completed' natural experiments (Diamond and Robinson 2010; Riede 2014) with which to test the impacts of environmental hazards and climate variability on a range of different societies, each with their different cultural capacities and limitations (Dugmore et al. 2012; Dow et al. 2013; Spielmann et al. 2016; Nelson et al. 2016). It is our argument, however, that attempts to develop a more

K. Smiarowski · T. H. McGovern
Hunter Zooarchaeology Laboratory, Department of Anthropology, Hunter College, City University of New York, 695 Park Ave, New York City 10021, USA

robust, practical framework for archaeological 'relevance' have been undermined by the deterministic nature of collapse discourse itself when considered in light of its most popular tropes, their uses and abuses. While the discourse has appeal to funding bodies its utility in helping to identify lessons from the past relevant to contemporary climate adaptation is not sufficiently established, nor particularly promising when envisaged in terms of case-to-case correspondence.

In this chapter, we recontextualise the Norse Greenland case study in its wider North Atlantic context to highlight the opportunities and limitations of using archaeology and other palaeo-sciences to find solutions to contemporary environmental and social challenges. We stress the utility of "qualitative scenario storylines" (Rounsevell and Metzger 2010)—an approach used in predictive modelling—to provide a structured method for archaeologists to construct possibilistic futures, while avoiding environmental determinism or reductive historical analogy (Riede 2019; Jackson et al. 2018a).

The Archaeological Science-Policy Interface

Recently, archaeological and historical researchers have highlighted the need to contribute datasets that link human and environmental processes and contribute to our understanding of global change (Riede 2014; Hartman et al. 2017; Hambrecht et al. 2018; Haldon et al. 2018; Ortman 2019). This goes beyond narratives that highlight lessons through broad literature reviews (for example Diamond 2005; Tainter 1988; Middleton 2012). Instead, researchers have adopted the language of global change research—particularly the human dimensions of global change (i.e. adaptation, vulnerability, resilience)—and have drawn parallels between contemporary challenges and evidence derived from historical and archaeological datasets (see for example ARCUS 1997; Sigurðardóttir et al. 2019). However, to gain purchase in policy arenas, historical sciences (including archaeology) need to draw clearer attention to relationships between evidence and specific policies—in other words, avoiding generalisation and delivering clear results (Cairney and Oliver 2020; Cairney 2016).

As Cairney and Oliver (2020) explain, there is a tendency across academic sciences to produce generalised advice that is of broad application. This generic advice has limited relevance to policy and policymaking, because it lacks contextualisation and a clear application (Boswell and Smith 2017). The challenge for historical sciences is, therefore, to show how historical data can be applied within a decision-making context while avoiding ideological biases (Cairney 2018), misconceptions or other problematic interpretations of history. Attention can be directed to historical information by problem-based research that targets specific challenges and supports policy solutions with historical evidence (Oliver and Cairney 2019). But to get here, archaeologists and historians will need to avoid the headline-grabbing narratives of collapse.

Problems of Collapse Discourse

Controversies arise from the identification of past social 'collapse' associated with different theoretical perspectives and whether a 'grand' unifying theory can explain why societies came to an end. Diamond (2005) frames collapse in terms of geographical and environmental determinism whereas Tainter (1988, 2006) frames collapse as a loss of socio-political complexity. Such theories have attracted vigorous academic debate and gained much wider attention for their focus on environmental and social determinants of population and institutional change (Butzer 2012) but they do not focus on the value of the archaeological record or expertise in the historical sciences, nor the (transient) role of archaeology in constructing the past (Lucas 2001, 2005). Archaeologists interweave multiple strands of evidence to produce narratives, often spanning extensive timescales (Gamble 2007). Archaeological information is derived from a combination of, or triangulation among, historical documents, archaeological contexts and environmental records (Lucas 2010). Although there is often a significant degree of uncertainty, the complex relationship between socio-culture and environmental change can be interpreted through the evaluation of multiple lines of physical evidence, written sources, ethnographic parallels and argument by analogy (Currie 2016). The relationship between human behaviour and the wider context of social, economic, political and environmental change can reveal multiple different scenarios that are unique to each case but can offer insights into overarching themes, such as the consequences of hierarchy, inequality, opportunity costs, and food insecurity (Vesteinsson et al. 2019; Hambrecht 2018; Nelson et al. 2016).

While this information has the potential to inform adaptation theory, the literatures that have received the greatest attention—including historical and archaeological literatures that feature in the IPCC reports—are those that fall within the 'collapse' genre (Kohler and Rockman 2020). Middleton (2017, 2018) has noted a burgeoning literature on what he terms 'collapsology.' He also notes the significant appeal of this literature across civil society and academia for the hyperbolic parallels that are often drawn between catastrophic climate impacts or other environmental hazards that face us today (Middleton 2012). However, the events described as collapse are rarely clear cut and the vague definitions used to explain all-encompassing social and environmental changes over extended periods of time cannot do justice to the nuance of human agency and decision-making when experiencing environmental change. Many archaeologists have critiqued the representation of collapse in the archaeological record, and question whether the term carries sufficient clarity as an academic concept (McAnany and Yoffee 2010; Vésteinsson 2013; Aimers 2007; Middleton 2012, 2017).

In light of these critiques, collapse can be understood as a metaphor that appeals to contemporary society's 'climate of fear' about future uncertainty but provides little evidence about 'the direction of history' (Popper 1957: 160) and humanity's destiny (cf. Hulme 2008, 2011). Collapse does not adequately describe the complex relationship between resource systems, demographic trends, and cultural adaptations, but operates as a heuristic for uncertainty, the end of a society or a sudden loss of socio-political complexity or population decline (Vogelaar et al. 2018; Middleton 2012, 2017; Tainter 1988). Vésteinsson (2013) has noted this recent trend in popular and academic discourse and adopts the polemical position that there is no such thing as collapse. For Vésteinsson, collapse as an idea fails to clarify what happened in the 'end game', in the years and decades preceding the putative transformative changes, and how humans attempted to respond to environmental challenges, which renders the concept unfit for purpose in historical disciplines. Recently, archaeologists have sought to distance themselves from notions of collapse—opting for concepts such as resilience and transformation to emphasise the human capacities to respond to change (McAnany and Yoffee 2010; Butzer and Enfield 2012; Strickland et al. 2018). Indeed, the lack of agency granted to human communities in deterministic narratives of environmental change is one of the most often-voiced critiques in takedowns of the collapse concept.

How Are Archaeological Data Useful Now?

Recent publications in archaeology and the palaeo-sciences have outlined the significant potential that long-term data have to improve our understanding of human impacts on ecosystems and how societies have responded to changing environmental conditions (Dunne et al. 2016; Boivin et al. 2016; van de Noort 2013). The application of these data can be divided into two categories associated with the (dis)continuity of data with the present day. First, archaeological data provide a long-term record of human–environment interaction and with it a continuous record of human impacts on the environment (Armstrong et al. 2017; d'Alpoim Guedes et al. 2016; Stephens et al. 2019). Archaeological and environmental data therefore provide an essential record of the human influences on ecosystem structures (Kwok 2017; Boivin et al. 2016) and impacts on the climate (Koch et al. 2019). In order to correlate human–environment interaction across regional-scales, and to compare the effects of human impacts and adaptation, Hambrecht and colleagues (2018) suggested compiling palaeo-environmental and palaeo-societal data to compare impacts on the environment (Fig. 1) with the most intimate association observed in site stratigraphies. Collectively, human and environmental datasets represent the long-term impacts humans have had on their environments, whether through vegetation clearance and cultural niche modification or the overhunting and extirpation of targeted species (Boivin et al. 2016; Ellis 2011). These so-called deep-time, spatially extensive datasets can be used to reconstruct human societies' long-term impact on biodiversity and ecosystem structures (Hambrecht et al. 2018). For example, long-term fishing data from west and northeast Iceland have been used to reconstruct human impact on Atlantic cod stocks (Edvardsson et al. 2019). These data can be used to reconstruct accurate environmental baselines and provide a key to the effective management and conservation of fish stocks and marine environments (Hambrecht et al. 2018; Kwok 2017).

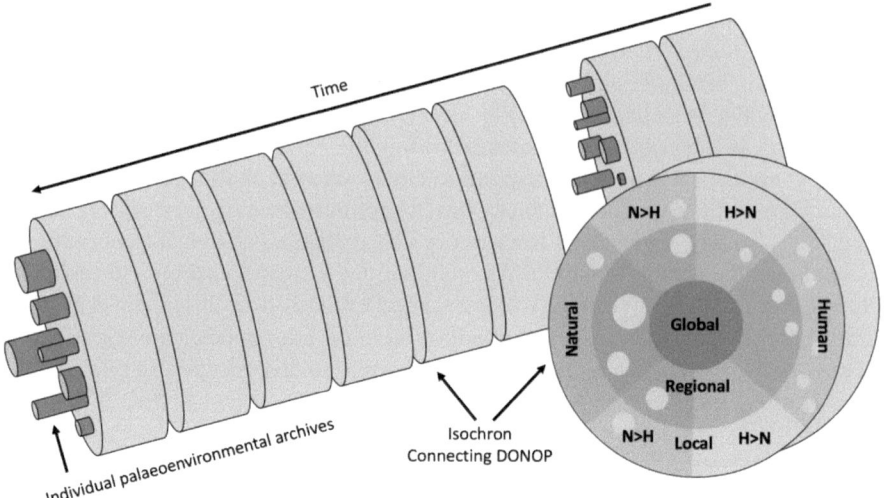

Fig. 1 Schematic representation of human and environmental datasets within Distributed Observing Networks of the Past (DONOP). Regional-scale datasets refer are more natural than human and local-scale datasets include evidence of human activity and human remains themselves (adapted from Hambrecht et al. 2018). The combination of human and environmental datasets, each with their own spatial reach, allow a reconstruction of human interaction with the environment using multiple stands of evidence

Human impacts on global environments and atmospheric chemistry have driven a rapid acceleration of warming in the Arctic and Sub-arctic, causing permafrost to thaw and records of material culture to degrade (Hollesen et al. 2015, 2016, 2017). In addition, there are now enhanced threats and impacts from flooding, fires and weathering across historical sites (e.g. Historic Environment Scotland 2019; McCaughie et al. 2020) and increased storm intensity and sea-level rise that is threatening and destroying ever more coastal heritage sites across the world (Dawson et al. 2017). Collectively, the destruction of cultural records has been termed 'burning libraries' (McGovern 2018), as ever greater proportions of limited, finite records of past environments and human history are lost forever. The destruction of archaeological heritage does not, however, represent solely a loss of data but also has a lasting impact on communities' sense of place and local identity (Harvey and Perry 2015).

A significant amount of archaeological information has no continuity with the present, making the interpretation of activities and beliefs more difficult—especially with the absence of historical records. Cases of collapse, by their nature, imply a discontinuity with preceding historical periods, which raises questions about what and how we, in the twenty-first century, can learn. One argument is that archaeological data can provide discrete records of putative limits to human adaptation and potentially identify critical degrees of vulnerability to impacts of social and ecological changes (Costanza et al. 2007; Dugmore et al. 2009). Because there is a lack of continuity with highly industrialised modern societies, these case studies cannot

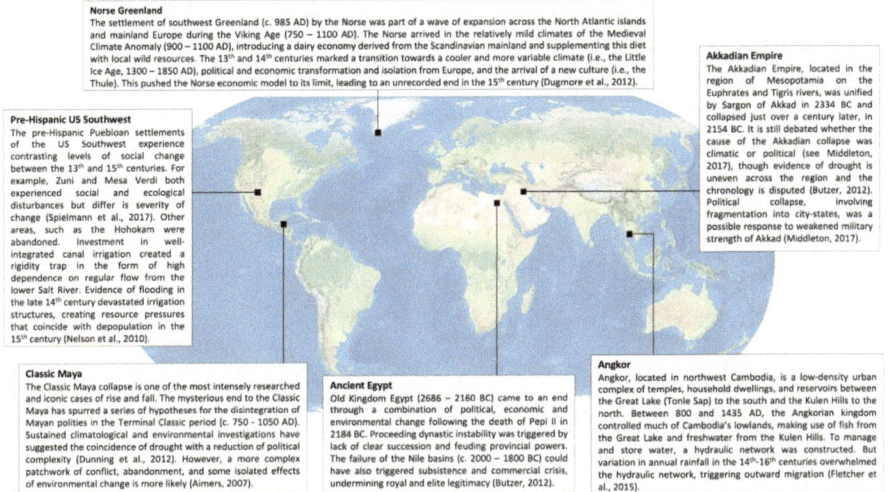

Norse Greenland
The settlement of southwest Greenland (c. 985 AD) by the Norse was part of a wave of expansion across the North Atlantic islands and mainland Europe during the Viking Age (750 – 1100 AD). The Norse arrived in the relatively mild climates of the Medieval Climate Anomaly (900 – 1100 AD), introducing a dairy economy derived from the Scandinavian mainland and supplementing this diet with local wild resources. The 13th and 14th centuries marked a transition towards a cooler and more variable climate (i.e., the Little Ice Age, 1300 – 1850 AD), political and economic transformation and isolation from Europe, and the arrival of a new culture (i.e., the Thule). This pushed the Norse economic model to its limit, leading to an unrecorded end in the 15th century (Dugmore et al., 2012).

Akkadian Empire
The Akkadian Empire, located in the region of Mesopotamia on the Euphrates and Tigris rivers, was unified by Sargon of Akkad in 2334 BC and collapsed just over a century later, in 2154 BC. It is still debated whether the cause of the Akkadian collapse was climatic or political (see Middleton, 2017), though evidence of drought is uneven across the region and the chronology is disputed (Butzer, 2012). Political collapse, involving fragmentation into city-states, was a possible response to weakened military strength of Akkad (Middleton, 2017).

Pre-Hispanic US Southwest
The pre-Hispanic Puebloan settlements of the US Southwest experience contrasting levels of social change between the 13th and 15th centuries. For example, Zuni and Mesa Verdi both experienced social and ecological disturbances but differ is severity of change (Spielmann et al., 2017). Other areas, such as the Hohokam were abandoned. Investment in well-integrated canal irrigation created a rigidity trap in the form of high dependence on regular flow from the lower Salt River. Evidence of flooding in the late 14th century devastated irrigation structures, creating resource pressures that coincide with depopulation in the 15th century (Nelson et al., 2010).

Classic Maya
The Classic Maya collapse is one of the most intensely researched and iconic cases of rise and fall. The mysterious end to the Classic Maya has spurred a series of hypotheses for the disintegration of Mayan polities in the Terminal Classic period (c. 750 - 1050 AD). Sustained climatological and environmental investigations have suggested the coincidence of drought with a reduction of political complexity (Dunning et al., 2012). However, a more complex patchwork of conflict, abandonment, and some isolated effects of environmental change is more likely (Aimers, 2007).

Ancient Egypt
Old Kingdom Egypt (2686 – 2160 BC) came to an end through a combination of political, economic and environmental change following the death of Pepi II in 2184 BC. Proceeding dynastic instability was triggered by lack of clear succession and feuding provincial powers. The failure of the Nile basins (c. 2000 – 1800 BC) could have also triggered subsistence and commercial crisis, undermining royal and elite legitimacy (Butzer, 2012).

Angkor
Angkor, located in northwest Cambodia, is a low-density urban complex of temples, household dwellings, and reservoirs between the Great Lake (Tonle Sap) to the south and the Kulen Hills to the north. Between 800 and 1435 AD, the Angkorian kingdom controlled much of Cambodia's lowlands, making use of fish from the Great Lake and freshwater from the Kulen Hills. To manage and store water, a hydraulic network was constructed. But variation in annual rainfall in the 14th-16th centuries overwhelmed the hydraulic network, triggering outward migration (Fletcher et al., 2015).

Fig. 2 Selected cases of so-called 'collapse'. These cases are selected for the influence they have had on collapse discourse and for featuring in major institutional reports, such as the IPCC Assessment Reports. Such cases provide records of human adaptation and vulnerability to environmental change (Adapted from Jackson et al. 2018a)

provide direct analogues for contemporary adaptation to environmental change. These cases do, however, provide experiments with which to understand processes, examine the dynamics of relationships between societies and environments, identify 'perfect storms' leading to collapse, learn lessons and test scenarios. Figure 2 presents six well-known cases of societal collapse (Jackson et al. 2018a). These cases feature prominently in the archaeological literature and have also played an increasing role in climate adaptation literature. Three of these case studies (Greenland and Iceland, Classic Maya, Ancient Egypt) featured in the Fifth Assessment Report of the IPCC (Working Group Two); however, it was not clear from these cases what 'relevance' historical and archaeological data has to the challenge of adapting to climate change (IPCC 2014: 920).

Climate change archaeology, a growing sub-discipline in archaeology, has developed into a broad literature that intersects with various challenges in global change and natural hazards research (Van de Noort 2011, 2013). Researchers have highlighted the potential for archaeology to contribute using its extended chronological scope on human adaptation (Riede 2014; Redman 2005) and to develop more holistic understanding of human–environment interaction over multi-generational to millennial timescales (Barnes et al. 2013; Hudson et al. 2012). However, what these studies have not addressed is how archaeology can contribute pre-modern data to social and environmental challenges that have no analogue in human history.

To resolve the discontinuity between pre-industrial and industrialised modern societies we argue that archaeological and historical records should be used as 'natural experiments' with which to examine the process of human adaptation and manifestation and implications of vulnerability. Examining the long-term human experience of social inequality, food insecurity, health inequality and numerous other social indicators can provide enhanced evidence of the potential synergistic social impacts of climate variability and environmental change (Hegmon 2016; Hegmon et al. 2018). Such information could be useful in scenario-building exercises, as longer timescales allow long-term processes to be examined across numerous case studies at multiple spatial scales over human history (Costanza et al. 2007; Fig. 2).

Getting Beyond Collapse?

In the last 5 years, future studies have grown significantly as a subdiscipline within the social sciences and humanities, exploring the potential impacts of technological change, natural hazards and climate change on society and the environment (Bostrom and Cirkovic 2008; Oreskes and Conway 2013; Urry 2015). New research centres dedicated to future studies and existential risk, including the Centre for Existential Risk (University of Cambridge), the Future of Humanity Institute (University of Oxford) and the Humboldt Forum (Berlin), have emerged with an explicit focus on major challenges facing humanity in the future (Bostrom 2013). The core research themes at these organisations encompass 'high impact/low probability' events, social interaction and disease, and catastrophic climate change and ecosystem collapse (https://www.cser.ac.uk/research/). Connection to other research communities focusing on the contemporary study and probability of societal risk offers significant ground for collaboration among academia, policy and industry (Urry 2011, 2015).

IHOPE (Integrated History and Future of People on Earth https://ihopenet.org/), a partner with Future Earth (https://futureearth.org/), has sought to highlight the importance of demonstrating the importance of the past to the future of human societies and ecosystems. Researchers in this network have demonstrated the relevance of historical and archaeological data for understanding global environmental change today, making use of the diverse tool kit of Historical Ecology (Meyer and Crumley 2011). To understand the environmental challenges today, so they argue, it is important to understand the role of domestication in land-use change (Boivin et al. 2016; Ellis 2011), uncovering long-term cultural traditions associated with food culture (Sykes 2014), reconstructing trade networks and the (over-)exploitation of marine and terrestrial species (Hambrecht et al. 2018) and understanding the long-term impacts of environmental change on the vulnerability of human communities (Nelson et al. 2016). Importantly, several IHOPE affiliated projects work directly to bring archaeological, historical, and traditional knowledge perspectives to bear on current and future land use and resource management issues (e.g. the *Herring School and Clam Garden Network* https://ihopenet.org/herring/, Thornton et al. 2010, also Peloponnese Project https://ihopenet.org/peloponnese/ Kaplan et al. 2012).

Human Ecodynamics Perspectives in the North Atlantic and Beyond

In the past 15 years North Atlantic and North Pacific researchers have increasingly adopted the perspectives of *human ecodynamics* to recognize and better articulate complex long-term interactions of humans, landscapes, and climate in the circumpolar north (Fitzhugh et al. 2019). This approach makes use of resilience discourse and the interdisciplinary tool kit of historical ecology, and from these perspective addresses questions of comparative social/environmental transformations (fast and slow, "painful" and advantageous: Hegmon et al. 2013). Particularly productive systematic human ecodynamics comparisons of contrasting cases from the American Southwest and the North Atlantic took place during 2012–18 through a series of workshops and joint meetings between teams from NABO (www.nabohome.org) and the *Long Term Vulnerability and Adaptations* project (https://ihopenet.org/southw est-us/) based at Arizona State University. These cross-regional comparisons of long-term outcomes (including 'full on collapse', 'painful transitions', and 'successful adaptation with costs') provided some valuable metrics on human vulnerability and adaptation to climate change in both desert and low arctic conditions (e.g. Nelson et al. 2016, 2017). These perspectives, as Nelson et al. (2017) argue, have the potential to inform food security planning in the IPCC reports and local adaptive management strategies. An increasing number of archaeologists and historians in different areas making explicit or implicit use of the human ecodynamics approach have sought to emphasise the explicit 'relevance' of archaeological and environmental records as deep-time experiments with which to test impacts and adaptation to climate change (Butzer and Enfield 2012; Hudson et al. 2012; Jackson et al. 2017; Riede 2017; Richer et al. 2019; Riede and Sheets 2020). The role of the North Atlantic cases (and especially the controversial collapse of Norse Greenland) in placing the perspective of long term human ecodynamics within the viewshed of groups like the IPCC represents real progress beyond the simple determinism of Diamond's 2005 account and reflects the power of interdisciplinary integration to tell stories that engage and inform on multiple levels. Can we do better in telling complex stories of humans, landscapes, and historical processes?

Qualitative Scenarios Storylines and Collaborative Conceptual Modelling

Collaborative conceptual modelling (CCM), a systems process used to map complex social-ecological feedbacks, is one such method allowing the consilience of archaeological data with existing methods of scenario planning (Newell and Proust 2019). CCM is a transdisciplinary approach that aims to bring together experts from within and outside the academy towards a common dialogue. By exploring potential threats using extended scenario storylines that incorporate historical datasets, the scope of scenarios and potential risks could be explored in further detail (Jackson et al. 2018a; Riede and Jackson 2020). Although the social impacts of environmental change and natural hazards is likely to be different to observations in pre-modern records, historical datasets provide natural experiments that can be contextualised with contemporary social-ecological systems.

The CCM process, as described by Newell and Proust (2019), involves six stages of dialogue to conceptualise possible futures (Table 1). In this six-stage dialogue, archaeologists and global change researchers together with policymakers and practitioners could work in dialogue to map the long-term effects of environmental challenges, such as climate change, food insecurity, seismic activity or volcanism, in a given regional context (Newell and Proust 2019; Riede 2017). By developing an effective inter- and transdisciplinary dialogue, there is greater potential to explore the strengths and limitations of historical data for understanding contemporary challenges. Effective scenarios are required if societies are to anticipate and adapt to the future impacts of climatic and environmental change. Representative concentration pathways (RCP) are one such way to project the possible outcomes of different emissions scenarios—with results showing the radiative forcing in W/m^2 (van Vuuren et al. 2011)—and are used, in turn, to project emissions scenario storylines (Rounsevell and Metzger 2010). The outcome of these scenarios is subject to further scenarios based on knowledge of the cause-effect relationship between biophysical

Table 1 Collaborative conceptual modelling process using historical evidence in Norse Greenland

What are the challenges?	Adapting to a cooler and more variable climate, adjusting to changing markets in continental Europe, coping with population decline, managing erosion and soil nutrient depletion, contact and potential hostilities with the Thule culture
What are the stories?	Evidence of successful colonisation, evolving adaptation to sub-arctic resource system and variable climate over multi-century timescales, but Norse society ultimately comes to an end in Greenland
Can I see how you think?	The dual challenge of recognising the relevance of environmental, historical and archaeological knowledge as an integrated evidence base that informs scenario development
What are the drivers of system behaviour?	Climate variability, political and economic change, human migration; system inheritance and inequalities between generations
Where are the leverage points?	Counterfactual strategies for negotiating environmental and social change
What future can we see?	Use of completed experiments to identify path dependency and potential rigidity traps, imagine alternative approaches to change, structure plausible scenarios and strategic responses to climatic, social, political and economic exposures in the future

and social systems. For example, the impacts of precipitation change on the French wine industry are then subject to understanding drivers, impacts/challenges and leverage points, if the industry is to adjust to different climatic conditions (Metzger and Rounsevell 2010).

Archaeologists and historians have potential to contribute a structured understanding of capacities and limitations created by a range of scenario storylines. As Dugmore and Vésteinsson (2012) have explained, historical context is essential information for understanding volcanic hazards. Discrete examples at the landscape-scale of both impacts and the absence of impacts of volcanic eruptions in Iceland, they argue, provide important lessons concerning processes of change, various forms of resilience, complex feedbacks and perceptions of risk in response to a range of eruption events. Such examples are essential to understanding volcanic hazard-risk and planning response strategies in areas prone to volcanic events (Donnovan and Oppenheimer 2010). However, historical information has been less influential in contemporary climate risk planning—a subject that has more recently gained attention in global change research (see, for example, Ford et al. 2018; Thomas et al. 2019).

For lessons about climate variability and food security, the Norse Greenland case study can provide important information about how vulnerability is assessed in the long-term (Nelson et al. 2016). Ford et al. (2018: 198) have acknowledged the need for climate change adaptation literatures to expand its scope to "focus on the long-term historic processes creating vulnerability [which] is essential for developing a richer understanding of the dynamics that shape vulnerability, representing a set of completed historical experiments on climate-society interactions". In what follows, we use the CCM approach, described by Newell and Proust (2019), to identify lessons that could have a bearing on future climate risk scenarios—bringing together a richer understanding of the long-term processes shaping vulnerability to climate change.

Norse Greenland

Using the CCM framework, we can start by identifying the challenges facing Norse society in Greenland between the tenth and fifteenth centuries. An evaluation of publications on Norse Greenland since the 1980s reveals three challenges facing Norse society: increased climate variability from the mid-thirteenth century (Dugmore et al. 2007b); European economic and world systems change from the fourteenth century (McGovern 1985, 1994); and cultural contact and potential hostility with Thule cultures (Gulløv 2008). These are similar challenges to those facing societies today, with the dual threats of climate variability and economic stagnation facing numerous societies in tandem (O'Brien and Liechenko 2000; Liechenko and O'Brien 2019). Expansionist activity, conflict and hostility are further challenges to human security and can be exacerbated by economic stagnation and climate variability (Barnet and Adger 2007; Hsiang et al. 2013).

The mysterious, unrecorded end of Norse society in Greenland, and the persistence of Icelandic society, has been discussed at great length in both academic and mainstream literatures (Buckland et al. 1996; Barlow et al. 1997; Ogilvie 1998, 2016, 2019). As discussed earlier in this chapter, common criteria for collapse, such as that set out in Jared Diamond's *Collapse,* are reflected in a mainstream discourse of fear about climate (Hulme 2008). Collapse has become a panacea for societal challenges today (Diamond 2005), where a nuanced narrative that exposes adaptive success and failure would be more appropriate (McAnany and Yoffee 2010; Dugmore et al. 2009). Focusing on why certain decisions were taken (*retrofactuals*) and why others were not taken (*counterfactuals*) is more revealing of the interplay between capacities and limits of past societies, such as Norse Greenland.

The Norse established two settlements in Greenland—the Eastern and Western Settlements—in the late tenth century. Recent interdisciplinary research has indicated that Norse settlement of Iceland ca. 880 involved intensive walrus hunting for low bulk/high value that may have depleted the native Icelandic walrus population, leading to hunters and their patrons moving to Greenland ca. 985 (Frei et al. 2015; Keighley et al. 2019). Available pastureland would have been an opportunity for chieftains to claim and award land, establishing status in the process; and walrus and other arctic exotica offered high-value commodities to trade with Europe (Vésteinsson et al. 2002; Roesdahl 2008). In the past decade, systematic GPS high resolution survey combined with a large-scale radiocarbon dating program has allowed for the development of a three-phase settlement model (Madsen 2014, 2019). The *settlement and expansion* phase (ca. 985-1160 CE) saw the initial establishment of farms and shielings in the inner fjords of the Eastern Settlement in the southwest (modern Kujalleq district) and in the smaller Western Settlement further north (modern Nuuk district). Settlement stimulated immediate localised erosion and initiated hydrological responses through vegetation cover change on what was a stable landscape. Subsequent landscape stabilization during this early phase indicates recognition and adaptation to these human mediated changes to local environments (Ogilvie et al. 2009; Golding et al. 2015). While Greenland apparently did not attract the large and rapid influx of settlers seen in Iceland (Lynnerup 1998, 2014, Vésteinsson and McGovern 2012) population growth expanded settlement into the highlands and outer fjord areas, eventually filling in virtually all areas with patches of vegetation suitable for pasture. Christianity arrived with first settlement, and a pattern of small turf and stone churches and churchyards was established. The distinctive Greenlandic subsistence economy combining sheep, goat, and cattle husbandry sustained over winter by fodder gathering and production of higher quality fodder from small fertilized and irrigated homefields (Adderley and Simpson 2006). This was integrated with caribou hunting and extensive sealing, fully established early on, as was the hunting of walrus for ivory and hide that attracted transatlantic trade (Smiarowski et al. 2017). By ca 1123 Greenlandic chieftains felt prosperous enough to exchange a polar bear for their own bishop from the King of Norway.

The following *consolidation* phase (ca. 1160–1300 CE) saw the establishment of the Episcopal seat at Garðar (modern Igaliku) with the stone cathedral as the largest manor in Greenland and political absorption of Greenland along with Iceland into the Norwegian Atlantic Realm (1262–4). This period saw the apparent "market dominance" of Greenlandic walrus products in western Europe (Star et al. 2020; Barrett et al. 2020) as the walrus hunt intensified and apparently expanded northwards into Disko Bay and perhaps beyond. This period also saw a pattern of the closing of small local turf and stone churches and the consolidation around a few larger two-cell Romanesque churches at high status farm locations, later (ca. 1250–1300) to be replaced by the large and impressive stone churches furnished with imported stained glass and church bells (some probably constructed with imported architects, perhaps as a mark of royal favour). Jette Arneborg has described this as a pattern of takeovers by a few great families, possibly paralleling a better documented competition and consolidation process in Iceland that led to a civil war among the elites.

By 1300 Norse Greenland had become a fully medieval society with a bishop, royal agents, a monastery and nunnery and tax and tithe obligations to Nidaros (Trondheim) and Rome. Comparative research (Vésteinsson et al. 2019) has suggested that Green-land ca. 1300 may have been more sharply stratified politically and economically than contemporary Iceland, with the Bishop's manor at Garðar and a few chieftain farms dominating a landscape filled with smallholders and tenants. Figure 3 illus-

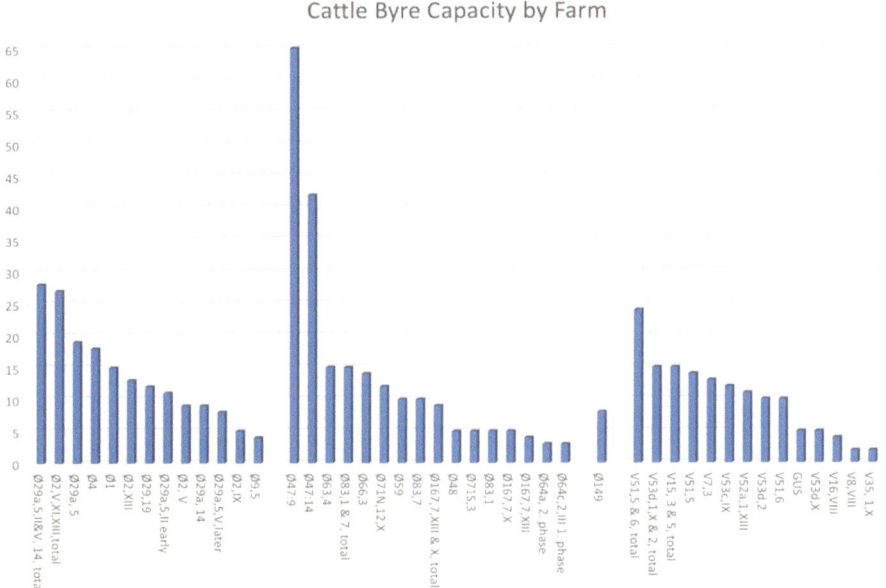

Fig. 3 Cattle byre stall capacity by farm. The first cluster of farms from Ø29a to Ø9 are around Tunulliarfik fjord with the best pastures, followed by the Igaliku fjord and Vatnahverfi (Ø47–Ø64c). Ø47 is the Bishop's farm Gardar, the other farms are from Vatnahverfi where the economy were based on sheep and goats. Ø149 is the farm of the Benedictine nunnery and the last cluster (V51–V35) represents farms in the Western Settlement. Note that the Bishop's farm (Ø47) and the chieftains´ farms (V51, Ø29a). Other high status farms with church are Ø1, Ø83, Ø66 and V7. are very atypical of the majority of smallholders such as V35, V8

trates the relative cattle byre capacities of excavated Norse farms with the two byres at Garðar far in the lead with a combined capacity of nearly 100 cattle.

Perhaps significantly, the Western Settlement chieftain's farm W51 Sandnes also has the greatest concentration of walrus ivory extraction debris, suggesting elite investment in the northern walrus hunt and the connection to European markets and society that generated church bells and ecclesiastical authority (McGovern et al. 1996; Roesdahl 2008).

The period also saw first contacts between Norse hunters in the northern hunting grounds with Thule-culture Inuit arriving ca. 1200 in their migration from Alaska. On the other side of the planet, a massive volcanic eruption at Samalas in modern Indonesia in 1257 was to trigger major climate change and a period of increased summer drift ice between Iceland and Greenland around ca. 1300 (Miller et al. 2012; Zhong et al. 2011

Severe sea-ice years at this time are also noted in documentary records from Iceland (Ogilvie 1982, 1991). Volcanic activity would have reduced solar insolation, causing increased ice delivery to the East Greenland Current (Zhong et al. 2011). On the inner fjord areas of the Eastern Settlement, increased levels of summer sea ice, correlates with increased ice-cover in the summer months (Jensen et al. 2004). The increased presence of summer sea ice and enhanced climate variability and storm activity is reflected in changes to the dietary and household economies in the same period (Arneborg et al. 2012).

The *contraction* phase (ca. 1300–1450 CE) saw the abandonment of many coastal and upland settlements and the whole of the Western Settlement. This was accompanied by a decline and sometimes rapid ending to land management for livestock fodder production. The community contracted into the early settled (and most protected) inner fjord zone of the Eastern Settlement concentrated around Tunulliarfik and Igaliku fjords, and several researchers have modelled dangerous population loss from emigration (Lynnerup 1999, 2014). The few contemporary documentary records that are extant concerning Norse Greenland were mainly written in Iceland. One of the last of these describes the solemnization of a marriage in 1408 at the site of the impressive stone at Hvalsey near modern-day Qaqortoq. Possible reasons for the lack of records after this time have been discussed elsewhere (Ogilvie 2016, 2020) but radiocarbon and paleoenvironmental data suggest an end game period lasting until ca. 1450. Thule contact seems to have intensified, with winter settlements extending progressively further south after ca. 1300 (Lund et al. 2011). Both zooarchaeology and human stable isotope research document a significant change in Norse diet, expanding sealing (with seal bone making up to 80% of some samples) after 1300 with an apparently successful intensification of communal seal drives and pooling of labor for the annual hunt (Ogilvie et al. 2009).

It is likely that a combination of factors involving the wider North Atlantic political and economic sphere and a period of relatively cold climate around 1300 reduced pasture productivity and also hindered both transatlantic contact and local travel. Nevertheless, the Norse Greenlanders showed significant resilience and adaptive capacity to intensify their existing pattern of sealing and changed diet away from dairy and domestic meat to the marine food web (Ogilvie et al. 2009; Arneborg et al.

2012; Smiarowski et al. 2017). It must be strongly emphasised that this was by no means a static society that "chose to fail" (Diamond 2005). Nonetheless, it is evident that, in one way or another, they lost their battle for survival.

Lessons from the Past

Norse Greenland can provide multiple insights into climate change adaptation and vulnerability, whether by contributing to theory (see Ford et al. 2018) or providing examples of limits and barriers to adaptation (Dugmore et al. 2012; Dow et al. 2013). Critically, the Norse settlers responded well to the Arctic climate of Greenland (very different to the climate of Iceland) by supplementing domestic shortfalls with wild resources. However, the transportation of domesticated landscapes and associated animal husbandry practices from Norway to Iceland and Greenland also transferred cultural and institutional path dependencies (Jackson et al. 2018b). Animal husbandry, iron tools, wood boats and fixed dwellings, and, from around AD 1000, Christianity, were central to the Norse way of life, but the accumulated skills and behaviours that were learned and operationalised in mainland Scandinavia and the eastern North Atlantic were not sustainable in Greenland in the long-term (Jackson et al. 2018b), and indeed of themselves may have contributed to community demise. Managing livestock would have become increasingly difficult as the climate became more variable in the fourteenth and fifteenth centuries (Ogilvie 1998; Dugmore et al. 2007a, b, 2009) modifying the critical growing season and with land management and organisation designed to increase fodder productivity instead causing its reduction. Because the Norse were not prepared to abandon farming, hunting activities would have been tethered to pastureland on the sheltered inner fjord, making hunting less efficient (Hambrecht et al. 2018). During the consolidation phase (c.1160–1300) the Norse experienced increasing social inequality and concentration of economic and religious power in the Garðar manor and the surviving major church farms. The significant investment in church architecture by this small community hints at the key role of belief and ideology in coordinating the many communal subsistence tasks and the increasingly costly northern walrus hunt, as well as representing sunk costs in immobile landesque capital investments (Lynnerup 2014; Janssen et al. 2003). When Norse managers faced the multiple challenges of the contraction phase after 1300 (declining contact with Europe, climate impacts, increased Thule competition) they did so with over 300 years of experience in living in Greenland, and with a functioning cultural landscape shaped and dominated by elites with a strong stake in the walrus hunt and the connection to Europe it maintained.

This accumulated local and traditional knowledge (LTK) and established cultural landscape and social structure and ideology provided both useful tools for adaptation but also boundaries to resilience. Path dependency in managers using existing tools for social control and mobilization may have critically limited the range of acceptable alternative choices just as some key elements in LTK (seasonal change, safe seafaring, winter travel conditions, length of winter fodder support) became increasingly devalued by environmental change (Jackson et al. 2018b). Increased social hierarchy and inequality may have empowered elite managers to enforce adaptive choices, but at the expense of closing options and (perhaps) eventually undermining their own authority and control.

This case study provides a scenario storyline that tests how modern communities and institutions respond to environmental stress. Considering cultural path dependence, environmental knowledge and limits to adaptation in contemporary communities could provide a basis with which to test climate scenarios and how adaptively communities respond (Riede and Jackson 2020). However, in order to gain greater recognition in mainstream research on global environmental change, archaeologists need to increase their visibility by writing in a format and venue that is accessible to global change researchers.

Scenarios and Counterfactuals

While a direct analogy cannot be drawn between Norse Greenland and modern societies, an assessment of adaptive capacities and limits can be translated into contemporary scenario exercises. By visualising the Norse adaptive pathway as a set of decisions that are constrained by the changing social and economic context, as shown in Fig. 4, we see that some decisions remain adaptive, such as marine hunting, but others become increasingly maladaptive, such as farming. The links between these subsistence strategies represent the interplay between path dependent strategies, that are associated with cultural limits (Dugmore et al. 2012), and innovations that are the outcome of social learning—or, perhaps more appropriately, landscape learning (Rockman 2003).

Archaeological and historical studies offer the opportunity not only to examine completed experiments, but also the interplay between continuity and change in social-ecological systems (Redman 2005; Costanza et al. 2007). A resolved and nuanced picture of the past can therefore reveal not simply cautionary tales of failure, but rather a far more vivid and revealing picture that tells of a combination of adaptive and maladaptive strategies that, when considered over discrete segments of time, show a series of alternative outcomes. Consider, for example, Norse society in Greenland between the late 10th and late thirteenth centuries: population was at its peak, walrus ivory production sustained contact and trade with the European continent, and subsistence was organised within an efficient seasonal schedule that organised production between individual farm households and communal hunting from specialised marine shielings (McGovern 1980; McGovern et al. 2014; Madsen 2014). Fast-forward to the fourteenth and fifteenth centuries: ivory trade had declined together with the decline of Norwegian political and economic power (now concentrated in the German Hansa towns), the climate has become increasingly variable and stormy, and the Western Settlement has been abandoned (Dugmore et al. 2007a, b).

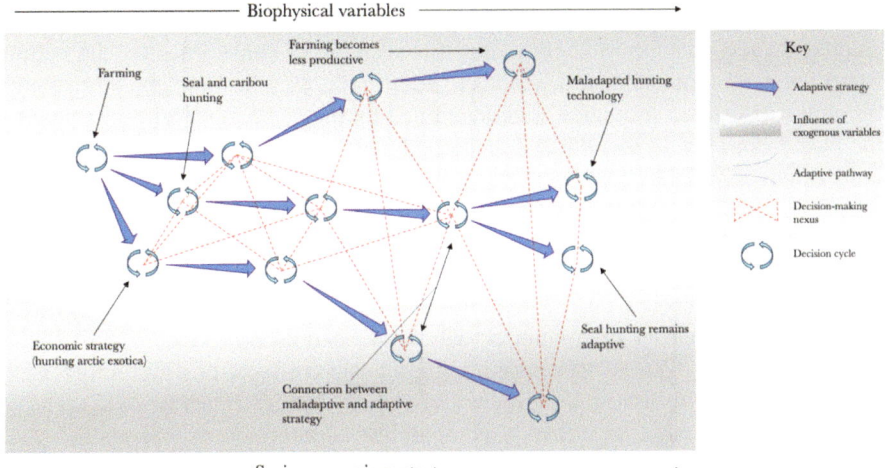

Fig. 4 Norse decision-making constrained by environmental and social-economic context (adapted from Wise et al. 2014). Above, the Norse adaptive strategy is constrained by the impacts of climate variability and landscape organisation on fodder and livestock productivity. Increased sea-ice presence also makes seal hunting less predictable and increasingly dangerous. Below, changes to European political power and markets transformation devalue walrus ivory and other arctic exotica. In the middle, the adaptive pathway that the Norse follow shows evidence of innovation (i.e. the adoption of hunting as a subsistence and market strategy) before becoming increasingly constrained by their existing decision-making strategies (path dependence) and the nexus of choices that they make (i.e. continued reliance on walrus ivory, taking up valuable time during the short, intensive summer season)

Did the Norse choose to fail, as Diamond described (and forewarned)? Or, alternatively, did the Norse settlers adapt successfully within the bounds of what was considered culturally acceptable or possible? The Norse were not passive to the changing climate, nor were they unwilling to adapt, but perhaps a more chilling conclusion is that the Norse adapted well, overcoming what Moser and Ekrstrom (2010) would term 'barriers' to adaptation, until they reached the cultural limits of their flexible subsistence strategy (Dow et al. 2013). Cultural limits are significant because culture is responsible for the intergenerational transmission of knowledge concerning the environment and the beliefs and values that deliver a sense of identity and place (Berkes 2017; Brace and Geoghegan 2011).

The utility and relevance of history and archaeology is in the interplay of success and failure, continuity and change. From the local to the regional scale, historical research uncovers what is likely to have happened and also asks counterfactual 'if-and-then' questions about why certain decisions were taken, at what cost, for whom, and why other decisions were not (see Baron-Cohen 2020). These counterfactuals are not purely hypothetical as they can still be constrained within the social and environmental context of the North Atlantic and the cultural barriers and limits to decision-making (Dugmore et al. 2012). For example, *if* the Norse abandoned agriculture *and* adopted semi-mobile seal-hunting *then* they would have had sufficient seal meat to support society. But this scenario is constrained by the cultural limits created by antecedent decisions to adopt sedentary farming and Christianity; rational subsistence strategies are constrained by the social and cultural context of the Norse diaspora in the North Atlantic. An agricultural system supplemented by wild resources could support the peak Norse populations in southwest Greenland whereas a nomadic system might have struggled to do so. This initial necessity could so reinforce a particular mind-set that by the time the population declined to a level where a nomadic system could support all, this crucial change was not perceived as appropriate even though it might have been the only route to survival.

"What if" scenarios, or counterfactuals, have been used increasingly in political methodology and history, to determine the plausible effects of different decisions (Levy 2008) and to assess the efficacy of policy choices (Harvey 2015). Harvey (2015) has explained, the utility of comparative counterfactual analysis is located ability to assess the *comparative plausibility* of different historical experiments. More broadly, as Carr (1964) has explained, counterfactuals are unavoidable in historical (and archaeological) studies: "the study of history is a study of causes" (87; cited Levy 2008). In this sense, what could have happened, but did not, can be gleaned from consideration of the range of possible decisions and outcomes. This, as Walter Scheidel (2017) has argued, helps historians and archaeologists identify, more confidently, the factors that were responsible for observed outcomes, and can, in turn, inform the range of possibilities in the future. As Fig. 4 illustrates, individual decisions could be considered adaptive, but, in combination with other constraints, become maladaptive. CCM and qualitative scenario storylines might benefit from plausible counterfactuals to develop effective adaptive strategies.

Conclusion

Archaeologists and historians are now uniquely positioned to provide relevant data for global change research and inform environmental policy and planning. As this chapter has argued, archaeologists and historians should depart from the broad narratives of collapse and focus instead on informing theory using evidence of different adaptive capacities and limits that can be observed using natural experiments. The nuance of archaeological and historical interpretation requires the explicit consideration of a range of counterfactuals to interpret human activities in the past, which undermines environmental determinism, to consider a range of potential adaptive pathways. But far from undermining the relevance of archaeological and historical information, the range of plausible and implausible scenarios, retrofactuals and counterfactuals, provide an evidence base that can inform climate change adaptation frameworks. CCM and qualitative scenario storylines require are two existing frameworks that could incorporate archaeological-historical information to consider the interplay between different climate impacts and adaptive capacities in different geographical contexts. The Norse Greenland case study illustrates an important lesson for climate change adaptation scenarios; even a highly adaptive society can, over the course of several centuries, reach limits to adaptation when exposed to unanticipated social and environmental change. Furthermore, a high adaptive capacity does not always translate as an ability (or willingness) to adapt.

References

Adderley WP, Simpson IA (2006) Soils and palaeo-climate based evidence for irrigation requirements in Norse Greenland. J Archaeol Sci 33:1666–1679

Aimers JJ (2007) What Maya Collapse? Terminal classic variation in the Maya Lowlands. J Archaeol Res 15(4):329–377

Arneborg J, Lynnerup N, Heinemeier J (2012) Human Diet and Subsistence Patterns in Norse Greenland AD 1000–AD 1450: Archaeological interpretations. Journal of the North Atlantic 3:119–133

ARCUS (1997) People and the arctic. a prospectus for research on the human dimensions of the arctic system. ARCUS, Fairbanks, Alaska

Armstrong CG, Shoemaker AC, McKechnie I, Ekblom A, Szabó P, Lane PJ, McAlvay AC, Boles OJ, Walshaw S, Petek N, Gibbons KS, Morales EQ, Anderson EN, Ibragimow A, Podruczny G, Vamosi JC, Marks-Block T, LeCompte JK, Awâsis S, Nabess C, Sinclair P, Crumley CL (2017) Anthropological contributions to historical ecology: 50 questions, infinite prospects. PLoS One 12(2):1–26

Barlow LK, Sadler JP, Ogilvie AEJ, Buckland PC, Amorosi T, Ingimundarson JH, Skidmore P, Dugmore AJ, McGovern TH (1997) Interdisciplinary investigations of the end of the Norse Western settlement in Greenland. Holocene 7(4):489–499

Barnes J, Dove M, Lahsen M, Mathews A, McElwee P, McIntosh R, Moore F, O'reilly J, Orlove B, Puri R, Weiss H (2013) Contribution of anthropology to the study of climate change. Nat Clim Chang 3(6):541–544

Brace C, Geoghegan H (2011) Human geographies of climate change: Landscape, temporality and lay knowledge. Progress in Human Geography 35(3):284–302

Barrett JH, Boessenkool S, Kneale CJ, O'Connell TC, Star B (2020) Ecological globalisation, serial depletion and the medieval trade of walrus rostra. Reviews 229 Quaternary Science: 106122

Bostrom N, Cirkovic MM eds (2011) Global catastrophic risks. Oxford University Press

Barnett J, Adger WN (2007) Climate change, human security and violent conflict. Polit Geogr 26(6):639–655

Baron-Cohen S (2020) The pattern seekers: how autism drives human invention. Penguin, London

Berkes F (2017) Sacred ecology, 4th edn. Routledge, London

Boivin NL, Zeder MA, Fuller DQ, Crowther A, Larson G, Erlandson JM, Denham T, Petraglia MD (2016) Ecological consequences of human niche construction: examining long-term anthropogenic shaping of global species distributions. Proc Natl Acad Sci 113(23):6388–6396

Bostrom N, Cirkovic MM (2011) Global catastrophic risks. Oxford University Press

Bostrom N (2013) Existential risk prevention as global priority. Global Pol 4(1):15–31

Boswell C, Smith K (2017) Rethinking policy 'impact': four models of research-policy relations. Palgr Commun 3(1):44

Brace C, Geoghegan H (2011) Human geographies of climate change: landscape, temporality, and lay knowledges. Prog Hum Geogr 35(3):284–302

Buckland PC, Amorosi T, Barlow LK, Dugmore AJ, Mayewski P, McGovern TH, Ogilvie AEJ, Sadler JP, Skidmore P (1996) Bioarchaeological and climatological evidence for the fate of Norse farmers in medieval Greenland. Antiquity 70:88–96

Butzer KW (2012) Collapse, environment, and society. Proc Natl Acad Sci 109(10):3632–3639

Butzer KW, Endfield GH (2012) Critical perspectives on historical collapse. Proc Natl Acad Sci 109(10):3628–3631

Cairney P (2018) The UK government's imaginative use of evidence to make public policy. Br Polit 14:1–22

Cairney P, Oliver K (2020) How should academics engage with policymaking to achieve impact. Polit Stud Rev 18(2):228–244

Cairney P (2016) The politics of evidence-based policy making. Springer

Carr EH (1964) What is history? Penguin, Hammondsworth

Ceballos G, Ehrlich PR, Dirzo R (2017) Biological annihilation via the ongoing sixth mass extinction signaled by vertebrate population losses and declines. Proc Natl Acad Sci 114(30):E6089–E6096

Ceballos G, Ehrlich PR, Barnosky AD, García A, Pringle RM Palmer TM (2015) Accelerated modern human–induced species losses: entering the sixth mass extinction. Sci Adv 1(5):e1400253

Costanza R, Graumlich L, Steffen W, Crumley C, Dearing J, Hibbard K, Leemans R, Redman C, Schimel D (2007) Sustainability or collapse: what can we learn from integrating the history of humans and the rest of nature? AMBIO: J Hum Environ 36(7): 522–527.

Cronon W (1992) A place for stories: nature, history, and narrative. J Am Hist 78(4):1347–1376

Crumley C (1994) Historical ecology. School of American Research Press, Santa Fe

Currie A (2016) Ethnographic analogy, the comparative method, and archaeological special pleading. Stud Hist Philos Sci Part A 55:84–94

d'Alpoim Guedes J, Crabtree SA, Bocinsky RK, Kohler TA (2016) Twenty-first century approaches to ancient problems: climate and society. Proc Natl Acad Sci 113(51):14483–14491

Dawdy SL (2009) Millennial archaeology: locating the discipline in the age of insecurity/doomsday confessions. Archaeol Dial 16(2):131–142

Dawson T, Nimura C, López-Romero E, Daire MY (eds) (2017) Public archaeology and climate change. Oxbow Books, Oxford, p 185

Donovan A, Oppenheimer C (2012) Governing the lithosphere: Insights from Eyjafjallajökull concerning the role of scientists in supporting decision-making on active volcanoes. Journal of Geophysical Research 117: B03214

Diamond J (2005) Collapse: how societies choose to fail or survive. Penguin, London

Diamond J, Robinson JA (2010) Natural experiments of history. Harvard University Press

Dow K, Berkhout F, Preston BL, Klein RJT, Midgley G, Shaw MR (2013) Limits to adaptation. Nat Clim Change 3(4):305–307

Dugmore AJ, Vésteinsson O (2012) Black sun, high flame, and flood: volcanic hazards in Iceland In: Cooper J, Sheets P (eds) Surviving sudden environmental change: answers from archaeology. University of Colorado Press, Boulder

Dugmore AJ, Keller C, McGovern TH (2007a) The Norse Greenland settlement: reflections on climate change, trade and the contrasting fates of human settlements in the Atlantic islands. Arct Anthropol 44(1):12–37

Dugmore AJ, Borthwick DM, Buckland PC, Church M, Dawson A, Edwards KJ, Keller C, Mayewski P, McGovern TH, Mairs KA, Sveinbjarnardóttir G (2007b) The role of climate in settlement and landscape change in the North Atlantic Islands: an assessment of cumulative deviations in high-resolution proxy climate records. Hum Ecol 35:169–178

Dugmore AJ, Keller C, McGovern TH, Casely A, Smiarowski K (2009) Norse Greenland settlement and limits to adaptation. In: Adger NW, Lorenzoni I, O'Brien KL (eds) Adapting to climate change. Cambridge University Press, Cambridge

Dugmore AJ, McGovern TH, Vésteinsson O, Arneborg J, Streeter R, Keller C (2012) Cultural adaptation, compounding vulnerabilities and conjunctures in Norse Greenland. Proc Natl Acad Sci 109(10):3658–3663

Dunne JA, Maschner H, Betts MW, Huntly N, Russell R, Williams RJ, Wood SA (2016) The roles and impacts of human hunter-gatherers in North Pacific marine food webs. Sci Rep 6(1):1–9

Edvardsson R, Patterson WP, Bárðarson H, Timsic S, Ólafsdóttir GÁ (2019) Change in Atlantic cod migrations and adaptability of early land-based fishers to severe climate variation in the North Atlantic. Quat Res 1–11

Ellis EC (2011) Anthropogenic transformation of the terrestrial biosphere. Philos Trans Royal Soc A: Math Phys Eng Sci 369(1938):1010–1035

Fitzhugh B, Butler V, Bovy K, Etnier M (2019) Human ecodynamics: a perspective for the study of long-term change in socioecological systems. J Archaeol Sci Rep 23:1077–1094

Ford JD, Pearce T, McDowell G, Berrang-Ford L, Sayles JS, Belfer E (2018) Vulnerability and its discontents: the past, present, and future of climate change vulnerability research. Clim Change 151(2):189–203

Frei KM, Coutu AN, Smiarowski K, Harrison R, Madsen CK, Arneborg J, Frei R, Gudmundsson G, Sindbæk SM, Woollett J, Hartman S, Hicks M, McGovern TH (2015) Was it for walrus? Viking age settlement and medieval walrus ivory trade in Iceland and Greenland. World Archaeology 47(3):439–466

Gamble C (2007) Archaeology: the basics, 2nd edn. Routledge, London

Gulløv HC (2008) The nature of contact between native Greenlanders and Norse. J North Atlant 1(1):16–24

Haldon J, Mordechai L, Newfield TP, Chase AF, Izdebski A, Guzowski P, Labuhn I, Roberts N (2018) History meets palaeoscience: consilience and collaboration in studying past societal responses to environmental change. Proc Natl Acad Sci 115(13):3210–3218

Hambrecht G, Anderung C, Brewington S, Dugmore A, Edvardsson R, Feeley F, Gibbons K, Harrison R, Hicks M, Jackson R, Ólafsdóttir GÁ, Rockman M, Smiarowski K, Streeter R, Szabo V, McGovern T (2018) Archaeological sites as distributed long-term observing networks of the past (DONOP). Quatern Int. https://doi.org/10.1016/j.quaint.2018.04.016

Hartman S, Ogilvie AEJ, Ingimundarson JH, Dugmore AJ, Hambrecht G, McGovern TH (2017) Medieval Iceland, Greenland and the New Human Condition: a case study in integrated environmental humanities. Glob Planet Change 156:123–139. https://doi.org/10.1016/j.gloplacha.2017.04.007

Harvey DC, Perry J (2015) The future of heritage as climate change: loss, adaptation and creativity. Routledge, London

Hegmon M (2016) Archaeology of the human experience: an introduction. Archaeol Pap Am Anthropol Assoc 27(1):7–21

Hegmon M, Arneborg J, Dugmore AJ, Hambrecht G, Ingram S, Kintigh K, McGovern TH, Nelson M, Peeples MA, Simpson A, Spielmann K, Streeter R, Vésteinsson O (2013) The human experience of social change and continuity: the Southwest and North Atlantic in "Interesting Times" ca. 1300. In: Lacey S, Tremain C, Sawyer M (eds) Climates of change: the shifting environments of archaeology. Proceedings of the 44th annual Chacmool conference. University of Calgary

Hegmon M, Peeples MA, LTVTP-NABO (2018) The human experience of social transformation: Insights from comparative archaeology. PLOS One 13(11):e0208060

Hollesen J, Mattiesen H, Elberling B (2017) The Impact of climate change on an archaeological site in the Arctic. Archaeometry 59(6):1175–1189

Hollesen J, Mattiesen H, Moller AB, Elberling B (2015) Permafrost thawing in organic soils accelerated by ground heat production. Nat Clim Chang 5(6):574–578

Hollesen J, Mattiesen H, Moller AB, Westergaard-Nielsen A, Elberling B (2016) Climate change and the loss of organic archaeological deposits in the Arctic. Sci Rep 6(9):28690

Hsiang SM, Burke M, Miguel E (2013) Quantifying the influence of climate on human conflict. Science 341(6151):1235367

Hudson MJ, Aoyama M, Hoover KC, Uchiyama J (2012) Prospects and challenges for an archaeology of global climate change. Wiley Interdisc Rev: Clim Change 3:313–328

Hulme M (2008) The conquering of climate: discourses of fear and their dissolution. Geogr J 174(1):5–16

Hulme M (2016) Weathered: cultures of climate. Sage, London

Golding KA, Simpson IA, Wilson CA, Low EC, Schofield JE (2015) Europeanization of sub-Arctic environments: Perspectives from Norse Greenland's outer fjords. Human Ecology 43(1):61–77

Harvey F (2015) "What If" History Matters? Comparative Counterfactual Analysis and Policy Relevance, Security Studies, 24(3):413–424, https://doi.org/10.1080/09636412.2015.1070606

Historic Environment Scotland (2019) A Guide to Climate Change Impacts: On Scotland's Historic Environment. Technical Advice and Guidance.https://www.historicenvironment.scot/archives-and-research/publications/publication/?publicationId=843d0c97-d3f4-4510-acd3-aadf0118bf82 [Accessed: 09/03/2022]

Hulme M (2011) Reducing the future to climate: a story of climate determinism and reductionism. Osiris, 26(1):245–266

IPCC (2014) Climate Change 2014: impacts, adaptation, and vulnerability. Part A: global and sectoral aspects. Contribution of Working Group II to the fifth assessment report of the intergovernmental panel on climate change [Field CB, Barros VR, Dokken DJ, Mach KJ, Mastrandrea MD, Bilir TE, Chatterjee M, Ebi KL, Estrada YO, Genova RC, Girma B, Kissel ES, Levy AN, MacCracken S, Mastrandrea PR, White LL (eds)]. Cambridge University Press, Cambridge, United Kingdom and New York, NY, USA, 1132 pp

Jackson RC, Dugmore AJ, Riede F (2017) Towards a new social contract in archaeology and climate change adaptation. Archaeological Reviews from Cambridge 32(2):197–221

Jackson R, Arneborg J, Dugmore A, Madsen C, McGovern T, Smiarowski K, Streeter R (2018a) Disequilibrium, adaptation, and the Norse settlement of Greenland. Hum Ecol 46(5):665–684

Jackson RC, Dugmore AJ, Riede F (2018b) Rediscovering lessons of adaptation from the past. Glob Environ Chang 52:58–65

Jensen KG, Kuijpers A, Koç N, Heinemeier J (2004) Diatom evidence of hydrographic changes and ice conditions in Igaliku Fjord, South Greenland, during the past 1500 years. The Holocene 14(2):152–164

Janssen MA, Kohler TA, Scheffer M (2003) Sunk-cost effects and vulnerability to collapse in ancient societies. Curr Anthropol 44(5):722–728

Kaplan JO, Krumhardt KM, Zimmermann NE (2012) The effects of human land use and climate change over the past 500 years on the carbon cycle of Europe. Glob Change Biol 18:902–914

Keighley X, Pálsson S, Einarsson BF, Petersen A, Fernández-Coll M, Jordan P, Tange Olsen M, Malmquist HJ (2019) Disappearance of Icelandic Walruses coincided with Norse Settlement. Mol Biol Evol 36(12):2656–2667

Koch A, Brierley C, Maslin MM, Lewis SL (2019) Earth system impacts of the European arrival and Great Dying in the Americas after 1492. Quatern Sci Rev 207:13–36

Kohler TA, Rockman M (2020) The IPCC: a primer for archaeologists. Am Antiq 85(4):627–651

Kwok R (2017) Historical data: hidden in the past. Nature 549(7672):419–421

Levy JS (2008) Counterfactuals and case studies. In: Box-Steffensmeier JM, Brady HE, Collier D (eds) The Oxford handbook of political methodology. Oxford University Press, Oxford

Liechenko R, O'Brien K (2019) Climate and society: transforming the future. Polity, Cambridge

Lucas G (2001) Critical approaches to fieldwork: contemporary and historical archaeological practice. Routledge, London

Lucas G (2005) The archaeology of time. Routledge, London

Lucas G (2010) Triangulating absence: exploring the fault lines between archaeology and anthropology. In: Garrow D, Yarrow T (eds) Archaeology and anthropology: understanding similarity, exploring difference. Oxbow Books, London

Lucas G (2019) Writing the past: knowledge production and literary production in archaeology. Routledge, London

Lynnerup N (1998) The Greenland Norse: a biological-anthropological study. Man and society 24. Copenhagen

Lynnerup N (2014) Endperiod demographics of the Greenland Norse. J North Atlantic 7:18–24

Madsen CK (2014) Pastoral settlement, farming, and hierarchy in Norse Vatnahverfi, South Greenland. Ph.D. Thesis, University of Copenhagen

Madsen CK (2019) Marine shielings in Medieval Norse Greenland. Arct Anthropol 56(1):119–159

McAnany PA, Yoffee N (2010) Questioning collapse: human resilience, ecological vulnerability, and the aftermath of empire. Cambridge University Press, Cambridge

McCaughie D, Simpson IA, Hyslop E, Graham C, Turmal A (2020) Baselining sandstone heritage for conservation in a climate change(d) future. In: Seigesmund S, Middendorf B (eds) Monument future: decay and conservation of stone. Proceedings of the 14th international congress on the deterioration and conservation of stone, vols I and II. Mitteldeutscher Verlag, pp 717–722

McGovern T (2018) Burning libraries: a community response. Conserv Manag Archaeol Sites 20(4):165–174

McGovern TH (1980) Cows, harp seals, and churchbells: adaptation and extinction in Norse Greenland. Hum Ecol 8(3):245–275

McGovern TH (1985) The Arctic frontier of Norse Greenland." In: Green S, Perlman S (eds) The archaeology of frontiers and boundaries. Academic Press, New York

McGovern TH, Harrison R, Smiarowski K (2014) Sorting sheep & goats in medieval Iceland and Greenland: local subsistence or world system. In: Harrison R, Maher RA (eds) Long-term human ecodynamics in the North Atlantic: an archaeological study. Lexington Publishers, Lanham

Meyer WJ, Crumley CL (2011) Historical ecology: using what works to cross the divide. In: Moore T, Armada X-L (eds) Atlantic Europe in the first millennium BC: crossing the divide. Oxford University Press, Oxford & New York, pp 109–134

Lund KA, Benediktsson K (2011) Inhabiting a risky Earth: The Eyjafjallajökull eruption in 2010 and its impacts. Anthropology Today 27(1):6–9

McGovern TH (1994) Management for extinction in Norse Greenland. In Crumley CL (ed.), Historical Ecology: Cultural Knowledge and Changing Landscapes, University of Washington Press, Washington

McGovern TH, Amorosi T, Perdikaris S, Woollett J (1996) Vertebrate Zooarchaeology of Sandnes V51: Economic Change at a Chieftain's Farm in West Greenland. Arctic Anthropology 33(2):94–121

Miller F, Osbahr H, Boyd E, Thomalla F, Bharwani S, Ziervogel G, Walker B, Birkmann J, van der Leeuw S, Rockström J, Hinkel J, Downing T, Folke C Nelson D (2010) Resilience and Vulnerability: Complementary or Conflicting Concepts? Ecology and Society 15(3):11

Middleton G (2017) Understanding collapse: ancient history and modern myths. Cambridge University Press, Cambridge

Middleton GD (2018) This is the end of the world as we know it: Narratives of collapse and transformation. In: Vogelaar AE (ed) The discourses of environmental collapse. Routledge, Abingdon

Middleton GD (2012) Nothing lasts forever: environmental discourses on the collapse of past societies. J Archaeol Res 20(3):257–307

Miller GH, Geirsdóttir Á, Zhong Y, Larsen D, Otto-Bliesner B, Holland MM, Bailey DA, Refsnider KA, Lehman SJ, Southon JR, Anderson C, Björnsson H, Thordarson T (2012) Abrupt onset of the little ice age triggered by volcanism and sustained by sea-ice/ocean feedbacks. Geophys Res Lett 39(2):1–5

Moser SC, Ekstrom JA (2010) A framework to diagnose barriers to climate change adaptation. Proc Natl Acad Sci 107(51):22026–22031

Nelson MC, Ingram SE, Dugmore AJ, Streeter R, Peeples MA, McGovern TH, Hegmon M, Spielmann KA, Simpson IA, Strawhacker C, Comeau LE, Torvinen A, Madsen CK, Hambrecht G, Smiarowski K (2016) Climate changes, vulnerabilities, and food security. Proc Natl Acad Sci 113(2):298–303

Nelson MC, Kintigh KW, Arneborg J, Streeter R, Ingram SE (2017) Vulnerability to food insecurity: tradeoffs and their consequences. In: Hegmon M (ed) The give and take of sustainability: archaeological and anthropological perspectives on tradeoffs. Cambridge University Press, Cambridge

Newell B, Proust P (2019) Introduction to collaborative conceptual modelling. Working Paper, ANU Open Access Research. https://digitalcollections.anu.edu.au/handle/1885/9386. Accessed 11 Nov 2020

O'Brien K, Liechenko R (2000) Double exposure: assessing the impacts of climate change within the context of economic globalization. Glob Environ Chang 10(3):221–232

Ogilvie AEJ (1982) Climate and society in Iceland from the medieval period to the late eighteenth century. Unpublished Ph.D. thesis, School of Environmental Sciences, University of East Anglia, Norwich, UK

Ogilvie AEJ (1991) Climatic changes in Iceland A. D. c. 865 to 1598. In: The Norse of the North Atlantic (Presented by G.F. Bigelow). Acta Archaeologica, vol 61-1900. Munksgaard, Copenhagen, pp 233–251

Ogilvie AEJ (1998) Historical accounts of weather events and related matters in Iceland and Greenland, A.D. c. 1250 to 1430. In: Frenzel B (ed) Documentary Climatic Evidence for 1750–1850 and the 14th century. Paläoklimaforschung 23. Special Issue: ESF Project "European Palaeoclimate and Man" 15. European Science Foundation, Strasbourg, Akademie der Wissenschaften und der Literatur, Mainz, 25–43

Ogilvie AEJ (2016) The Norse in Greenland. In: Stewart W, A Fortunate land. Offsetdruckerei Karl Grammlich GmbH, Pliezhausen, Germany, pp 114–126

Ogilvie, A.E.J. (2019). Opportunities and Challenges for Nordic Arctic and Subarctic Regions: A Case Study Approach. In: Nilsson, K., Karlsdóttir, A. and Refsgaard, K. (eds.) *Nordic Working Papers Opportunities and Challenges for Future Regional Development*, Nordisk ministerråd, Copenhagen. http://urn.kb.se/resolve?urn=urn:nbn:se:norden:org:diva-5802

Ogilvie AEJ (2020) Famines, mortality, livestock deaths and scholarship: environmental stress in Iceland ca. 1500–1700. In: Kiss A, Prybil K (eds) The dance of death. Environmental stress, mortality and social response in late Medieval and Renaissance Europe, London, Routledge, pp 9–24, Online 2019. https://doi.org/10.4324/9780429491085

Ogilvie AEJ, Woollett JM, Smiarowski K, Arneborg J, Troelstra S, Pálsdóttir A, McGovern TH (2009) Seals and sea ice in medieval Greenland. J North Atlantic 2:60–80

Oreskes N, Conway EM (2013) The collapse of Western civilization: a view from the future. Daedalus 142(1):40–58

Ortman L (2019) A new kind of relevance for archaeology. Front Digit Human 6(16):1–13

Oliver K, Cairney P (2019) The dos and don'ts of influencing policy: a systematic review of advice to academics. Palgrave Communications, 5(1):21. https://doi.org/10.1057/s41599-019-0232-y

Popper K (1956) The Poverty of Historicism. London: Routledge

Redman CL (2005) Resilience theory in archaeology. Am Anthropol 107:70–77

Richer S, Stump D, Marchant R (2019) Archaeology has no relevance. Internet Archaeol 53. https://doi.org/10.11141/ia.53.2

Riede F (2014) Climate models: use in archaeology. Nature 513(7518):315

Riede F (2017) Past-forwarding ancient calamities. Pathways for making archaeology relevant in disaster risk reduction research. Humanities 6(4):79

Riede F, Jackson RC (2020) Do deep-time disasters hold lessons for contemporary understandings of resilience and vulnerability?: the case of the Laacher see volcanic eruption. In: Sheets P, Riede F (eds) Going forward by looking back: archaeological perspectives on socio-ecological crisis, response, and collapse. Berghahn Books, New York

Riede F (2019) Environmental determinism and archaeology. Red flag, red herring. Archaeol Dialog 26(1):17–19

Rockman M (2003) Knowledge and learning in the archaeology of colonization. In: Rockman M, Steele J (eds) Colonization of Unfamiliar Landscapes: the archaeology of adaptation. Routledge, London

Roesdahl E (2005) Walrus ivory—demand, supply, workshops, and Greenland. In Viking and Norse in the North Atlantic: select papers from the proceedings of the fourteenth Viking Congress, Tórshavn, 19–30 July 2001 (pp 182-191)

Rounsevell MD, Metzger MJ (2010) Developing qualitative scenario storylines for environmental change assessment. Wiley Interdisc Rev: Clim Change 1(4):606–619

Scheidel W (2017) The great leveler: violence and the history of inequality from the stone age to the twentieth century. Princeton, Princeton University Press

Sigurðardóttir R, Newton A, Hicks MT, Dugmore A, Hreinsson V, Ogilvie AEJ, Júlíusson ÁD, Einarsson Á, Hartman S, Simpson IA, Vésteinsson O, McGovern TH (2019) Trolls, water, time, and community: resource management in the Mývatn District of Northeast Iceland. In: Lozny LR, McGovern T (eds) Studies in human ecology, Vol. 11: Global perspectives on long term community resource management. Studies in human ecology and adaptation book series (STHE). vol. 11, Springer International Publishing. https://doi.org/10.1007/978-3-030-15800-2

Spielmann K, Peeples MA, Glowacki DM, Dugmore A (2016) Early warning signals of social transformation: a case study from the US Southwest. PLoS One 11(10):1–18

Stephens L et al, ArchaeoGLOBE Project Authors (2019) Archaeological assessment reveals earth's early transformation through land use. Science 365(6456):897–902

Strickland KM, Coningham R, Gunawardhana P, Simpson I (2018) Hydraulic complexities: collapse and resilience in Sri Lanka. In: Sulak F, Pikirayi I (eds) Water and society: from ancient times to the present. Routledge, pp 259–281

Sykes N (2014) Beastly questions: animal answers to archaeological issues. Bloomsbury Publishing

Star B, Barrett JH, Gondek AT, Boessenkool S (2018) Ancient DNA reveals the chronology of walrus ivory trade from Norse Greenland. Proceedings of the Royal Society B, 285:(1884) 20180978

Smiarowski K, Harrison R, Brewington S, Hicks M, Hérbert FJ, Prehal B, Hambrecht G, Woollett J, McGovern TH (2017) Zooarchaeology of the Scandinavian settlements in Iceland and Greenland: Divergent pathways. In Albarella U (ed.), Oxford handbook of Zooarchaeology, OUP, Oxford

Tainter JA (1988) The collapse of complex societies. Cambridge University Press, Cambridge

Tainter JA (2006) Social complexity and sustainability. Ecol Complex 3(2):91–103

Thomas K, Hardy RD, Lazrus H, Mendez M, Orlove B, Rivera-Collazo I, Roberts JT, Rockman M, Warner BP, Winthrop R (2019) Explaining differential vulnerability to climate change: a social science review. Wiley Interdisc Rev: Clim Change 10(2):e565

Thornton TF, Moss ML, Butler VL, Herbert J, Funk F (2010) Local and traditional knowledge and the historical ecology of Pacific herring in Alaska. J Ecol Anthropol 14(1):81–88

Urry J (2011) Climate change and society. Polity Press, Cambridge

Urry J (2015) What is the Future? Polity Press, Cambridge

Van de Noort R (2011) Conceptualising climate change archaeology. Antiquity 85:1039–1048

Van de Noort R (2013) Climate change archaeology: building resilience from research in the World's Coastal Wetlands. Oxford University Press, Oxford

Vésteinsson O (2013) Collapse or resilience? Archaeology, metaphor and global warming. In: Sabatini S, Bergerbrant S (eds) Counterpoint: essays in archaeology and heritage studies in Honour of Professor Kristian Kristiansen. Bar Publishing

Vésteinsson O, Hegmon M, Arneborg J, Riche G, Russell WG (2019) Dimensions of inequality. Comparing the North Atlantic and the US Southwest. J Anthropol Archaeol 54:172–191

Vésteinsson O, McGovern TH, Keller C (2002) Enduring impacts: social and environmental aspects of Viking Age settlement in Iceland and Greenland. Archaeologia Islandica 2:98–136

Vogelaar AE, Peat A, Hale B (2018) Introducing the end. In: Vogelaar AE, Hale B, Peat A (eds) The discourses of environmental collapse: imagining the end. Routledge, Abingdon

Van Vuuren DP, Edmonds J, Kainuma M, Riahi K, Thomson A, Hibbard K, Hurtt GC, Kram T, Krey V, Lamarque JF, Masui T (2011) The representative concentration pathways: an overview. Climatic change, 109(1):5–31

Vésteinsson O, McGovern TH (2012) The peopling of Iceland. Norwegian Archaeological Review 45(2):206–218

Wise RM, Fazey I, Smith MS, Park SE, Eakin HC, Van Garderen ERMA, Campbell B (2014) Reconceptualising adaptation to climate change as part of pathways of change and response. Global Environmental Change 28:325–336

Zhong Y, Miller GH, Otto-Bliesner BL, Holland MM, Bailey DA, Schneider DP, Geirsdottir A (2011) Centennial-scale climate change from decadally-paced explosive volcanism: a coupled sea ice-ocean mechanism. Climate Dynamics, 37(11):2373–2387

Resilience of Coupled Socio-Ecological Systems: Historic Rice Fields of the U.S. South

Edda L. Fields-Black, R. Daniel Hanks, Travis F. Folk, Rob Baldwin, Ernie P. Wiggers, Andrew Agha, Daniel D. Richter, and Richard H. Coen

Abstract While resilience is defined differentially by social scientists and ecologists, sustainability is possible where resilient social and ecological systems meet and interact, and sustainable resilient systems promote societal use of ecosystem services supporting contemporary societal needs without risk to future generations. Yet it is possible for seemingly appropriate and rational decisions from individuals, and society at large, to be counter to long-term sustainable solutions. Historic rice field cultivation in the wetlands of the Carolinas and Florida provides an example of various forms of resilience and sustainability within the theoretical framework of alternate stable states, whereby a resilient system can exist in more than one state and where stability is achieved when disruptive variables are not so disruptive as to generate tipping points from one state to another. This contribution examines the changing role and political as well as environmental impacts of rice agriculture in the region with particular emphasis on the contingent processes of environmental and cultural transformation that took place between the seventeenth and twenty-first centuries.

E. L. Fields-Black (✉)
Carnegie Mellon University, Pittsburgh, USA
e-mail: fieldsblack@cmu.edu

R. Baldwin · R. H. Coen
Clemson University, Clemson, USA

T. F. Folk
Folk Land Management, Inc., Green Pond, USA

R. D. Hanks
Weyerhaeuser Company, Clemson, USA

D. D. Richter
Duke University, Durham, USA

A. Agha
New South Associates, Stone Mountain, USA

E. P. Wiggers
Nemours Wildlife Foundation, Beaufort, USA

© The Author(s) 2022   273
A. Izdebski et al. (eds.), *Perspectives on Public Policy in Societal-Environmental Crises*,
Risk, Systems and Decisions, https://doi.org/10.1007/978-3-030-94137-6_18

Defining Resilience

The anthropogenic modification of Earth and Earth systems has, in recent years, garnered attention from a variety of academic disciplines resulting in increased awareness and interest in resiliency across multiple system types that include ecological and social systems. Generally, resilience is the amount of disturbance a system can absorb without changing state (Gunderson 2000, 426). However, resilience can be defined differentially by various disciplines: social scientists define social resilience as comprising three dimensions—coping capacity, adaptive capacity, and transformative capacity (Keck and Sakdapolrak 2013, 5–7). Ecologists measure ecological resilience by the magnitude of disturbance that can be absorbed by a system before its variables and processes controlling the system's behavior are changed such that the system redefines its structure or settles in an alternate stable state (Gunderson 2000, 426). Ecologists also identify systems designed with a single operating objective and a global optimum where alternate states should be avoided as engineering resilience (Gunderson 2000, 426).

There is potential for sustainability where resilient social and ecological systems meet and shape one another through various interactions. Sustainable resilient systems promote societal use of ecosystem services supporting contemporary societal needs without risk to future generations (Adger and Hodbod 2013). It is possible for seemingly appropriate and rational decisions from individuals, and society at large, to be counter to long-term sustainable solutions, even to the point where the resiliency of the support systems that promote such decisions is disrupted (Kahn 1966, 23, 24, 27). The consequences of known and unknown socio-ecological risks to the present and the future are diminished where resilient systems support sustainability (Adger and Hodbod 2013, 79–81). Risks that may be mitigated include changing climate, natural disasters, and wholesale socio-political systems change (e.g., abolition of slavery).

Alternate stable states theory suggests a resilient system can exist in more than one state. A state is considered stable when the disruptive variables influencing it are not so disruptive that it shifts state or regimes. The resiliency of a system determines how large of a perturbation the system can experience without the system shifting states. Stable state theory is commonly visualized with a cup-and-ball diagram (Fig. 1), where the "cup" represents the system, and the ball represents the state of the system. The parameters maintaining the system state are represented by the walls of the cup; the position of the ball in the cup represents the state of the system. When the ball is at the bottom of the cup (i.e., the basin of attraction) the system is most stable. The steepness of the walls determines the return time of the system after disturbance, whereas the width of the cup pertains to the system's resilience. Negative feedback mechanisms tend to push the system back towards the stable bottom of the cup after a perturbation. For the system to change to an alternate state, the perturbation must be large enough for the ball to breach the precipice of the cup and enter another stable state. Once an alternative stable state has been entered it is unlikely that the previous stable state can be reclaimed.

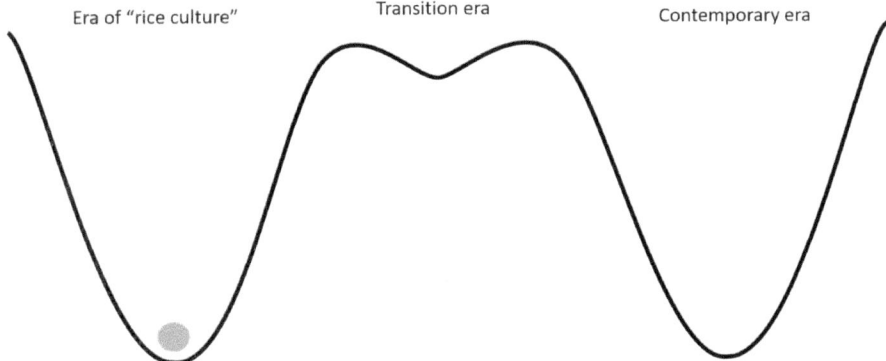

Era of "rice culture"   Transition era   Contemporary era

Fig. 1 Characteristics of socio-ecological stable states are represented by the various components of the cup-and-ball diagram. The state of the system is represented by the ball, while the parameters that control the system's state are represented by the walls of the cup. In the figure presented there is a transition era where the precipice of the original state (i.e., Era of "rice culture") has been breached and the system is gaining momentum towards transitioning to the contemporary era but there are a variety of forces attempting to push the system's state back into the stability of the era of "rice culture". We refer to this era as a transition era

Resilience on Lowcountry Wetlands and in Lowcountry Rice Fields

The economic, cultural, and environmental significance of southeastern rice is almost completely obscured by that of King Cotton, the export crop that came to dominate southern agriculture by the mid-1800s. Even still, the role of rice agriculture is almost completely unknown to the American consciousness as the driving national force it once was, one that created the wealthiest planters and largest slaveholders in British North America before the American Revolution. South Carolina rice planters claimed Nathaniel Heyward, the wealthiest slave holder with the largest slave holdings, as one of their own until his death in 1851 (Heyward 1993, xi, liv, 3–6, 12–13, 18–19). The value of rice to global society has been and continues to be immense.

Across time and space, Queen Rice can demonstrate how human labor and wetlands can be harnessed for economic ends, and how such systems can be resilient or collapse with political change such as that of the 13th Amendment of the US Constitution. Wetlands are valuable natural resources that provide high biological diversity. They filter and transform nutrients and pollutants and act to buffer the impacts of human activities on downstream aquatic systems (Hook 1993, 2157, 2160). Wetlands have been transformed in West Africa's Upper Guinea Coast since c. 1500 CE (Fields-Black 2008, 135–160) and in North America since the 1720s to grow rice as a staple and commodity crop. In North America, the modification of wetlands continued into the 1920s (Heyward 1993, x, xxii–xxx, xxxiii–xxxiv, liii–liv, 14, 33–34, 45, 48, 50, 54, 56, 60, 62, 63–65, 67–69, 74–75, 81–82, 90, 213–216, 39). During this period, some 234,354 acres of wetlands in South Carolina were modified

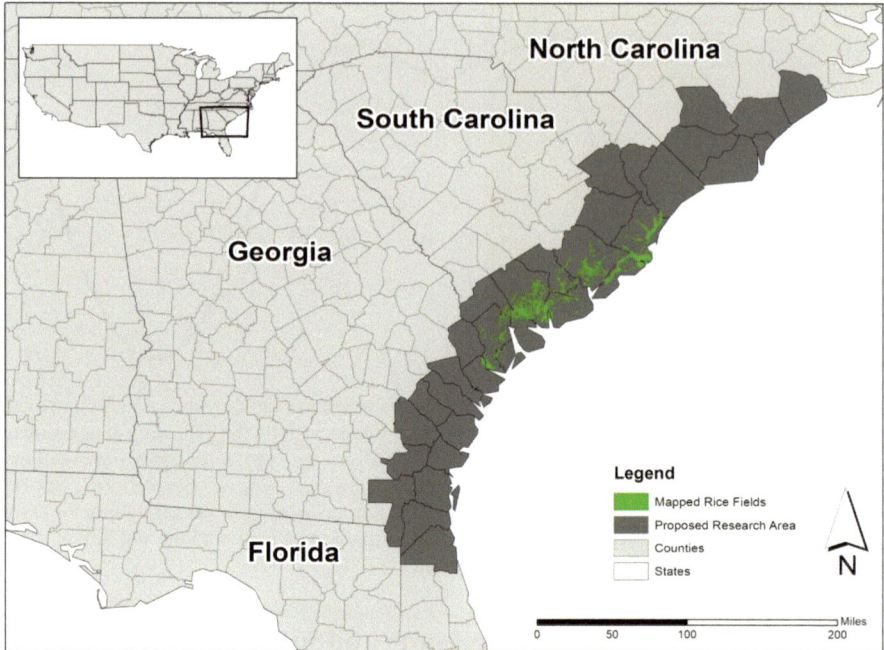

Fig. 2 Historic rice fields of the coastal plains of the southeastern U.S.A. extend from southeastern North Carolina to northeastern Florida. In green are the results of a mapping project using high resolution LiDAR to identify the historic rice fields of South Carolina

for rice production (Hanks et al. 2021, 7) (Fig. 2). Historic rice fields provide an interesting case study of the various forms of resilience and sustainability within the theoretical framework of alternate stable states.

Wetland clearing, tree stump removal, land leveling, and construction of water retention devices resulted in the first foray in the coastal plains of southeastern North Carolina to northeastern Florida, U.S.A. into the culture of rice as a staple crop to be followed with its use as a commodity crop. Ecosystems over extensive areas were substantially transformed by the rice economy and altered soils, vegetation, and hydrologies present invaluable opportunities for habitat conservation, specifically for economically important and migratory waterfowl. Queen Rice also provides important insights into resilience, in that ongoing and ecological and human dynamics produce new and invaluable stable states despite previous collapses.

The Queen Rice economy was entirely dependent on enslaved people to help planters identify microenvironments where rice could be successfully cultivated and perform the arduous labor of converting coastal plain marshes and forests to rice agriculture from the late 1600s to mid-1800s. Enslaved laborers levelled the primeval forest, felling cypress trees with axes–trunks of the largest cypress trees measured up to 12 ft in circumference–and deracinated the cypress knees with hand tools (Harper 1998, 59). Enslaved laborers made the earth as "level as a billiard table" with shovels, baskets, and planks of wood and moved 500,000 cubic feet of river muck per mile on South Carolina's Cooper River alone to construct the hydraulic irrigation system (Rosengarten 1998, 40; Ferguson 1992, 147). Then, enslaved laborers cultivated, harvested, and processed the rice crops and maintained the hydraulic irrigation system. American and European slave traders imported an estimated 198,314 Africa captives into South Carolina and Georgia colonies between 1701 and 1858 (www.slavevoyages.com). It is likely that the majority of these enslaved people were forced to labor in the swamps on Lowcounty South Carolina and Georgia rice plantations. South Carolina rice planters and congressmen led the legislative lobby to keep the trans-Atlantic slave trade open and legal. The Compromise of 1796 resulted in the abolition of the trans-Atlantic slave trade in US colonies in 1808. The illegal slave trade continued into the Lowcountry. South Carolina and Georgia's share of slave importation equalled more than half (53%) of the entire number of African captives imported into the US mainland. (www.slavevoyages.com). Dependence on an unfree and unpaid labor system set up a social system and an economic system, a plantation economy, which were not resilient.

The earliest rice culture was not resilient. It used low-lying, inland freshwater environments, valley bottoms of cypress-hardwood streams, and wetlands at the heads of creeks and rivers. Colonial naturalist Mark Catesby described the land that planters converted as "impregnated by the washings from the higher lands" over a period of years to become "deep soyl [sic]...a sandy loam of a dark brown colour" (Catesby 1771, iii). Early planters adapted rice cultivation techniques to the Lowcountry environment. Inland, freshwater swamp rice plantations depended on the natural features of watersheds to better irrigate rice fields. By hand, enslaved laborers cleared hardwood forests of cypress, tupelo, and sweet gum trees, exposing soils with slow water permeability and high water-holding capacity into which rice was sowed.

The fields were located on low-lying nearly level terrain with very slight downward-sloping grades (0–2%) that allowed drainage of freshwater into the rice fields. Locations of rice fields were chosen to take advantage of water from higher elevations flowing down into the fields from surface and groundwater sources (Smith 2019, 3–6, 7–8, 29–35, 38, 96–97, 109–112, 116–118, 120–121). Enslaved laborers built clayey embankments that retained water in reservoirs and quarter-drains then released it.

The 1700s saw demand and market share for rice from the Carolinas increase and concurrently, the development of more effective technologies and infrastructures for managing water. By 1710, rice had become South Carolina's most successfully cultivated staple crop and by the 1730s demand for rice exports was exploding. Annual exports from Charleston harbor increased five-fold between 1720 to 1740, from about 6 to 30 million pounds (Coclanis 1989, 83). Growth in the 1730s is partly attributed to the relaxation of the British Navigation Acts, enabling Carolina planters to export rice directly to Southern Europe on British ships.

Planters in the early eighteenth century were focused on economic growth. But, with increased demand and market share, planters established greater management control over the flow of water into the rice fields by constructing a complex of irrigation channels, trunks, and gates. Early trunks consisted of hollowed out tree trunks with plugs on each end. The irrigation resembled technologies used in West Africa's rice-growing region (Carney 1996, 27; Doar 1970, 12). Planters further mechanized water control systems with valves, rectangular boxes that were open to incoming water on one end, and sliding gates to trap and release freshwater into the rice fields. Simultaneously, the valves resembled technologies English farmers were using to drain fens (Edelson 2006, 6, 34, 48, 73–74, 76, 197).

Despite technological improvements, inland rice cultivation presented serious limits to the agricultural system's potential for growth. Planters were limited by the unidirectional flow of stream and river water, and gestating rice was vulnerable to both droughts and floods. Subtropical storm runoff readily breached the earthen embankments and washed out inland rice fields. On the other hand, the burgeoning rice export trade demanded more efficient, reliable, and predictable irrigation systems that would give planters more control over water.

South Carolina rice planter, Nathaniel Hayward's, primary inheritance was an inland rice fields on a tributary of the Broad River in Beaufort County, South Carolina and tidal rice fields only partially reclaimed near the Combahee River. He made a good faith effort planting the inland rice fields, which took advantage of water on higher elevations in the marsh where evergreen loblolly pine and oak trees grew in dry, sandy, clay soils without altering the environment. Nathaniel Heyward planted his inland rice fields for only one season. When the rice was still in the fields and almost ready to be harvested, a "freshet" destroyed the entire crop. Legend has it that he drove out to his inland rice field in his buggy the next morning after the storm to survey the damage. His fine rice crop was submerged. Heyward took one look at the water covering it, got back in the buggy, and never went back to or planted the inland rice field again. From this point, he focused on managing his brothers' inheritances, draining the swamps on his partially reclaimed tidal ride fields and his brothers' larger tidal rice fields, and subsequently acquiring and building his own empire, 17 "gold mines," tidal rice fields on the Combahee River in South Carolina's Colleton County (Heyward 1993, 11–12, 14, 50, 65–67).

Inland rice culture provided means to store water in reserve systems as a way to provide some mitigation against unpredictable water supply, which was the primary threat to a successful crop (Smith 2019, 3, 9, 81–82, 88, 99, 195; Coclanis 1989, 78–83). There remained some uncertainty for successful crop production depending on the severity of the drought where upstream reserves could not store enough water to supply the downstream rice field. It was not until the development of the tidal field system that large scale and consistent rice crops could be produced and shipped to market. Unlike Nathaniel Heyward, not all rice planters gave up inland rice cultivation entirely. Middling planters who were not uber-wealthy like the Heywards, continued to grow inland rice primarily for subsistence. After the collapse of the Queen Rice economy in the early twentieth century, African American farmers on heirs' property, property owned by the descendants of landowners who died intestate, well into the twentieth century (Heyward 1993, xiii–xiv; Hilliard 1975, 57–58; Merrens 1972, 187, 192; Smith 2019, 3, 9, 81–82, 88, 99, 195; Vernon 1993, 248, n.9, 274 n.7).

Inland swamp rice production was not perfect. It did not allow planters to control water and often led to too much water, in torrential deluges called freshets, or too little water in the rice fields. Perfecting water control would require more technology, more labor, even more knowledge of the coastal wetlands, and much more capital to reclaim the swamps and establish tidal rice production. It would take another few decades for all of these factors to come together in the mid-eighteenth century.

Tidal rice plantations were based on a simple principle of water control; they used the fresh water in river estuaries, which rose and fell with the tides, to flood rice fields. Tidal rice fields are located in an inland zone that is close enough to the coast to experience tidal fluctuations but far enough upstream to be freshwater. This geographic location allowed them to take advantage of the predictable tidal fluctuations and the associated freshwater wedge, thereby allowing the management of freshwater into and out of the fields.

Tidal rice plantations used water-control structures to inundate fields with the freshwater of estuaries that floated on top of the salt water and that rose and fell with the tides. With predictable environmental conditions (i.e., predictable supply of freshwater) ensured, the physical structure of the former wetlands as tidal rice fields became engineered and managed. As the process of rice culture in tidal fields was fine tuned and overall yield per acre increased, the tidal fields system was focused on a singularly objective, the production of rice for its market value. The immensity of the change to the structure and function of the wetlands across such a large geography resulted in a state shift from natural wetlands to engineered and managed wetlands, or rice fields. This combination made for the only naturally irrigated soils in the US. When management control was established over the hydrology of tidal wetlands, the area of rice cultivation greatly expanded.

Planters and enslaved laborers selected level land along rivers with tidal estuaries, impacted by the tides, accessing freshwater on either side. Planters and enslaved laborers converted level floodplains and swamplands along tidal rivers to rice fields, managing both fresh- and saltwater. Canals with floodgates were even constructed to channel freshwater from tidal estuaries to remote rice fields.

Slave labor was used throughout the time where the methodology of growing rice in the southern coastal plains of the United States was fine-tuned and increasingly focused on the tidal rice system. This societal stable state allowed forced and enslaved labor to transform a vast geography of wetlands across these coastal plains (more than 234, 354 acres in South Carolina, USA alone) (Hanks et. al. 2021, 7). Enslaved laborers converted tidal swamps to rice fields with water control infrastructures by building ditches and embankments around an entire floodplain. Next, they cleared, burned, and removed trees and stumps. Embankments that surrounded fields had wooden "trunks" installed for flooding and drainage. Trunks passed through the dikes and allowed movement between rice fields or between a tidal rice field and the tidal river. Initially, cypress trunks were used after being hollowed from top to bottom and installing a large plug to control water flow. Trunks eventually evolved into sophisticated wooden structures with gates at each end that opened and shut depending on the tide. Individual fields had cross ditches and channels excavated and check banks installed for highly precise water level control (Fig. 3). In the final stage, enslaved laborers planted rice on the floodplains.

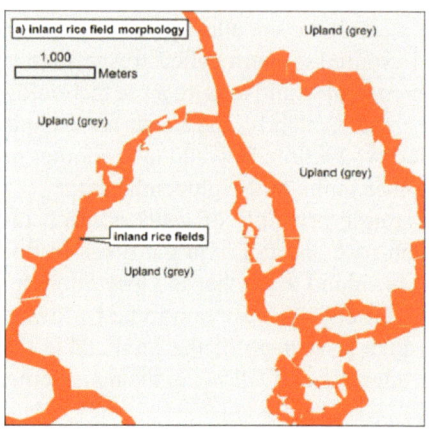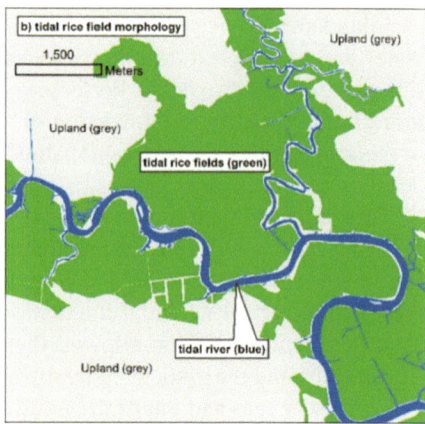

Fig. 3 Comparison and description of inland (**a**) and tidal rice field (**b**) morphology. Inland fields are linear, gravity fed wetlands. Widths of these wetlands can range from over 1000' wide to just 100' wide. Inland fields are bordered by uplands. Tidal fields were constructed in floodplain areas between sections of tidal, freshwater rivers and uplands. Inland fields tend to be narrow and long which result in smaller acreage impoundments. Tidal fields are not as constrained by adjacent uplands and individual impoundments on the average are larger acreages than those in inland rice fields

Conversion of a tidal swamp was a "great undertaking," one considerably more difficult than that of the inland swamps. It required significant engineering knowledge and knowledge of the natural landscape, as well as intensive capital and enslaved labor inputs over many years. In addition to clearing and leveling the land, constructing the hydraulic irrigation system, enslaved laborers maintained the irrigation system against the threat of breaks in the embankments and the rush of water. Because of the capital, labor, and time demands of tidal rice culture, only wealthy planters with multiple investments like Nathaniel Heyward, not small farmers, could undertake swamp reclamation. Then, they reinvested profits from one tidal rice plantation into acquiring more labor and reclaiming more tidal lands. The intensification of capital in the new American economy was vital to tidal rice culture (Heyward 1993, 18). By the mid-eighteenth century, tidal rice culture was extensive. By 1770, rice exports exceeded 60 million pounds per year (Fig. 1), largely due to tidal rice cultivation (Coclanis 1989, 83).

The system of slave labor became entangled with the environmental system of engineered and managed rice fields so intensively that the environmental system of engineered and managed rice fields could not exist were it not for the social system. With the environmental system of rice production now being stable and consistent (i.e., in an alternate stable state) across such a large geography and the singular focus of the rice fields being to produce marketable rice, slave labor became an integral cog in the socio-environmental system of rice culture in the southeastern United States. In other words, to maintain the stability of the environmental system (i.e., rice fields) required the stability of the social system where slave labor was foundational.

The full tragedy of Queen Rice is that African–Americans were at high risk of disease and survival rates were notably low in New World rice plantations. Though yields of tidal rice were incredibly productive and financially lucrative and though enslaved laborers contributed labor as well as critical skills, tidal rice culture came at a high cost for enslaved laborers. The saturated and inundated environments of the inland and tidal rice fields were extremely unhealthy environments. Rice plantations in the South Carolina and Georgia lowcountry had among the highest death rates among enslaved children and adults in the US, and second only to sugar plantations in the entire New World (Forster and Smith 2011, 909, 923). Whereas death rates for slaves on cotton and tobacco plantations dropped to approximately 33% by the 1780s, double the number, approximately two-thirds, of enslaved people still died on lowcountry rice plantations up until the Civil War. Along the Savannah River, 90% of children enslaved on rice plantations died before 15 years of age. In the winter of 1839 when most of the children died on Pierce Butler's Butler Island rice plantation on the Altamaha River near Darien, GA, the mother of a child who recently died described the plight of many women enslaved on Lowcountry rice plantations saying over and over again to Fannie Kemble, Butler's English wife: "I've lost a many; they all goes so" (Dusinberre 1996, 2–3, 49–51, 55–56, 58, 61, 74, 80–81, 84, 103, 121, 188, 199, 237–238, 240–241, 245, 325, 412–413, 415–416; Kemble 1984, 130).

The tight coupling of the social and environmental systems in the production of rice as a commodity crop through the engineered tidal field system operated within an engineering resilience framework. Production of rice for market value was the single objective of the tidal rice field system and any alternate state, for either the environmental or social system, was potentially not acknowledged but certainly to be avoided at all costs. Maintaining the system in an engineering resilient framework required constant input of energy which was provided by slave labor. Southern planters were unable to assess potential risks to the coupled socio-environmental system they had developed and fostered because they were operating in an engineering resiliency state-of-mind, where no other potential alternate stable states were envisaged. In this framework, the environmental stability of the tidal rice fields was a requisite that had to be maintained to produce rice and the only way this was possible was through slave labor.

The external forces that tested the resilience of the environmental system (i.e., engineered and managed tidal rice fields) still existed in both chronic (e.g., the erosion of dikes along riverbanks) and episodic (e.g., hurricanes) but this was overcome with the immense human power of forced free labor provided by a social system that enabled slave labor. Any threats to the social system (i.e., slavery) was quashed by the system itself. That is until the American Civil War put an end to slavery. At this point, both environmental and social systems underwent interlinked perturbations that allowed both to realize various trajectories previously unacknowledged and likely beyond contemporaneous comprehension. People who were once enslaved were now "free" and those who enslaved them now had to envision new ways to manage their engineered rice fields without slave labor. This resulted in a period after the American Civil War where planters attempted to continue to produce and manage rice as a commodity crop within the engineered rice fields up until the last commercial rice crop in South Carolina was produced by ss. In Fig. 1 this is represented by the small (and temporary) alternate stable state (termed "transition era"), which was fragile at best, requiring only small perturbations to push the system into a completely different stable state. These perturbations came in the form of several hurricanes and disruptions in market supply chains that affected planters in the southeastern coastal plains.

This transition era reflects the point made in Haldon et al. (2020, 1) that systems seldom collapse in the dramatic fashion so often popularized in modern culture. The rice culture that began in the 1670s and saw its ultimate demise beginning just prior to the American Civil War did not change the social nor the environmental systems in rapid fashion. Still today American society struggles with racism, some 155 years after the end of the American Civil War. The environmental system continues to change and may require more time to reach the stable bottom of the cup than that of the social system.

The postbellum social system disrupted the production of rice as a commodity crop in the southeastern U.S. due to the abolishment of slavery, but it also affected the engineered rice fields where, after some time, it was no longer seen as a tenable option to manage for rice production. These managed wetlands now experienced various ecological trajectories. While some previously managed fields were unmanaged and allowed to "re-wild," still others were managed for another objective–waterfowl hunting. The utilization of rice fields for waterfowl hunting beginning in the 1910 and 1920s has continued to evolve as a conservation ethic has percolated through this region. In recent years management of engineered tidal rice fields has become more diverse in the set of objectives that now includes, along with waterfowl, ecosystem services such as wildlife habitat (e.g., black rail, wood storks, roseate spoonbill, American alligator, etc.). These "new" objectives and their socio-environmental stable states are represented by the third cup in Fig. 1 and have yet to be fully established.

Resilience, Climate Change, and the 21st Century Rice Field Infrastructure

Given that the rice fields in South Carolina have significant historical, cultural and ecological significance, it is prudent to consider the permanence of these man made systems. Over the course of rice history in South Carolina, there have been several shifts in locations and objectives of rice fields. Initially the potential to reap financial profits started the rice economy in South Carolina and planters developed and used the inland field system. Ultimately tidal rice fields provided a financially successful alternative to inland fields and planters shifted their development and management in this area. These tidal fields (and inland fields to a lesser degree) transitioned from a financial to an ecological objective with the end of rice as a commodity crop and the rise of sporting pursuits on the rice plantations. From the early twentieth century this ecological objective has persisted and become more diverse with recognition that the management of these tidal and inland rice fields benefits waterfowl and many other non-game species like bald eagles, wood storks, black rails, etc.

There is increasing awareness by the owners and managers of the rice fields that the permanence of the existing system is being influenced negatively. Federal, state, and private owners of rice fields are faced with changing forces on this infrastructure. Decision making about how, when and how much to invest in these man-made systems is complex and has been ongoing since their creation.

Rice fields are composed of raised dikes that prevent water from entering or exiting a particular area where rice was grown and now is managed for wildlife habitat (Fig. 3). Canals were excavated to enable water flow throughout the system. Water control structures were placed in dikes to permit water retention or dewatering in these rice fields. Other than the water control structures (which are typically wooden), this system is constructed from onsite mud and soil. The properties of this mud and soil have influenced the maintenance regimes of the earthen infrastructure from the very inception of rice field construction in the early eighteenth century. As an example, some plantations may have been established along sections of rivers where the local material had a high sand content. Other rice fields were established along particular rivers where the available soil material was higher in loam or clay. When building dikes that stand 4–6' above the wetland grade, soils with higher clay content are able to persist longer with the normal erosion forces of rain, travel across the dikes, etc. These constant erosion forces require these dikes to have material placed back on top and the interim period between these efforts can vary from 5 to 10 years for areas of sand versus 10–15 years for areas of material with higher clay content.

Rice planters initially and wildlife habitat managers today have dealt with this constant degradation of the built rice field system. Retopping dikes, installing new water control structures, excavating canals for improved water circulation were all acts that have been required from the very beginning of rice culture and were incorporated into the mental calculus of determining whether it was worth the financial investment to perform these remedial acts. Maintenance and repair issues have always been a part of the rice field landscape and until recent decades were costs nearly all were willing to bear to maintain the system.

The costs and efforts, though, of maintaining rice fields, especially tidal ones, has become more costly and more frequent in recent decades. Environmental forces that have been influenced by climate change are causing increased degradation and loss of the rice field infrastructure. These forces include increases in tidal amplitude as driven by sea level rise and increase in the frequency of tropical storm events as a result of climate change. We discuss each of these factors below and how they negatively impact rice field infrastructure.

Tidal rice fields were placed along the tidal rivers in creeks to take advantage of the tidal, freshwater. Dikes and water control structures allowed the intentional flooding and dewatering of the rice fields depending on different times of the rice growing season. Since the transition to management of wildlife habitat, the intentional flooding and dewatering of the tidal rice fields enables management for native plant communities (e.g., widgeongrass in brackish and smartweed and panic grasses in freshwater rice fields) that do not develop under normal tidal regimes in free tidal flow marshes.

At a mechanistic level, tidal fields worked because the dike heights were built to a grade above mean high water. The bed of the rice field (i.e., the broad acreage where rice initially and not native plants were and are grown) was at a level above mean low water (Fig. 4). This enabled full dewatering of the rice field because as the tidal river reached low tide, i.e. below the bed level, water from the tidal rice field passed through water control structures from the rice field and into the free tidal flow water bodies.

Sea level rise impacts are most visible in coastal communities. The impacts of more frequent flooding on tourist and beach communities are commonly reported in the media (Tibbetts and Mooney 2018; Cappucci 2020). Additionally, sea level rise also influences tidal amplitude inland along tidal rivers. In South Carolina, these river systems are typified by the Combahee, Ashepoo, Edisto and Santee Rivers. Coastal South Carolina has seen a significant increase in tide levels. Since 1950, tidal waters have increased 10 inches in Charleston, which is central along the coast to rice growing areas of SC (Anonymous 2019, 6). Additionally, this increase appears to be increasing to a point where tidal increases of 1 inch every 2 years is anticipated.

These dramatic statistics within the tidal rice field landscape are forcing landowners to think of how increasing mean high water marks interact with the grade of perimeter dikes. Indeed, increasing numbers of spring tides, i.e. the strongest tide of the lunar cycle, are now overtopping these perimeter dikes. Efforts to raise them

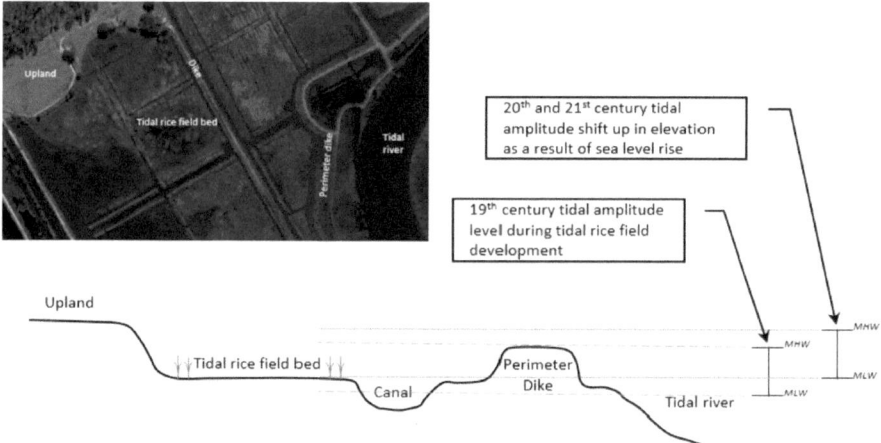

Fig. 4 Cross section sketch of a tidal rice field bed, canal and perimeter and dike. Tidal amplitudes during the era of rice field development (i.e., late 1700s until mid 1800s). MLW level is at a point below the rice field bed such that water could easily pass from the tidal rice field into the tidal rivers. Sea level rise in the late twentieth and current twenty-first century have shifted this tidal amplitude up such that MLW levels are just below or at bed level and thereby making it difficult to fully dewater tidal rice fields for wildlife habitat management objectives

are ongoing on rice plantations. In fact, because of the increasing mean high water, efforts to raise dike levels over the last decade are ongoing in all areas of tidal rice fields across South Carolina.

However, there is another more difficult dynamic associated with the increased tidal amplitude. Increasingly managers of tidal rice fields have difficulty fully dewatering the fields during low tide periods because the mean low water mark has also increased over time. Thus, the time and potential of water to pass from the tidal rice field and into the tidal river is limited. Because these tidal fields are managed for wildlife habitat and because many of the native plants require periods of drying to the bed, the management focus for wildlife is becoming more difficult. Some plantations have installed mechanical pumping stations to ameliorate the impact of increasing mean low water marks, but this cost is additive to the cost of dike maintenance and retopping.

Climate change has also impacted the frequency of tropical storm events (Marsooli et al. 2019, 5). Coastal South Carolina has seen these impacts recently with nine tropical storms events impacting coastal South Carolina since 2016 (Mizzell et al. 2020, 12). Luckily, there is a trend for the increased tropical storm events to be storms with less intensity than in the past; however, these events lead to excessive flooding from rainfall and tidal surge. Both sources of excess water in tidal rivers cause excessive dike erosion, dike breaching, and loss of water control structures. In recent years, some plantations have performed these dike repairs between hurricane seasons only to be forced into the same repair the next year caused by another storm.

Tidal rice fields are more susceptible to these impacts than inland fields for two reasons. First, tidal fields are directly adjacent to tidal rivers and creeks. Therefore they are immediately impacted by increases in tidal amplitude and storm surge as a result of tropical storm events. In contrast, inland fields are insulated by geography from these impacts; however, intense rain events can cause excessive flooding through the system, which degrades dikes and water control structures. In general, the cost of maintenance of tidal fields versus inland fields is 1.5–2 times higher (per acre of managed wetland). One complicating factor though is that more tidal fields than inland fields are still actively maintained and managed for wildlife habitat.

There are few absolutes when dealing with environmental forces. One is that man is rarely successful in engineering solutions that will completely ameliorate the impacts on an anthropogenic landscape. So why then do owners and managers of tidal rice fields continue to repair and maintain these systems in the face of increasing pressures? What are the alternatives to the tidal rice field system for wildlife habitat management? What are the societal forces that would entrain owners and managers to continue to repair tidal rice fields versus move back to inland fields where restoring dikes, canals, and water control structures is easier and less costly per acre? All of these questions are being considered informally across the coast of South Carolina.

Historically, real estate values have been the motivators for persistence in maintaining and repairing tidal fields through the recent impacts of increases in tidal amplitude and storm frequency. Tidal rice fields, when well managed for wildlife, can be valued at $8000 to $10,000 per acre. These values are significant when considering some tidal fields are hundreds of acres. The value of tidal fields where the dikes not intact and management not possible would diminish to approximately $250 per acre. Furthermore, the overall real estate of many of these plantations exists because of the presence of functional tidal rice fields on the property. Would land owners be willing to pay these prices if the tidal fields were not functional (i.e., dikes are breached, water control structures are not present)?

The attractiveness of the habitat to wildlife itself must also be considered in the question of tidal versus inland rice fields. In general, many water birds (waterfowl and other species) typically do not use the area of habitat in close proximity to the dike infrastructure. There is a band that surrounds these dikes where the habitat may be suitable, but wildlife do not use it with as great a frequency as the interior of these wetland units. This dynamic has not received much research, but it is hypothesized that water birds avoid dikes due to the disturbance of travel across the dikes.

Many acres are still accessible to water birds in a squarish-configuration of hundreds of acres of tidal rice fields. This is in contrast to inland fields that tend to be narrow and long. Might many inland fields be too narrow to allow full habitat utilization by wildlife without the hypothesized perimeter disturbance effect? Research on this has not been conducted but would be useful in better understanding how transitioning financial and logistical resources away from tidal and into inland fields might succeed.

Conclusion

Change is certain, whether it be social or ecological and often the two co-mingle and affect one another. Unlike the white rice field planters of the antebellum southern U.S., who were unable to envisage any other stable state, contemporary society must attempt to project future changes and evaluate mitigating solutions to possible risks. Historic rice fields, as a social-environmental system, are facing multiple changes due to climate change. As such, they are a prime case for evaluating society's ability to consider alternative scenarios where current social and environmental functions may change but large scale changes, or even functional loss, may be ameliorated. In this light, many stakeholder groups who are part invested in the socio-environmental system of historic rice fields are confronted with deciding how and if they should deal with increased sea levels and tropical storm activity. Understanding the magnitude of impacts and the nature of the resource lost to climate change is a wise first step in evaluating the type of mitigating response.

In the case of rice fields, we know their historical significance, their place in the cultural fabric of coastal South Carolina, and their importance as valuable wildlife habitat. How can these components of historic rice fields, or potentially a subset of these components, be maintained given the impending threats? Several questions come to mind: (1) What, of the socio-environmental attributes of the rice fields, should be the focus of conservation efforts?, (2) What are the possible implications of conserving one attribute while forgoing conserving another (i.e., what does a new alternate stable state look like?), (3) Who decides what attribute(s) to conserve?, (4)

What effect will possible changes in the environmental system have on the social system and vice-versa?, (5) What potential means are available to achieve conservation of the decided upon attributes to be conserved? These are not easy questions to answer given the complexity of either the social or environmental systems as separate entities, let alone when attempting to consider them as a single co-mingled and interacting system.

As a hypothetical example, it may be possible to shift the ecological function of the threatened tidal fields (due to sea level rise and increased tropical storm activity) to inland fields. In this example, inland fields could be cleared of timber, tree trunks removed, the land leveled, and water retention and control structures and devices installed to reclaim the forested land to its former rice field structure. As the structure of tidal fields changes and presumably functional loss occurs, the hope would be for these reclaimed inland fields to take on the structure and function of tidal rice fields. If this were to be successful, it would increase the resilience of the ecological system; however, there are likely cascading effects to the social system that must be considered. The loss of structure and function of the tidal fields, where the primary focus is on waterfowl hunting and their contemporary maintenance is provided by a relatively small group of landowners, will likely result in changes in the structure of monetary support for conservation efforts in the region. In other words, the structure and function of today's tidal fields and the resulting large scale conservation activities are primarily maintained by a small group of landowners. Without the available environmental structure to support waterfowl hunting, these landowners may forgo financial support for conservation across the region. Shifting the ecological structure and function of the tidal fields to the reclaimed inland fields will result in a different spatial arrangement and potentially different land ownership of these landscape features. The owners of these newly reclaimed rice fields may have a different conservation ethic than the owners of the "lost" tidal fields. What might this mean for the ecological structure and function? How might this impact the societal system within the region?

With inevitable change to both environmental and societal systems, some of which is foreseeable while others may not be, society should consider what the value of the various attributes of any given social-environmental system are, what maintains the system's structure and function, what are the possible threats, however likely, to the various attributes of the system, and what the structure and function of alternate stable states might be. Such efforts should provide various scenarios where the resilience of socio-environmental systems is increased. In doing so, society will hopefully be able to promote wise use of environmental resources that meet the needs of both contemporary and future societies.

References

Adger, WN, Hodbod J. (2013) Ecological and social resilience. In: Handbook of sustainable development, pp 91–102. https://doi.org/10.4337/9781782544708.00014

Anonymous (2019) Flooding and sea level rise strategy. Charleston SC

Cappucci M (2020) The week started with major coastal flooding in Charleston. The weather was beautiful. Washington Post, September 22, 2020

Carney JA (1996) Landscapes of technology transfer: rice cultivation and African continuities. Technol Cult 37(1):5–35

Catesby M. The natural history of Carolina, Florida and the Bahama Islands: containing the figures of birds, beasts, fishes, serpents, insects, and plants: particularly the forest-trees, shrubs, and other plants, not hitherto described, or very incorrectly figured by authors. together with their descriptions in english and french. to which, are added observations on the air, soil, and waters: with remarks upon agriculture, grain, pulse, roots, & c. To the whole, is prefixed a new and correct map of the countries treated of London: Printed at the expence of the author and sold by W. Innys, R. Manby, Mr. Nauksbee and the author, 1743

Coclanis PA (1989) The shadow of a dream: economic life and death in the South Carolina low country, 1670–1920. Oxford University, New York

Doar D (1970) Rice and rice planting in the South Carolina low country. Charleston Museum, Charleston, SC

Dusinberre W (1996) Them dark days: slavery in the American rice swamps. Oxford University, Oxford

Edelson SM (2006) Plantation enterprise in Colonial South Carolina. Harvard University, Cambridge

Ferguson L (1992) Uncommon ground: archaeology and early African America. DC, Smithsonian Institution Press, Washington

Fields-Black EL (2008) Deep roots: rice farmers in West Africa and the African diaspora. Indiana University Press

Forster M, Smith SD (2011) Surviving slavery: mortality at Mesopotamia, a Jamaican sugar estate, 1762–1832. J Roy Stat Soc 174(4):907–929

Gunderson LH (2000) Ecological resilience—in theory and application. Ann Rev Ecol Syst 31(1):425–439. https://doi.org/10.1146/annurev.ecolsys.31.1.425

Haldon J, Chase AF, Eastwood W, Medina-Elizalde M, Izdebski A, Ludlow F, Middleton G, Mordechai L, Nesbitt J, Turner BL (2020) Demystifying collapse: climate, environment, and social agency in pre-modern societies. Millennium 17(1):1–33. https://doi.org/10.1515/mill-2020-0002

Hanks RD, Baldwin RF, Folk TH, Wiggers EP, Coen RH, Gouin ML, Agha A, Richter DD, Fields-Black EL (2021) Mapping antebellum rice fields as a basis for understanding human and ecological consequences of the era of slavery. Land 10:831. https://doi.org/10.3390/land10 080831

Harper F (1998) The travels of William Bartram, Naturalist. University of Georgia, Athens

Heyward DC (1993) Seed from Madagascar Columbia. University of South Carolina

Hilliard SB (1975) The tidewater rice plantation: an ingenious adaptation to nature. Geosci Man 12:57–66

Hook DD (1993) Wetlands: history, current status, and future. Environ Toxicol Chem 12(12):2157–2166. https://doi.org/10.1002/etc.5620121202

Keck M, Sakdapolrak P (2013) What is social resilience? Lessons learned and ways forward. Erdkunde 67(1):5–19. https://doi.org/10.3112/erdkunde.2013.01.02

Kemble FA (1984) Journal of a residence on a georgian plantation in 1838–1839. University of Georgia, Athens

Marsooli R, Lin N, Emanuel K, Feng K (2019) Climate change exacerbates hurricane flood hazards along US Atlantic and Gulf Coasts in spatially varying patterns. Nat Commun 10:3785

Merrens HR (1972) A view of coastal South Carolina in 1778: the journal of Ebenezer Hazard. South Carolina Histor Mag 73(4):177–193

Mizzell H, Griffin M, Malsick M (2020) South Carolina hurricanes comprehensive summary. South Carolina Department of Natural Resources

Rosengarten T (1998) In the master's garden. Art and landscape in Charleston and the Lowcountry. Spacemaker Press, Washington, pp 1–21

Smith HR (2019) Carolina's golden fields: Inland rice cultivation in the South Carolina lowcountry, 1670–1860. Cambridge University Press

Tibbetts J, Mooney C (2018) Sea level rise is eroding home value, and owners might not even know it. Washington Post, August 20

The trans-Atlantic and Intra-American slave trade database. www.slavevoyages.org

Vernon AW (1993) African Americans at Mars Bluff, South Carolina. Baton Rouge, Louisiana State University

The Short- and Long-Term Effects of an Early Medieval Pandemic

Merle Eisenberg and Lee Mordechai

Abstract This article examines short- and long-term governmental policy responses to the effects of the Justinianic Plague (c. 541–750 CE). While many studies have linked the Justinianic Plague to significant changes across all sectors of life, they overlook how states responded to the pandemic's impact at different temporal scales—from immediate reactions to medium term politics. First, we discuss the immediate state responses to the initial outbreak in Constantinople in 542 at a micro-scale, which included measures to bury large numbers of dead. Second, we investigate the effects over a five-year time frame following the first outbreak to understand how the state responded to potential impacts through fiscal and economic policies. And, third, we reflect upon the post-five year changes scholars often connect to the plague outbreak to reveal the deep difficulties in making in such linkages.

Keywords Justinianic plague · Pandemics · Disease · Late antiquity · Early middle ages · Roman empire

Introduction

As a near endless mine of events, causes, and effects, the past is an attractive source for lessons to learn from and to apply to ongoing situations today. It is an obvious truism that less 'standard' written evidence tends to exist the deeper we look into the past, but paradoxically the stories told using what scant evidence we have grows more sweeping on every conceivable scale. The effects of disease are no different. The 1918 Influenza Pandemic killed perhaps 50–100 million people worldwide, but until the 1990s few people were even aware of its devastating demographic effects, since it was soon forgotten (Crosby 1989; Beiner 2021). Even today its political and economic effects are considered minimal and instead scholarship has focused on the

M. Eisenberg (✉)
Oklahoma State University, Stillwater, OK, USA
e-mail: merle.eisenberg@okstate.edu

L. Mordechai
Hebrew University in Jerusalem, Jerusalem, USA

© The Author(s) 2022
A. Izdebski et al. (eds.), *Perspectives on Public Policy in Societal-Environmental Crises*, Risk, Systems and Decisions, https://doi.org/10.1007/978-3-030-94137-6_19

localized experience of the pandemic (Bristow 2012). In contrast, the sixth-century Justinianic Plague (c. 541–750 CE) has only a tiny fraction of the evidence, but is nonetheless said to have caused massive political, economic, and cultural effects that eventually resulted in major historical events such as the fall of the Roman Empire or the rise of Islam (e.g. Harper 2017).

These supposedly drastic effects have drawn significant attention, especially in the context of the Covid-19 pandemic. Although popular articles in the media have used earlier pandemics such as the Justinianic Plague to contextualize Covid, the vast majority of these are superficial and repeat known truisms that do not align with historical facts. In other words, the past—or actually an exaggerated version of it—is used to frame discussions in the present. Such approaches, however, hardly offer useful ideas for contemporary policy since they tend to reinforce the writer's pre-existing ideological goals. More broadly, the very assumption that pandemics in the present somehow follow past pandemics (often in ways that are never explicitly stated) should be openly questioned. The social and ecological contexts of how and why a disease spreads are pivotal to understanding its impact. This is even true when pandemics are caused by the same pathogen, or happen within a similar ecological, technological or social context. For example, the first plague pandemic (mid-6th to mid-eighth centuries) shared little in common with the second plague pandemic (fourteenth-eighteenth centuries), and both were distinct from the third plague pandemic at the turn of the twentieth century (Eisenberg and Mordechai 2020).

This chapter focuses its discussion on a specific past pandemic to examine its implications for the present. As with any other case study, this early medieval pandemic cannot offer direct lessons for our own situation because the states and societies of the sixth century are fundamentally different from those of the twenty-first century. A careful analysis of this case study, however, reveals several similarities with the Covid pandemic. It also demonstrates the methodological issues in drawing conclusions from the medium and long term effects, and the difficulties of applying these conclusions to our present.

Setting the Scene: The Roman Empire and the Outbreak of Plague

The Justinianic Plague was part of the first plague pandemic, a general concept that encompasses epidemic disease outbreaks that occurred between the mid-sixth and mid-eighth centuries in western Eurasia. Some, but not all, of these outbreaks were related, and some, but likely not all, were caused by the pathogen *Yersinia pestis*, which causes plague. Although earlier research has claimed that as many as 100 million people died in this pandemic, several recent studies have argued for a far lower death toll (Mordechai et al. 2019; Mordechai & Eisenberg 2019).

Practitioners today define three major manifestations of plague in humans, all caused by the *Y. pestis* bacterium. The most common—and best known—is bubonic plague, named after its buboes—large swellings near the lymph nodes often in the armpit or the groin. Mortality occurs on average just a few days after the symptoms appear in 30% or more of untreated cases, although antibiotics today can drastically reduce mortality. The two other types of plague are pneumonic plague and septicemic plague. Although both cause mortality much faster and their case fatality rates often reach 100%, only a small fraction of humans develop pneumonic plague, which is the only way in which people can infect each other directly. Septicemic plague is rarely discussed in the context of the Justinianic Plague because of its rarity and lack of discernible symptoms in the historical sources (CDC 2020).

The Justinianic Plague (c. 541–544) reached and affected multiple states, but most of our evidence comes from its most severe outbreak in the Roman Empire. The empire's political and social institutions had originally developed in Rome and Italy centuries earlier, but following the Roman conquests of the Mediterranean world, they continued to operate in what is today the Balkans, Turkey, and the Middle East. By the mid-sixth century, the empire's capital was permanently located in Constantinople (modern Istanbul).

The emperor Justinian (r. 527–565), during whose reign the pandemic reached the empire, pursued multiple large-scale projects to solidify his authority before the outbreak of the plague. Within a few years of gaining the throne, he led a major reform of Roman law and rebuilt the great church of Hagia Sophia in Constantinople, making it the largest building in Western Eurasia for centuries. Hoping to restore Roman territories in the central Mediterranean that had been lost 50 years earlier, he sent armies to reconquer North Africa, Sicily, and Italy. His generals conquered the first two quickly, while the initial phase of the war in Italy concluded favorably. Yet at the same time as his armies expanded the empire's frontiers, a long-standing conflict with the Persian Empire (centered in modern-day Iraq and Iran) reignited on its eastern frontier. Although the conflict was usually limited to border regions, in 540 a large Persian raid sacked and burned Antioch, the Roman Empire's third-largest city (situated at the northeast corner of the Mediterranean), deep within Roman territory (Heather 2018).

Then, the plague hit the empire. While the origin of the plague did not occupy the minds of contemporaries, modern scholars have focused an inordinate amount of attention on this question and now believe it came from central Asia through Indian Ocean trade networks (see Eisenberg and Mordechai 2019 for the literature). Over the next three years, plague spread around the Mediterranean. Few specifics are known about its effects in most of the empire's territory, although it appears to have hit areas in Egypt and Syria. The most significant outbreak was in Constantinople (Mordechai and Eisenberg 2019).

Similar to other pre-modern states, the social contract binding together the Roman Empire was inherently different from present-day developed countries. The state was obliged to provide only minimal services, most importantly security from foreign enemies in the form of a standing army, which it funded using taxes. Centralized expenditure on infrastructure was minimal, although the state maintained a public road system and a state-only postal system. The state might help build city walls, aqueducts or churches in specific locations, but these investments were exceptional and the burden of maintaining them would generally fall on local elites. The state did not offer any institutionalized and centralized health, education, or welfare system.

Constantinople was a partial exception to these rules. Emperors and the central government maintained the public structures of the capital—defense systems, water infrastructure, and cultural institutions such as churches. The state also supported, at considerable expense, the *annona* system by which it supplied free (basic) food to the city. As the hallmark of the system, the government bought huge amounts of grain in Egypt (an imperial province), moved it by sea to Constantinople, and distributed it among its citizens. In this manner, the state maintained a much higher population in its capital than could be sustained by the resources in and around it.

As a consequence of the state's variegated system, the experiences and expectations of the people in the empire would have been radically different, depending on where they were. Rural provincial inhabitants normally had minimal interaction with the state. The state was more visible in provincial urban settlements, but had a substantial presence only in the empire's largest cities. At the same time, even in the capital—filled with monuments, ceremonies, and officials—expectations from state were far more limited compared to the present.

As a result of the different social contract between the state and its citizens, there is no indication that the state directed a coordinated response to the onset of plague nor was it even expected to mount any type of response. The state had information about the plague, its spread, and its effects but does not appear to have done anything with this information to reduce the potential impact on its citizens (Malalas 481). Although the outbreak in Constantinople occurred about a year after the early outbreaks in Egypt and southern Palestine, and although the slow movement of the pandemic allowed observers in the capital to trace its advancement, the imperial government did not prepare for the pandemic (John of Ephesus 93). There is no evidence that it attempted to slow traffic, come up with a response plan to plague's arrival in the capital or alert its citizens. At least some of these must be attributable to the novelty of the event, alongside that epidemic disease response was not within the purview of an early medieval state.

Within this context, we divide the impacts of plague on the early medieval societies it affected into short- and long-term effects. The short-term section focuses on immediate changes that can be directly attributed to plague, particularly the public reactions to the government's efforts to bury the increased number of dead in Constantinople. The long-term section focuses on the potential secondary and tertiary effects of the first outbreak. They include, among others, economic, and political impacts.

Short-Term Responses to the First Plague Outbreak in 542 C.E

Public Reaction

Once the plague appeared in Constantinople in the spring of 542, the inhabitants of the city tried to make sense of what was happening through their own beliefs and knowledge of the time, which seem strange and irrational from a 21st century perspective. Yet, these ideas are, in many respects, no more or less rational than contemporary responses to disease outbreaks, such as HIV or Covid (for HIV, McKay 2017; for Covid, Gillespie 2020; van Doremalen 2020). To use today's terms, these late antique beliefs are equivalent to conspiracy theories or misinformation, all of which flourish in contexts where reliable information is partial and ambiguous. Such beliefs have an adaptive function—whether by offering believers agency (or the illusion of agency), or by helping them make sense of the world around them.

Observers of the plague outbreak tried to explain the ways in which an individual was infected. Beliefs included demons in human form that infected humans by touching them in the real world or in their dreams. Another common idea in Constantinople identified monks and clerics—who were identifiable by their appearance—with death. Rather than seeking solace from these religious authorities, the local population fled from them. In response, monks and clerics walked around the streets less often (John of Ephesus 108–109). Although the state and church were concerned with correct religious belief and in other contexts were willing to commit significant resources to enforcing it, we hear of no attempts to correct this unorthodox belief, which persisted for two years.

The underlying cause of the pandemic was debated inconclusively at the time. Many attributed it to the divine, but others argued the plague was the result of malignant demonic activity. Both beliefs circulated together, and some authors of the time suggest both were at fault. For example, one author described plague as "[divine] chastisement" but also accepted the testimony of others who claimed to have seen fearful ghost ships of bronze—complete with headless corpses—which spread plague along the Mediterranean coasts (John of Ephesus 82–83).

Compared to experiences with Covid, which focused on its geographical origins, the origins (whether proximate or ultimate) of plague did not particularly interest late antique contemporaries. One sixth century writer (Evagrius 4.29) mentions Ethiopia, a country stereotypically associated with diseases, while a few noted only the first place it affected within the empire—Pelusium, a port town in northeastern Egypt (e.g. Procopius, Wars 2.22). Other authors remain silent, since they either do not know or think it was important.

Somewhat surprisingly, historical sources refer to only a few beliefs that were meant to cure or get rid of plague during the initial outbreak (Mulhall, 2021). One such belief was the conviction that if the city's population broke ceramic pitchers by throwing them down into the street, the noise would scare death (i.e. plague) away from the city. This rumor allegedly convinced women in particular (John of

Ephesus 108). Attempts to prevent infection are similarly few and concentrated on staying away from other people. Some people in Constantinople appear to have locked themselves at home—whether as an attempt to escape demons or contagion (John of Ephesus 101). When plague reached southern Asia Minor, the local farmers refused to bring their goods into the city of Myra since they feared infection, likely preserving themselves but causing a food shortage in the city (Life of St. Nicholas of Sion 52). Institutional response was rare. In Roman-controlled Italy during a later outbreak, for example, Pope Gregory the Great instituted a religious procession to atone for sins that caused the plague outbreak and prevent future infection (Gregory of Tours 10.1). From a modern biomedical perspective, none of these attempts to stop plague worked. They did, however, offer some immediate comfort or perceived agency to people at the time, a practice not much different from immediate reactions to Covid, and some—such as de facto self-isolation—may have inadvertently worked.[1]

State Response in Constantinople

As with any pandemic, the larger and more diverse geographically a state is, the higher the variability of the disease's impact. Each location during the Justinianic Plague—from a regional to a neighborhood scale—was struck differently. Settlement patterns (urban vs. rural) and ecological factors (e.g. local species that could host and transmit plague) played a significant role in determining the more local effects of the pandemic, although these can rarely be known in detail. Crowded concentrations of humans in cities would typically increase the disease burden on urban populations even if plague's impact on individual cities would have varied considerably and some may not have been significantly affected at all. In some cases, a series of (mostly unknown) contingencies could cause a significant outbreak. Such a "perfect storm" appears to have happened in Constantinople in 542, drawing by far the most attention from contemporaries relative to any other outbreak during the first pandemic.

The plague outbreak in Constantinople led to an unusual, but significant central-ized governmental response (by late antique standards) to alleviate a major immediate problem—how to handle the increased number of corpses. Such burial problems frequently occur in pandemics or other mass mortality events (Crosby 1989; NYC 2020). As the death toll increased, observers were dismayed that even the corpses of the elite were not buried (Procopius, Wars 2.23.5). Emperor Justinian, therefore, appointed an administrator named Theodoros to manage the process of disposing of corpses. This decision broke with tradition. Burial in the empire was the responsi-bility of a person's relatives, communal organizations such as burial associations, or the church, who buried the poor (Rautman 2006: 10–12, 78). The state, however, did not usually interfere in burial practices and normally had no capacity to complete this task. Justinian's immediate decision, therefore, was a necessary adaptation.

---

[1] Note that self-isolation would have been less effective than during e.g. the Covid pandemic if the vector were rodent fleas.

Theodoros oversaw the operation in person and upon request, he would send his men to collect corpses. His presence on the streets also enabled him to act quickly. On one occasion, he and his staff came upon a locked house that gave off a strong stench. Breaking in, they found approximately twenty decomposing corpses, which they had removed by generously paying day laborers (John of Ephesus 101). The corpse disposal service Theodoros offered to the city's inhabitants, and the economic incentive he offered to his workers resulted in the successful collection of corpses, but the more pressing problem of what to do with the corpses once they were massed together remained.

Theodoros first tried to appropriate existing graves that belonged to the city's citizens. This was another break with tradition since the state did not, under normal circumstances, infringe upon its citizens' property. It became quickly clear, however, that the existing graves were insufficient for the task. At this point, officials likely realized that providing services (i.e. burying corpses) was not enough, and that they would have to adapt by constructing the necessary infrastructure to deal with this problem. While Justinian ordered the production of new litters to carry corpses and dig new graves, Theodoros instructed his teams to dig large ditches. The patterns of public participation in the project suggest social cohesion, and when combined with financial incentives the city gained the necessary labor to dispose of bodies (John of Ephesus 100).

Unfortunately, as a first solution Theodoros' initiative was inadequate—digging new graves took too long while the corpses continued to accumulate. Theodoros therefore decided to send the corpses elsewhere. After dumping some corpses into the surrounding water, he decided to send most of them to the nearby suburbs, alleviating some of the problem. They were first placed in fortifications in the city's suburbs, but problematically the changing wind carried the stench back into the city (Procopius 2.23.9) forcing Theodoros to find another solution. Soon afterwards, Theodoros decreed the digging of huge graves—our sources claim these could hold 70,000 corpses, an exaggeration—which finally resolved the burial problem successfully (John of Ephesus 100–101).

Theodoros employed some individuals on a more permanent basis based on their skills and his needs, while others were hired as non-specialized labor. There does not appear to have been any attempt of the state to coerce the population into working in exploitative conditions. In fact, it appears that the pay for corpse removal—ranging from carrying corpses to digging graves—was generous and immediate, since Theodoros placed paymasters who disbursed a standard payment to anyone who arrived with a load of corpses (John of Ephesus 100–101). Supposedly, many workers preferred to work for less remuneration as they scorned receiving pay for their labor in this context and some went as far as to reject all payment (John of Ephesus 104; 103).

A final point regarding the burial issue in Constantinople and cooperation it facilitated concerns the two main rival factions in the city. Each was well organized and could mobilize its followers to show support, for example in context of the popular local chariot races, and sometimes to conduct organized or semi-organized violence, often against each other. Despite their history of past (and future) strife, both factions cooperated with the state in the burial efforts (Procopius 2.23.13; Cameron 1976). Whether the initiative came from the factions themselves or the state is unclear, but it demonstrates that domestic differences and rivalries were temporarily shelved until some measure of stability might be achieved.

Annual to Decadal-Scale Responses

State Effects in the First Five Years: The 540s

A bureaucratic and relatively centralized state by premodern standards, the Roman Empire maintained itself through assuring the continuity of its government and administration, collecting taxes, and fielding armies (Haldon 2016). None of these core functions were severely impacted by the Justinianic Plague and the government continued operating as before. We know of no administrators who died during the Justinianic Plague, and although the emperor Justinian himself fell ill, raising some concern in the military, he recovered and continued to govern (Procopius, *Secret History* 4.1–3).

There is no concrete indication that the state's ability to collect taxes was affected as a result of the plague. After other disasters such as earthquakes, the Roman government sometimes offered afflicted regions limited tax relief or other forms of support. Their targeted use suggests the government maintained, or at least could repurpose, existing surpluses (Mordechai and Pickett 2018). After the first plague outbreak, however, there is no evidence for any change of governmental policy, or any indication of tax relief, to meet a seemingly more significant disaster. The single exception is a law that attempted to ensure laborers were not receiving more money than they had previously received for their work (Meier 2016). Scholars have argued that the plague could have reduced the amount of taxes raised overall and therefore may have halted the empire's military campaign against Persia during the initial plague outbreak (Kislinger and Stathakopoulos 1999). However, the empire's two active major wars in Italy and Persia soon resumed and continued for over another decade. Moreover, less than a decade after the first plague outbreak, the empire sent a new military expedition to Spain, on the far side of the Mediterranean, where it conquered the southern part of the peninsula.

The onset of plague did coincide with the disruption and reduction of the state-managed grain supply that fed Constantinople. Whether this was due to less demand because of mortality or emigration, a supply problem from Egypt, or a temporary reduction to conserve imperial resources is uncertain, since the sources only provide numbers but do not explain the fluctuation (Zuckerman 2004). Regardless, as with the military efforts, the supply of grain to the capital returned within a few years to its normal amount and, within five years, had surpassed pre-plague amounts. Whatever the cause of the reduction, the state was able to fulfill its grain requirements without significant difficulties after a few years. The economic impact of the plague was small enough that the state could maintain a key tenet of its governing ideology.

Another adjustment at the state level appears to have been monetary. The empire used a tri-metallic monetary system, which included gold, silver, and bronze coins. In 538, Justinian instituted a monetary reform, which increased the size and weight of the empire's bronze coins—which were predominantly used for commerce. When the plague reached Constantinople four years later the imperial fisc partially rolled back the reform. New coins were smaller and lighter, although still larger and heavier than their pre-reform counterparts. The value of the coins did not change during the reform or after the plague—their denomination and face value remained constant (Stahl 2021). This suggests that difficulties in acquiring, mining or moving enough bullion to maintain the larger, heavier reform weight standards—any of which may have been an effect of the plague—could have been the reason for not pursuing the reform further. This coinage change was centrally ordered and controlled, and was probably an attempt to alleviate worse economic damage.

From a macro-governance perspective, the state continued to function while meeting its immediate needs during the worst years of the plague outbreak. The Roman state withstood the stress of the pandemic and managed the immediate crisis without much difficulty. Even during a severe outbreak in the empire's densely popu-lated capital and without any biomedical response, the threat to the state ideology, system, and its structures was minimal. Based on the three examples above—the mili-tary budget, the grain supply, and the monetary reform—it appears that the central government could mitigate the most pressing problems through minor governance changes, additional outlays of resources, and economic fixes.

Effects on a Decadal Scale: Causality and Caution

Plague's impact is far less clear on the decadal scale across the broader Eurasian world from the 550s onward. This is partially due to the fewer written sources and the less data that exist on the topic outside Constantinople. But it is also due to how much we can plausibly link plague—or any pandemic—with long-term changes in society. Since plague's effects differed based on ecological and human context, outbreaks varied to a great extent—with some severe and some negligible, and none of the reasons behind this impact have yet been disentangled (Echenberg 2007 for the Third Pandemic as one example; Eisenberg and Mordechai 2020 for all three pandemics). Overall, human decisions based on power and authority played an overlooked but crucial role.

Of the spectrum of large-scale changes that certainly occurred from the mid-sixth century onward across the broader Mediterranean world, some may have been related to early medieval plague outbreaks. In some cases, the pandemic continued and perhaps accelerated existing trends such as the decline in political, economic, and cultural power that started in the late fifth century due to the fragmentation of the Roman Empire in the western Mediterranean. However, the causal connection between the initial outbreak of plague in the 540s and these changes is impossible to prove one way or the other. The city of Arles in southern France, for example, had long been a key regional center because of its location at the mouth of the Rhone River, a vital trade conduit from the Mediterranean northward into the rest of France. While Arles certainly falls off the map in importance from the 550s onward, its decline as a political, economic, and cultural hub had begun decades before the plague. The immediate effects of plague on the city are unclear. Mortality in Arles was probably higher compared to locations further northward such as Clermont, but we hear of no implementation of temporary burial measures as in Constantinople (Gregory of Tours 4.5). By the late sixth century, power continued to shift northward to what is now the Paris region, while France's commercial links to the Mediterranean continued to decline (Klingshirn 1994; Wood 1994). Plague may have exacerbated the slow breakdown of trade links northward or further made the city less politically vital, but none of these changes can be attributed to the plague alone or even the plague as a significant factor. It hardly caused a systemic collapse (McCormick 2021). Ultimately, the causal connection between plague and a transformation of a city in its broader regional importance—if even extant—was a slow process that took over a century.

The most basic problem for outbreaks of plague on a regional and long-term scale is that no contemporary writer directly linked a single outbreak to transformation, resilience or decline. This contrasts with the way the same authors described other disasters. Earthquakes or fires, for example, were used as causes for the decline or destruction of cities (e.g. Malalas 419–422). Significant disease outbreaks can certainly accelerate existing trends—such as in Arles—but they do so as part of a longer, gradual process for social, economic, and cultural reasons in which plague may play a role, while political power, will, and agency remain key factors. It is undoubtedly true that many regions in the post-Roman west became less populated and less economically productive over the sixth century, but to link these changes causally to plague is problematic at best, and impossible at worst (Wickham 2005 for an alternative explanation).

Conclusion

As perhaps might be expected in evaluating institutional response to complex events, the response of the Roman government to plague was mixed. Although contemporaries had experienced disease and epidemics on multiple occasions, the original plague outbreak in 541 was unprecedented in its scale and mortality. There is no indication that the central government used its resources and capabilities to prepare for plague or mitigate its impact in advance. While we should not expect a modern biomedical response, the absence of significant religious or cultural responses appropriate for the times is puzzling. Other, smaller disasters in the mid-sixth century did in fact result in notable public, imperial responses. For example, Justinian publicly mourned and did not wear his crown for a service after an earthquake hit the important city of Antioch, while after another earthquake hit Constantinople he removed his crown and did not wear it for 30 days (Malalas 421, 489). It appears therefore that the empire did not utilize its relative advantages when responding to the pandemic. The lack of a formal top-down administrative response to plague contributed to the diversity of local responses, only a fraction of which are preserved in our sources.

It is only when glimpsed from our view 1,500 years later with the ability to retrospectively connect causes and effects centuries in the future that the Justinianic Plague is linked to long-term changes, such as the rise of Islam almost a century after the outbreak of the plague. Yet as a general rule, the further in time the perceived effects of plague are from the first and most significant outbreak itself, the less robust the causal connection is. To make up for the absence of explicit societal effects of outbreaks, scholars piece together different types of evidence from written sources to archaeological remains and ancient DNA, but pinning the causal explanation on plague alone is simply impossible given that dating via non-written sources cannot be linked directly to an annual outbreak or even a single decade. In the case of the initial outbreak, it is perhaps easiest and most accurate to argue that early medieval states were resilient to the pandemic, since none of them collapsed or underwent any transformational changes in the decade immediately after they experienced the worst outbreak in recorded history by that point in time.

As we have demonstrated above, the early medieval pandemic may have seemed like the end of the world to many contemporaries, who encountered an unknown situation and had understandable difficulties in comprehending it. Individual responses varied—some were more rational, others perhaps less so—but none of them threatened the existing socio-political order. The state returned to its pre-pandemic functions even if some prioritizing was necessary to adapt to the new reality, for example by giving up on an overly ambitious monetary reform. Elites do not appear to have lost their status. There were surely problems—some created by the government—and challenges—some overcome by its administrators. However, the Romans recognized there were times when social cohesion and support was more important than internal rivalries. The social contract between ruler and ruled remained strong—the fact that the central government reinstituted the grain shipments and maintained the distribution of free food in the capital supported the Roman economic-ideological system. As a result, the impact of these problems and challenges was limited on the state scale, even if their consequences would have been felt far more acutely on a local level, where administrative decisions could result in unnecessary deaths or unexpected survival. Communities and individuals likely suffered over the short-term, but in a manner reminiscent of the Influenza pandemic of 1918, these effects were soon forgotten in the Roman world.

References

Beiner G (ed) (2021) Pandemic re-awakenings: the forgotten and unforgotten flu of 1918–1919. Oxford University Press, Oxford

Bristow NK (2012) American pandemic: the lost worlds of the 1918 influenza epidemic. Oxford University Press, New York

Cameron A (1976) Circus factions. Clarendon Press, Oxford

CDC (2020) Plague home' CDC: centers for disease control and prevention. https://www.cdc.gov/plague/index.html Accessed 4 Mar 21

Crosby AW (1989) America's forgotten pandemic the influenza of 1918. Cambridge University Press, New York

Echenberg M (2007) Plague ports: the global urban impact of Bubonic plague, 1894–1901. New York University Press, New York

Eisenberg M, Mordechai L (2019) The Justinianic Plague: an interdisciplinary review. Byzantine and Modern Greek Studies 43(2):156–180

Eisenberg M, Mordechai L (2020) The Justinianic plague and global pandemics: the making of the plague concept. Am Hist Rev 125(5):1632–1667

Gillespie E (2020) Sanitize groceries, discard takeout containers immediately: Doctor demonstrates "sterile technique', FOX TV Digital Team. https://www.fox5dc.com/news/sanitize-groceries-discard-takeout-containers-immediately-doctor-demonstrates-sterile-technique. Accessed 4 Mar 21

Haldon JF (2016) The empire that would not die: the paradox of eastern Roman survival, 640–740. Harvard University Press, Cambridge, Massachusetts

Harper K (2017) The fate of Rome: climate, disease, and the end of an Empire. Princeton University Press, Princeton

Heather P (2018) Rome resurgent: war and empire in the age of Justinian. Oxford University Press, New York

John of Ephesus=Witakowski W (1987) The Syriac chronicle of Pseudo-Dionysius of Tel-Maḥrē: a study in the history of historiography. Almqvist & Wiksell International, Stockholm, Sweden

Kilgannon C (2020) As Morgues Fill, N.Y.C. to Bury some virus victims in Potter's Field. The New York Times. https://www.nytimes.com/2020/04/10/nyregion/coronavirus-deaths-hartisland-burial.html. Accessed 13 Apr 2020

Kislinger E, Stathakopoulos DC (1999) Pest und Perserkriege bei Prokop: chronologische Überlegungen zum Geschehen 540–545. Byzantion 69:76–98

Klingshirn WE (1994) Caesarius of Arles: the making of a Christian community in late antique Gaul. Cambridge University Press, New York

Krusch B, Levison W (eds) (1937) Gregorii episcopi Turonensis Historiarum libri X, in Monumenta Germaniae historica Scriptorum rerum Merovingicarum; t. 1, p 1. Hahn, Hannover

Malalas I (1986) Chronographia (published as The Chronicle of John Malalas), Byzantina Australiensia. Australian Association for Byzantine Studies, Melbourne

McCormick M (2021) Gregory of tours on sixth-century plague and other Epidemics. Speculum 96(1):38–96

McKay RA (2017) Patient zero and the making of the AIDS epidemic. The University of Chicago Press, Chicago

Mordechai L, Eisenberg M (2019) Rejecting catastrophe: the case of the Justinianic plague. Past and Present 244:3–50

Mordechai L, Eisenberg M, Newfield TP, Izdebski A, Kay JE, Poinar H (2019) The Justinianic Plague: an inconsequential pandemic? PNAS 116:25546–25554

Mordechai L, Pickett J (2018) Earthquakes as the quintessential SCE: methodology and societal resilience. Hum Ecol 46:335–348

Mulhall J (2021) Confronting pandemic in late antiquity: the medical response to the Justinianic Plague. J Late Antiquity 4(2):498–528

Procopius (1914) Wars. Loeb classical library. The Macmillan co., London and New York

Prokopios (2010) The secret history: with related texts. A. Kaldellis (trans.) Hackett Publishing, Indianapolis

Rautman ML (2006) Daily Life in the Byzantine Empire. Greenwood Publishing Group, Westport, Conn

Ševčenko I, Ševčenko NP (1984) The life of saint Nicholas of Sion. Hellenic College Press, Brookline, Mass

Stahl A (2021) Numismatic Evidence for the Pandemic of Justinian I', Presented at the 56th International Congress on Medieval Studies, Kalamazoo, MI

van Doremalen N, Bushmaker T, Morris DH, Holbrook MG, Gamble A, Williamson BN, Tamin A, Harcourt JL, Thornburg NJ, Gerber SI, Lloyd-Smith JO, de Wit E, Munster VJ (2020) Aerosol and Surface Stability of SARS-CoV-2 as compared with SARS-CoV-1. N Engl J Med 382:1564–1567

Wickham C (2005) Framing the early middle ages: europe and the Mediterranean 400–800. Oxford University Press, Oxford

Wood IN (1994) The Merovingian kingdoms, 450–751. Longman, New York

Zuckerman C (2004) Du village à l'empire: autour du registre fiscal d'Aphroditô (525/526). Association des amis du centre d'histoire et civilisation de Byzance, Paris

Migration and the Environment

The Integration of Settlers into Existing Socio-Environmental Settings: Reclaiming the Greek Lands After the Late Medieval Crisis

Georgios C. Liakopoulos

Abstract This chapter examines to what extent two late medieval nomadic groups in the southern Balkans adopted the economic practices of the areas they moved into, in order to achieve agricultural sustainability. In the fourteenth century, these two groups, Turk *yörük*s and transhumant Albanians, migrated to Greece in order to invigorate depopulated areas and reclaim lands in Thessaly and the Peloponnese respectively. Almost three generations after their establishment, Ottoman taxation cadastres cast light on their agricultural and pastoral activities. Even though these groups followed different trajectories in their sedentarisation—more or less dictated by their ethnocultural peculiarities—they both focused over time on farming basic, life-sustaining crops, such as cereals, which were complimentary to the manifold market-oriented farming activities of the long-settled local Greeks.

Keywords Nomads · Yörük · Albanians · Sedentarisation · Land-reclamation · Ottoman Empire · Greece

Introduction

The crisis of the Late Middle Ages that swept across Europe marked the fourteenth-century Balkans with political fragmentation, ranging from disputes between local lords or magnates to a larger scale of power shift among empires and various local polities. Almost constant unrest combined with the Black Death ravaged the demography and, consequently, the economy of the region (Fine 1996: 329–34; Varlık 2015: 107–12, 125; Kostis 2020: 303–20). This instability and shortage of agricultural labour caused an increasing demographic mobility, as local landlords tried to recruit landless people. In the big picture, entire vulnerable population groups fled their insecure war-torn homelands to seek a better life.

The menacing revenue shortfalls troubled the state authorities, which, in an organised fashion, invited or deported nomadic and transhumant groups to repopulate

G. C. Liakopoulos (✉)
Max Planck Institute for the Science of Human History, Jena, Germany
e-mail: liakopoulos@shh.mpg.de

© The Author(s) 2022
A. Izdebski et al. (eds.), *Perspectives on Public Policy in Societal-Environmental Crises*, Risk, Systems and Decisions, https://doi.org/10.1007/978-3-030-94137-6_20

regions and reclaim arable lands. This paper aims to explore the extent to which this was a wise practice and how two such groups, the nomads of Turkish origin (*yörük, tatar*) in Thessaly (central Greece) and the transhumant Albanians of the Peloponnese (southern Greece), exploited the natural resources of their new habitat, by analysing quantifiable fiscal data contained in contemporary Ottoman tax registers. The *yörük*s followed military chieftains and were then deported *en masse* by sultans from Anatolia into the Balkans, whereas the Albanians were invited to repopulate the countryside of southern Greece by the Venetians and the Byzantines. The Ottoman cadastres offer a cross-section of the two population groups after a period of between sixty and seventy years subsequent to their deportation or immigration to their new homelands, and thus furnish the historian with tools to investigate the level of sedentarisation required for undertaking agricultural activities. It should be noted that these groups were differentiated in the surveys only on the basis of their fiscal peculiarities (i.e. tax exemptions or reductions), not as distinct ethnic groups. The *yörük*s retained a more cohesive nomadic profile, while the Albanians seemed to have evolved to a sedentary level, as far as the circumstances allowed.

The Turkish Nomads in Thessaly

Around 1385, Thessaly witnessed the incursions of the Ottoman warlord Evrenos Beg, who appeared in the following winter of 1386–87 as the sovereign of the region (Beldiceanu and Năsturel 1983: 117–18; Savvides 1995: 38–40, 59–60; Kiel 1996: 114–15; Kiel 2013: 474). Under his command, nomadic groups of Anatolian origin infiltrated Thessaly, prior to the consolidation of Ottoman rule by Sultan Bayezid I in 1393. The topic of the use of nomadic tribes in the fields of colonisation and military institutionalisation by the sultanic authorities has been well addressed by the historians of the Ottoman Empire (Halaçoğlu 1991: 99–104; Yeni 2017: 188–91). One aspect that commonly gained currency is that the Ottoman state often resorted to the nomadic populations of Anatolia or farther eastern provinces to invigorate and safeguard newly conquered lands in the Balkans (İnalcık 1954: 125). Their areas of service included the military (auxiliary soldiers, raiders, guards of mountainous passes and bridges), transportation, rice cultivation, salt production, butter supplying and mining (İnalcık 1982: 104). Their contribution was so valued that the great economic historian of the period, H. İnalcık, considered them 'the backbone of the entire imperial organisation' (İnalcık 2014: 485). In the imperial chancery, these groups were referred to by the term *Türkmen* (Turcoman), for those present in eastern Anatolia; and *yörük* or *yürük* for those who had moved west of the Kızılırmak River, many of whom were funnelled at a later stage into the Balkans (Kasaba 2009: 21; Çetintürk 1943: 109; Barkan 1953–54: 209–13; Arıcanlı 1979: 30–31). On the basis of the registers of 1520–35, as many as 37,435 households (19.2% of the Muslim population) in the Balkans belonged to *yörük*s (Barkan 1957: 33).

In the first extant Ottoman taxation cadastre of central Greece (MM10 cadastre of Trikala) dated 1454/5, the *yörük*s are earmarked either by the note '*yörük*s inhabit' next to the place name or by the mention of *yörük* or Tatar cultivators of farmlands in the taxation section. Such entries are recorded in two fiefs: (a) one village and nine arable lands included in the fief of the governor of Trikala (MM10: 1b–60b), and (b) two villages and ten arable lands belonging to the fief of İbrāhīm and Yūsuf (MM10: 212a–221b) (Map 1). The village of Andriya Miḥal, belonging to the fief of Burāk, deserves our attention. Its headline reads as follows:

> This village was formerly inhabited; nomad Turcomans came and were settled (ordered by the authorities to settle); there are no infidels now; they did not even have grain sowed; it has a lot of lands; one could not reason with the *yörük*s. (MM10: 141b).

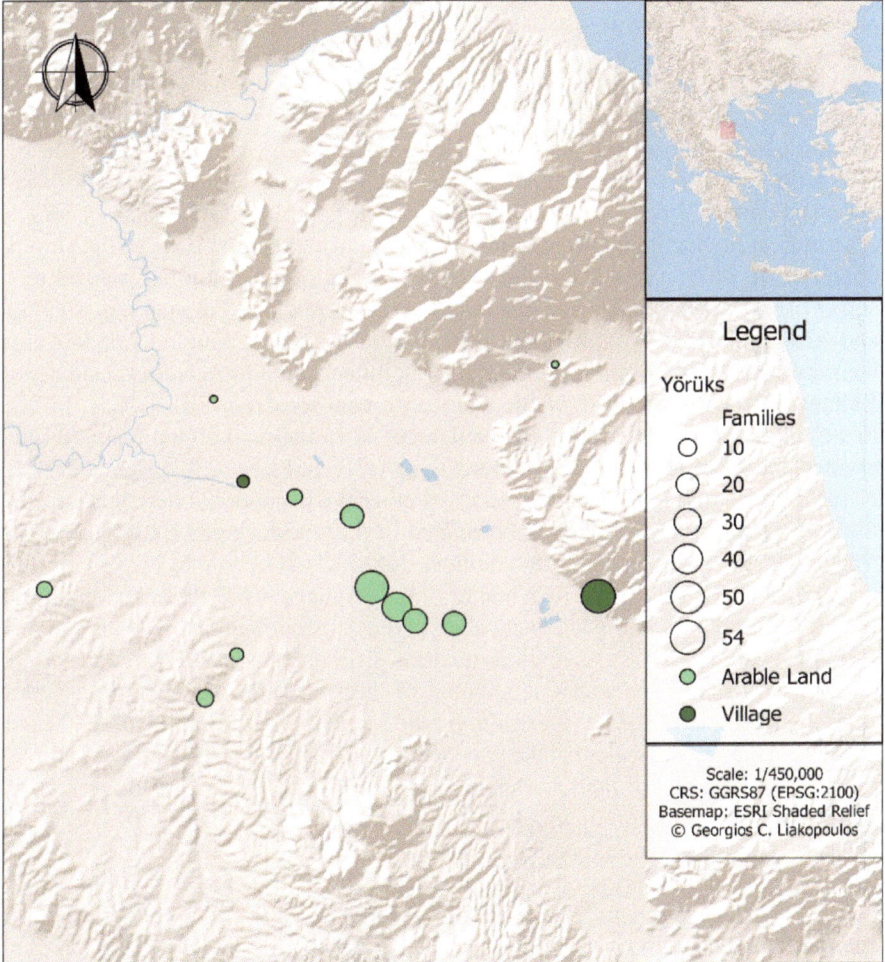

Map 1 *Yörük* settlements in Thessaly in 1454/5

Being on guard against anachronism, we should conceive the term *türk* as the equivalent of Turcoman or tribal, rather than an ethnonym (Beldiceanu and Năsturel 1983: 131; Ergul 2012: 634). The fields of this village were cultivated by outsiders, upon whom an annual lumpsum of one hundred silver coins (*akçe*s) was levied.[1] From this note, one can infer that the settlement of the nomadic tribes in the Balkans was not always carried out effortlessly; thorny issues arose both from the part of the *yörük*s and the local population. According to another cadastre (1530, TT167), the Christian villagers of Divlas in the Dojran district of central Macedonia had to resettle in neighbouring Dolna Ġırbas (Kato Sourmena) under pressure from *yörük* attacks (Coşkun et al. 2003: 276). The presence of nomads in an urban environment is, as expected, quite sporadic; one *yörük* is mentioned inhabiting the Bedrü'd-dīn Hoca quarter of Yeñişehir (Larissa) under the name Turhan the tailor (MM10: 55a).

Members of this group were by no means precluded from the fief-holders (timariots) class. Our list concludes with three *yörük* timariots: Muhammedī son of Altun, mentioned as a '*yörük* man' (MM10: 159b); and Mūsā and Nesīmī sons of Balcı, mentioned as members of the '*yörük* tribe' (MM10: 222b). These two cases illustrate a rather confined advancement of nomads to the military élite of local administration. The epithet 'from time immemorial' (*kadīmī*) employed for Muhammedī's fief is an indication that it had been initially granted during the first stage of the Ottoman conquest. In other cases, the cadastre clearly mentions that the fief-holders themselves or their forefathers came in Thessaly with Evrenos or Turahan Beg (Delilbaşı and Arıkan 2001: I, xx–xxiii). The Anatolian nomads who fought as raiders under the command of such military chieftains facilitated the Ottoman expansion in the Balkans in the second half of the fourteenth century (Yeni 2013: 185). In order to stimulate their immigration and settlement in Rumelia, the Ottoman authorities granted them arable lands, areas to set their tents and fiefs (Gökbilgin 1957: 15). N. Beldiceanu and P. Ş. Năsturel query whether the exceptional hereditary form of the early Thessalian fiefs should be related to these inducements (Beldiceanu and Năsturel 1983: 148; see also Moutafchieva 1988: 37–41).

Being a *yörük* was primarily a matter of tax privilege *vis-à-vis* the Ottoman state. It paid off to remain nomadic, even when perhaps engaging to some extent in cultivation of scattered fields. An obvious taxation difference which favoured the *yörük*s, as opposed to their sedentary neighbours of the same religion, was the exception from the payment of the *ra'iyyet kulluğu*, the regular agricultural tax of the Muslim subjects, a money equivalent of the *corvée* due to the fief holder, which was estimated on the basis of a plot of land workable by a pair of oxen, unless they were recorded as settled (*yerlü*) in a certain arable land (İnalcık 1959: 581; İnalcık 2014: 472). Our cadastre mentions nominally only ten such *yörük* families which evolved to a sedentary stage: seven in Sarıhānlu (Modestos) (MM10: 216b), one in Çullular (Melia) (MM10: 216b) and two in Sakallu (Melissa) (MM10: 217a). When it comes to transhumant nomads with no connection to the earth, the only mention given, as stated above, is the number of the cultivators. The Germiyanlu (Prinia) village is located,

---

[1] The Ottoman silver coin, *akçe* or *asper*, contained 1.01 grams of silver in the 1450s and 0.96 in the 1460s (Pamuk 2000: 46).

Table 1 Percentages of agricultural and pastoral taxes in Thessaly in 1454/5

Taxable asset	Yörüks (%)	Sedentary Muslims (%)	Greeks (%)
Cereals	94	75.72	38.93
Vineyards	1.04	0.31	6.85
Other cultivations	0.52	5.09	8.21
Animal husbandry	4.44	18.88	46.01

according to its heading, 'amid the sheep-breeders' (MM10: 215a). Despite the fact that the toponym refers to the homonymous principality in western Anatolia, a region that historically staffed with *yörük*s the Ottoman raider class in Rumelia, this village is not included in the enumeration of the *yörük* settlements, since no inhabitant is earmarked as such. The same holds true for Emīrhanlu *alias* Aydıñlı (MM10: 221a–221b); the anthroponymic study showed that at least one of its dwellers, Aṣlıḥan the tailor son of Tatar Maḥmūd, was of nomadic provenance. Finally, the rice-cultivators village Çeltükçi *alias* Ḥallāc Ḥamza (Neohori) (MM10: 215b) does not record any *yörük*s, even though the engagement of this group in risiculture is well documented (İnalcık 1982: 103–6).[2] The headman of the rice cultivators, ʿAlī Faḳīh, was most probably the chieftain of the neighbouring ʿAlī Faḳīlar (Kalamaki) village, where fifty-four *yörük* farmers are recorded (MM10: 215b–216a).

The nomads of Thessaly appear to combine agricultural and pastoral activities. The absolute majority of the taxes levied on the two hundred and forty-three families recorded in designated *yörük* localities, excluding the head taxes on those settled, belongs to the cultivation of cereals (94%) (Table 1; Appendix Table 5).[3] Nomadic groups with a rather loose engagement in farming tended to focus on staple crops. Within the context of self-consumption, they cultivated small plots in the vicinity of their pastures or abandoned lands to secure the supply of grain for their own diet and for fodder. Cereal production is almost equally divided between wheat (730.5 *kile*; 53.95%) and barley (615 *kile*; 45.42%) (Appendix Table 6). Their total agricultural income is eked out by a mere 1.04% of viticulture and 0.52% of cotton

---

[2] Cf. the nomads' contribution of 36.57% (7640-*akçe* tithe) to the total rice production (20,890-*akçe* tithe) of Adana's Kınık district in 1525 (Kurt 2004: xlv).

[3] The calculation of the exact cereal yield in kilograms is hindered by the fact that both the metric equivalent of the unit of measurement for weight (*kile*) and the actual percentage of the tithe (ʿöşr) remain uncertain for the specific province in the mid-fifteenth century. The cadastre mentions that two *kile*s equal one load (*yük*) (MM10: 7b); the latter may vary between 150 kg and 205.4 kg (Beldiceanu and Năsturel 1983: 106; Hinz 1955: 14, 36; İnalcık 1983: 330). Moreover, one cannot ascertain whether the *kile* employed refers to the one of Trikala (*kile-i Tırḥala*) or the one of Larissa (*kile-i Yeñişehir*). In 1506 the *kile* of Trikala equalled two *kile*s of Istanbul or 40 *okka*s, that is a total of 51.312 kg, and the one of Larissa equalled two *kile*s of Trikala (Beldiceanu and Năsturel 1983: 106; Taşkın 2005: 70). On the other hand, the tithe could fluctuate between 10.5% and 13.33% of the production, depending on the inclusion of the *sālāriyye* or *sālārlık* surtax (Alexander 1985: 490).

Table 2 Percentages of pastoral taxes in Thessaly in 1454/5

Taxable asset	Yörüks (%)	Sedentary Muslims (%)	Greeks (%)
Swine	0	0	23.47
Sheep	43.43	42.25	58.36
Apiculture	56.57	57.75	18.17

and madder (Table 1; Appendix Table 7). The latter two manufacturing productions should be construed in relation to the renowned nomadic kilim- and carpet-making. It is noteworthy that madder is not recorded among the taxable assets of either sedentary Muslims or Greeks. Both nomadic and sedentary Muslims present a similar ratio of apiculture to sheep breeding, namely 1:0.77 for the yörüks and 1:0.73 for the settled Muslims (Table 2). One would expect a more substantial contribution of animal husbandry to the yörüks activities than 4.44%. This low figure may be attributed to the 'relative invisibility' of certain yörüks; namely, those tribes who were not recorded in the cadastral surveys by virtue of pursuing exclusively pastoralism and not engaging in farming (Yeni 2013: 200). Their settled Muslim and Greek neighbours appear to flourish more in this sector, with 18.88% and 46.01% of agricultural and pastoral taxes, respectively (Table 1). Due to Islamic religious limitations, swine breeding is only recorded among the Greeks (Table 2).

This image is by no means particular to Thessaly. Across the Aegean, in the second half of the fifteenth century, the Meander Valley accommodated a substantial yörük population of nine communities numbering 294 households. On the basis of the cadastre of Aydın (1461–70, TT1), barley and wheat constituted 30.37% and 25.17%, respectively, of the total tithes (Erdoğru and Bıyık 2015: 23). Animal husbandry, on the other hand, contributed only 1.81% to the total revenue, which is interpreted by the editors as a marked preference of the inhabitants for agriculture, dictated by the fertility of the region, despite the nomadic presence (Erdoğru and Bıyık 2015: 36). In the previous century, the export of grain from the regions of Smyrna and the Meander proved to be of paramount importance for Venice (Zachariadou 1983: 163–65). We are thus witnessing parallel farming orientations and practices by yörüks in the major granaries of the southern Balkans and western Anatolia.

All in all, the yörüks of Thessaly enjoyed reduced taxation, which is mirrored in their average tax per family of 42.6 akçes, as opposed to the 67.5 akçes of the sedentary Muslims and the 66.4 akçes of the Greeks. The rates are evened out when one deducts the personal encumbrances: the yörüks paid, on average, 18.4 akçes less than a settled Muslim family and only 1.8 akçes less than a Greek one (Appendix Table 5). The overall impression is that to a large extent they adhered to the nomadic modus vivendi. Their engagement in farming should be viewed as a means of increasing the fief-holder's revenue through the reclamation of abandoned lands for agriculture. Out of their twenty-three recorded localities, one is a standing village (ʿAlī Fakīlar: 54 families), one is abandoned (Andriya Miḥal: 0 families) and the remaining twenty-two are cultivated lands (229 families). Finally, there are only two cases where former nomads earned a living as artisans, one in the local capital, Yeñişehir, and the other in Emīrḫanlu/Aydıñlı, as mentioned above.

The nomadic groups of Turkish origin who settled in Thessaly in the 1380s preferred hamlets to villages, where they engaged primarily in cereal cultivation and, to a lesser extent, in manufacturing production. In this aspect, they proved to be quite successful in reclaiming the fertile lands of Thessaly for staple crops. From parallel cases in Anatolia, we presume that the animal husbandry section of their economy must have been more significant than attested to in the tax register. Overall, they appear more destitute than their sedentary neighbours.

The Albanians in the Peloponnese

Less than a decade later and farther south, the first extant Ottoman taxation cadastre of the Peloponnese (TT10-1/14662), dated 1460–63, divides the recorded settlements into Greek and Albanian on the basis of different rates of taxation, which favoured the latter ethnic group.

Due to the persisting wars and the plague in the fourteenth century, the Peloponnesian population had suffered losses (Zakythinos 1949: 9–10; Panagiotopoulos 1987: 61–68). This demographic deficit was counterbalanced to some degree by the invitation and settlement of Albanian nomadic clans, this time Christian—as opposed to the Muslim *yörüks*—who formed populous groups consisting of families, or tribes. The Albanians, also known as Arvanites in the Greek lands, were first mentioned in the Peloponnese in the second half of the fourteenth century. By 1391 there had been an influx of Albanians that could be hired as mercenaries. The Venetians were in need of colonists and soldiers in their depopulated areas and hence offered plots of arable land, pastures and tax exemptions to the wandering Albanians in southern Greece (Thiriet 1959: 366; Chrysostomides 1995: 206, 291, 337, 339; Topping 1980: 261–71; Ducellier 1968: 47–64). A well-attested-to, more populous Albanian settlement took place during the rule of Theodore I Palaeologus (1384–1407), when ten thousand Albanians appeared before the Isthmus and asked Theodore for permission to settle in the Peloponnese (1394–95). A second wave of immigrants from southern Albania and western mainland Greece descended on the Peloponnese, perhaps in 1417–18. Their establishment was significant for the invigoration of the Albanian demography in the peninsula that led to the Albanian rebellion in 1453 (Zakythinos 1975: 247–56; Biris 1998: 133–40). In the first stage of their arrival, in the space of ten to thirty years, they were probably in search of appropriate land for animal husbandry. After mapping out the evidence contained in the Peloponnesian register, it becomes apparent that by the early 1460s the Albanians had established their settlements throughout the region regardless of land morphology and elevation (Map 2).

The main reason for occupying a different category in the cadastre is the 20% reduction in the poll tax (20 *akçe*s instead of the 25 the Greeks paid). This mirrors a late Byzantine (Vranoussi 1998: 293–305)—and even, as we saw, Venetian—practice of tax exemptions that the Ottomans adopted to control the intractable Albanians

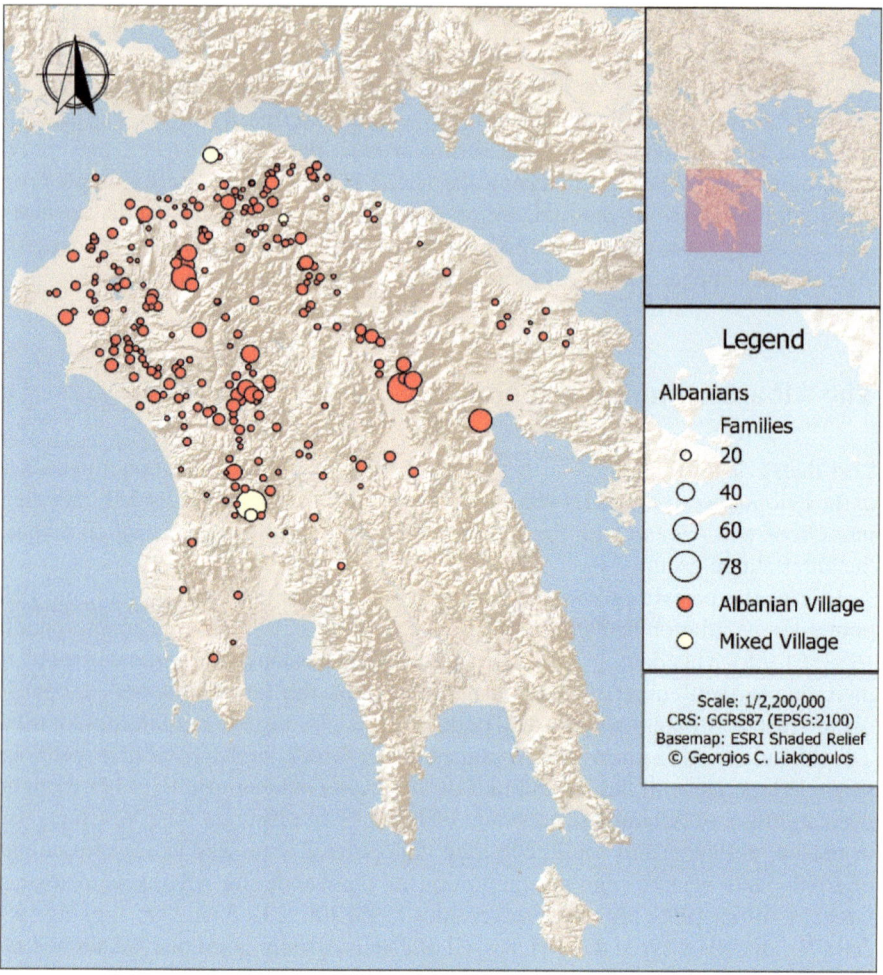

Map 2 Albanian settlements in the Peloponnese in 1460–63

and should be examined within the context of the 'continuity for stability' policy.
Their rebellions of 1423 and 1453, were reminiscent of their untamed nature (Chalcocondyles 1923: II, 16–17; Dilo 1969: 205–11). However, within half a century,
the favourable taxation terms granted to the Albanians had ceased to exist. The next
register of the Peloponnese (1514/5, TT80) recorded the same amount of poll tax
(25 *akçe*s annually) levied on both Greeks and Albanians and, for this reason, did
not earmark the Albanian villages. This shows that by the early sixteenth century
Ottoman rule in the peninsula had been consolidated. The TT10-1/14662, on the
other hand, clearly noted the Albanian villages with the heading 'of the Albanian
community'. Besides the villages marked as Albanian, there were inhabitants of

Albanian origin in Greek villages and towns, as their names indicate (Liakopoulos 2019, 214–16). These, however, like the rest, were due to pay a 25-*akçe* poll tax.

The analysis of the financial data contained in the Peloponnesian cadastre (Appendix Table 8) shows that an Albanian family paid, on average, a little less than three-fifths (58.85%) the amount of tax a Greek one did, which constitutes a first indication of the relative poverty of this ethnic group. It is plausible to suggest that such communities were mostly engaged in cultivation of life-sustaining crops (e.g. cereals), instead of agricultural activities that were geared towards securing a monetary surplus. An Albanian family appears to pay, on average, 4.7 *akçes* more wheat tithe and produce 238.6 kg more wheat than a Greek one.[4] Barley follows with a similar analogy of 0.8 *akçes* more tithe and 46.9 kg more production than the Greek equivalent (Appendix Tables 8 and 10). On the other hand, the contribution of the Albanians in other agricultural taxable assets ranges a little over one-tenth of the totals in the Peloponnese (11.9%).

Viticulture and wine production constitute two areas of agriculture and pre-industrial activity that require a certain amount of know-how and a closer connection to the earth; hence, they are more often performed by established sedentary societies. The transhumant Albanians, who must have continued being employed as mercenaries in armies of various Italian states (*stradioti*), show a very constrained, almost non-existent, contribution to the cultivation of the vine (3.25% of total viticulture and vinification taxes). This holds true for the cultivation of both taxed subjects' and fief-holders' personal demesne vineyards (Liakopoulos 2009: 202–3). The exclusive cultivation of resin trees by Greeks points to resin's use in the maintenance of wine jars and barrels. With the exception of oak trees in the Buryalisa village (TT10: 186), the Albanians did not engage in cultivating the fief-holders' fields to the same extent as their Greek neighbours did (Appendix Table 9). This, however, should be examined in the light of the main settlement pattern in the early Ottoman Peloponnese, which was the fortified large village or town dating back to the Franco-Byzantine era (thirteenth to fifteenth centuries). The largest fiefs were established around such a fortified centre, which functioned as the local administrative capital and market (Panagiotopoulos 1987: 45–49). The Albanian newcomers chose to inhabit a number of small satellite settlements in the periphery. This allowed them seclusion and loose relations with the local capitals, where the fief-holders had their mansions and fields.

On the other hand, the Albanians appear to thrive more in the sector of animal husbandry, different to the *yörük*s of Thessaly (Appendix Table 8). As determined by their semi-nomadic nature, they contributed to the tax levied on swine breeding and to sheep ownership, with 84.65% and 60.68%, respectively. However, the impact of pastoral activities on the overall economy is limited to 2.13% for the Albanians and 0.15% for the Greeks.

---

[4] The calculation of the actual cereal yield is based on the Adrianople *müdd* (*müdd-i Edrene*). One *müdd* of wheat cost 80 *akçe*s and a bushel (*keyl*) 4 *akçe*s, and one *müdd* of barley cost 60 *akçe*s and a bushel 3 *akçe*s (TT10: 26; İnalcık 1983: 324–25; Beldiceanu and Beldiceanu-Steinherr 1980: 57; Barkan 1964: 258). One *müdd* of wheat weighed 513.12 kg and one *müdd* of barley 445 kg (Hinz 1955: 47). The tithe on cereals is estimated at one-eighth of the total production (Beldiceanu and Beldiceanu-Steinherr 1980: 24).

From the above it is inferred that, even if we employ the rather optimistic family coefficient of 5, in most cases the cereal harvest, presumably produced by both ethnic groups, surpassed the level of domestic consumption. The yearly subsistence minimum per individual is estimated at 200 kg of cereals plus 59 kg for seed and 36 kg for tax, a total of 295 kg (Asdrachas 1999: 90; Kiel 2007: 41). The grain tax figures of the Peloponnesian cadastre are closely related to the population of each settlement (cereals in kg to families, R^2: 0.88; p-value: <0.001). The commercialisation of the most profitable cultivations' surplus, in our case the cereals and the vines, was obligatory for the sustainability of the household that was encumbered with the poll tax and the capitation (Balta 2015: 118). The complementarity of agrarian production constitutes a key characteristic of pre-industrial economies (Asdrachas 1999: 221–22; Asdrachas 1988: 15–17). However, the Albanians had less of a need to obtain the same surplus as the Greeks, due to their 20% reduction in the poll tax. Whereas they seem to have mainly focused on cereal cultivation (cereals: 89.36%; viticulture: 3.88%), the Greeks present a more balanced ratio (cereals: 42.97%; viticulture: 43.60%) (Table 3). In addition to that, Greeks are keener to engage in cotton and silk manufacturing. The latter should be connected with mulberry cultivation; the two combined reach an impressive 56.34% of the taxes levied on activities other than cereal cultivation and viticulture/vinification (Appendix Table 11). These findings tally with the image of destitute stockmen living in the countryside that the historical sources give about the Albanians:

> This entire race are nomads and they do not engage in any lasting activity. (Chalcocondyles 1923: II, 170).

> So, having acquired all these, without wanting to calm down anymore and being ungrateful to the beneficiaries, in the middle of winter after three years, he removes from the Illyrians in the Peloponnese all their herds, many horses, many oxen, several sheep and several swine. (Lambros 1926: 194).

The Albanians of the Peloponnese were settled throughout the region, inhabiting mostly small-sized villages in the periphery of the Greek towns. Their farming activities were clearly oriented towards cereal cultivation. As a matter of fact, their geographical distribution shows that they achieved high rates of cereal production regardless of elevation and soil type. On the basis of their impressive animal breeding scores, one can deduce that they must have been the main suppliers of animal products to the Greek towns. On the other hand, the sectors of viticulture and manufacturing lagged behind those of the Greek settlements.

Table 3 Percentages of agricultural and pastoral taxes in the Peloponnese in 1460–63

Taxable asset	Albanians (%)	Greeks (%)
Cereals	89.36	42.97
Vineyards	3.88	43.60
Other cultivations	4.63	13.28
Animal husbandry	2.13	0.15

Conclusions

The first Ottoman officials to arrive in a newly conquered land were the tax surveyors (Cvetkova 1983: 134). The Ottoman state, parallel to applying its own general taxation system in its core lands, incorporated, abolished or amended a number of local pre-existing taxation practices in a sense of pragmatism and flexibility (Dağlı 2013: 195–200). The Thessalian and the Peloponnesian cadastres possess, *mutatis mutandis*, the quality of a Doomsday Book in presenting a snapshot of the *yörük*s and the Albanians. After more than half a century since the two population groups under study had been introduced to their new homelands, they demonstrated different trajectories towards sedentarisation and achieved different standards of living (Table 4). Almost all (94.7%) of the *yörük*s of Thessaly retained their nomadic taxation status. On the other hand, all the 4900 Albanian families of the Peloponnese are recorded inhabiting villages, the majority of which (93.71%) numbered fewer than forty families, by contrast with the respective 27.03% of the 7103 Greek families. Eighty percent of the Albanian villages belonged to temporary transhumance settlements named after the clan chieftain (Liakopoulos 2019: 221, 223). This is an indication that the Albanians were, by the mid-fifteenth century, already advanced in pursuing a more sedentary livelihood than the *yörük*s. As a matter of fact, Albanians had only incidentally and occasionally been nomads in their history; their migration period in the southern Balkans is characterised by an acquired mobility necessitated by low living standards (Ducellier 1979: 35). Their relatively rapid transition into farmers is attested to by a similar episode in Attica. It was most probably the Florentine duke of Athens, Antonio Acciaiuoli, who resorted to their assistance between 1418–20, in an attempt to strengthen the defence of his duchy against the Venetians and the Ottomans, and to increase his revenues by reclaiming arable lands (Biris 1998: 108). Paleoenvironmental records from Brauron (Vravrona) in eastern Attica showed extensive soil erosion attributed to anthropogenic activities such as ploughing and herding in the first half of the fifteenth century, which coincides with the settlement of Albanians in the area (Triantaphyllou et al. 2010: 19–20; Kouli 2012: 273, 276).

The readily available manpower, in addition to the renowned military virtues of the *yörük*s and the Albanians, had been key determinants in their utilisation by Ottoman and pre-Ottoman authorities. These two population groups adopted at a different pace to the local practices and exploited natural resources to achieve sustainability and the necessary monetary surplus to cover their, no matter how reduced, personal taxes. Three generations after their immigration, no group seems to have achieved

Table 4 Average taxes per family in Thessaly and the Peloponnese in the mid-fifteenth century (in grams of silver *per annum*)

	Thessaly			Peloponnese	
	Sedentary Muslims	*Yörük*s	Greeks	Albanians	Greeks
Total	68	43	67	63	107
Excl. personal taxes	60	41	45	43	83.5

full integration with the local sedentary society; neither did they engage in the full breadth of agricultural opportunities the lands they settled in offered. Their high percentages of cereal cultivation indicate assiduity in the discharge of their land reclamation duties. On top of that, they met the needs of urban populations in animal products. The nomadic/transhumant and the sedentary populations complemented one another, perhaps even by the exploitation of the first. As attested to in posterior cadastres, the Peloponnesian Albanians gradually began to cover a wider gamut of agricultural activities and thus narrow the gap with the Greeks. It appears that in the long run the political choices of the authorities to mobilise these marginal groups were successful.

Appendix

See Tables 5, 6, 7, 8, 9, 10 and 11.

Table 5 Taxes in Thessaly in 1454/5 (in *akçes per annum*)

Taxable asset		*Yörük*s	Sedentary Muslims	Greeks[a]
Wheat	Total tithe	5904[b]	4940	5428
	Average tithe per family (*akçes*)	24.3	27.6	13.2
	Average tithe per family (*kile*)	3.01	3.45	1.65
Barley	Total tithe	3073	1740	1111
	Average tithe per family (*akçes*)	12.6	9.7	2.7
	Average tithe per family (*kile*)	2.53	1.94	0.54
Millet	Total tithe	32	100	160
	Average tithe per family (*akçes*)	0.1	0.6	0.4
	Average tithe per family (*kile*)	0.03	0.11	0.08
Vetch	Total tithe	10	–	175[c]
	Average tithe per family (*akçes*)	0.04	–	0.4
	Average tithe per family (*kile*)	0.01	–	0.08
Flax	Tithe	–	–	537

(continued)

Table 5 (continued)

Taxable asset		*Yörük*s	Sedentary Muslims	Greeks[a]
Walnuts	Tithe	–	–	67
Fruits	Tithe	–	149	15
Cotton	Tithe	40	120	365
Madder	Tithe	10	–	–
Broad beans	Tithe	–	–	96
Kitchen & vegetable gardens	Tithe	–	187	370
Viticulture & vinification	Tithe on vineyards & barrel tax[d]	100	28	1210
Animal husbandry	Tax on swine	–	–	1907
	Tax on sheep	185	714	4741
	Apiculture	241	976	1476
Average tax per family	Total	42.6	67.5	66.4
	Excl. personal taxes	40.9	59.3	44.4

[a] Entries of nine Greek villages included in the fief of İbrāhīm and Yūsuf
[b] In the arable land of Aynoġoli the tithe on grain is estimated at a lumpsum of 60 *akçe*s
[c] In Vaṣ irama (Vathyrrema) it also includes a broad bean tithe
[d] The tax on vinification or barrel tax is attested to only in the Greek villages of Kestric (Kastri) and Viseni/Büyük Göl (Aetolofos)

Table 6 Percentages of cereal production in Thessaly in 1454/5

Taxable asset	*Yörük*s (%)	Sedentary Muslims (%)	Greeks (%)
Wheat	53.95	62.66	70.25
Barley	45.42	35.31	22.97
Millet	0.48	2.03	3.31
Vetch	0.15	0	3.47

Table 7 Percentages of cultivations other than cereals and vines in Thessaly in 1454/5

Taxable asset	*Yörük*s (%)	Sedentary Muslims (%)	Greeks (%)
Flax	0	0	37.03
Cotton	80	26.32	25.17
Madder	20	0	0
Walnuts	0	0	4.62
Fruits	0	32.67	1.04
Broad beans	0	0	6.62
Kitchen & vegetable gardens	0	41.01	25.52

Table 8 Taxes in the Peloponnese in 1460–63 (in *akçes per annum*)

Taxable asset		Albanians	Greeks
Wheat	Total tithe	161,481	201,049
	Average tithe per family	33	28.3
	Average production per family (kg)	1691	1452.4
Barley	Total tithe	25,190	30,894
	Average tithe per family	5.1	4.3
	Average production per family (kg)	305	258.1
Flax	Tithe	6431	8986
Cotton	Tithe	463	5732
Silk	Tithe	628	26,920
Olive oil	Tithe	36	2144
Honey	Tithe	299	644
Resin	Tithe	1	10
Fruits	Tithe	84	2351
Mulberries	Tithe	–	13
Kitchen gardens	Tithe	–	299
Viticulture	Tithe on vineyards	7554	155,573
Vinification	Tax on wine	344	15,939
Animal husbandry	Tax on swine	4448.5	806.5
	Sheep (heads)	3885	2517
Average tax per family	Total	65.55	111.39
	Excl. poll tax	44.66	85.96

Table 9 Taxes of cultivations in timariots' personal demesne in the Peloponnese in 1460–63 (in *akçes per annum*)

Taxable asset	Albanians	Greeks
Vineyards	552	79,807
Fruit trees	142	4070
Olive trees	150	4404
Mulberry trees	398	13,460
Pomegranate trees	–	5
Resin trees	–	720
Oak trees	1000	50
Walnut trees	–	25
Pear trees	–	5
Bitter orange groves	–	450
Vegetable gardens	50	250
Kitchen gardens	–	1155

Table 10 Percentages of cereal production in the Peloponnese in 1460–63

Taxable asset	Albanians (%)	Greeks (%)
Wheat	84.71	84.84
Barley	15.29	15.16

Table 11 Percentages of cultivations other than cereals and vines in the Peloponnese in 1460–63

Taxable asset	Albanians (%)	Greeks (%)
Flax	66.42	12.53
Cotton	4.78	8
Silk	6.49	37.55
Oleiculture	1.92	9.13
Honey	3.09	0.9
Resin	0.01	1.02
Oaks	10.33	0.07
Walnuts	0	0.03
Fruits	2.33	8.95
Mulberries	4.11	18.79
Pomegranates	0	0.01
Pears	0	0.01
Bitter oranges	0	0.63
Kitchen & vegetable gardens	0.52	2.38

References

Alexander JC (1985) Toward a history of post-Byzantine Greece. The Ottoman kanunnames for the Greek lands, circa 1500–circa 1600. PhD thesis, Columbia University, reprint, Athens

Arıcanlı İ (1979) Osmanlı İmparatorluğu'nda Yörük ve Aşiret Ayrımı. Boğaziçi Üniversitesi Dergisi 7:27–34

Asdrachas S (1988) Οικονομία και νοοτροπίες. Ερμής, Athens

Asdrachas S (1999) Μηχανισμοί της αγροτικής οικονομίας στην τουρκοκρατία (ΙΕ΄-ΙΣΤ΄ αιώνας). Θεμέλιο, Athens

Balta E (2015) The testimony of the Ottoman tax registers on viticulture and wine-production in the Peloponnese (15th–18th c.). In Balta E (ed) Population and agricultural production in Ottoman Morea. Analecta Isisiana 132, The Isis Press, Istanbul, pp 115–122

Barkan ÖL, (1953–54) Osmanlı İmparatorluğu'nda Bir İskân ve Kolonizasyon Metodu Olarak Sürgünler. İstanbul Üniversitesi İktisat Fakültesi Mecmuası 15(1–4):209–237

Barkan ÖL (1957) Essai sur les données statistiques des registres de recensement dans l'Empire ottoman aux XVe et XVIe siècles. J Econ Soc Hist Orient 1(1):9–36

Barkan Ö (1964) Edirne ve Civarındaki Bazı İmâret Tesislerinin Yıllık Muhasebe Bilânçoları. Belgeler 1(2):235–377

Beldiceanu N, Beldiceanu-Steinherr I (1980) Recherches sur la Morée (1461–1512). Südost-Forschungen 39:17–74

Beldiceanu N, Năsturel PȘ (1983) La Thessalie entre 1454/55 et 1506. Byzantion 53(1):104–156

Biris KI (1998) Ἀρβανῖτες, οἱ Δωριεῖς τοῦ νεώτερου ἑλληνισμοῦ. Ἱστορία τῶν Ἑλλήνων Ἀρβανιτῶν. Μέλισσα, Athens

Çetintürk S (1943) Osmanlı İmparatorluğunda Yürük Sınıfı ve Hukuki Statüleri. Ankara Üniversitesi
    Dil Ve Tarih-Coğrafya Fakültesi Dergisi 2:107–116
Chalcocondyles L (1923) Historiarum demonstrationes. E. Darkó (ed). 2 vols. Editiones Criticæ
    Scriptorum Græcorum et Romanorum, Academia Litterarum Hungarica, Budapest
Chrysostomides J (1995) Monumenta Peloponnesiaca. Documents for the history of the Pelopon-
    nese in the 14th and 15th Centuries. Porphyrogenitus, Camberley
Coşkun A, Özkılınç A, Sivridağ A, Yüzbaşıoğlu M (2003) 167 Numaralı Muhâsebe-i Vilâyet-i
    Rûm-İli Defteri (937/1530) I. Paşa Livâsı Solkol Kazâları (Gümülcine, Yeñice-i Kara-su, Drama,
    Zihne, Nevrekop, Timur-hisârı, Siroz, Selanik, Sidre-kapsı, Avrat-hisârı, Yeñice-i Vardar, Kara-
    verye, Serfiçe, İştin, Kestorya, Bihlişte, Görice, Florina) ve Köstendil Livâsı. Dizin ve Tıpkıbasım,
    Defter-i Hâkânî Dizisi 9., T.C. Başbakanlık, Devlet Arşivleri Genel Müdürlüğü, Osmanlı Arşivi
    Daire Başkanlığı, Ankara
Cvetkova B (1983) Early Ottoman *tahrir defters* as a source for studies on the history of Bulgaria
    and the Balkans. Archivum Ottomanicum 8:133–213
Dağlı M (2013) The limits of Ottoman pragmatism. History and theory 52(2):194–213
Delilbaşı M, Arıkan M (2001) Hicrî 859 Tarihli Sûret-i Defter-i Sancak-ı Tırhala. 2 vols. Türk Tarih
    Kurumu, Ankara
Dilo T (1969) Luftërat e Shqiptarëve të Peloponezit kundër Turqve në shekullin XV. In: Konfer-
    enca e Dytë e Studimeve Albanologjike me rastin e 500-vjetorit të vdekjes së Gjergj Kastriotit-
    Skënderbeut, 12–18 Janar 1968. Universiteti Shtetëror, Instituti i Historisë dhe i Gjuhësisë, Tirana,
    pp 205–211
Ducellier A (1968) Les Albanais dans les colonies vénitiennes au XVᵉ siècle. Studi Veneziani
    10:47–64
Ducellier A (1979) Les Albanais du XIᵉ au XIIIᵉ siècle : nomades ou sédentaires ? Byzantinische
    Forschungen 7:23–36
Erdoğru MA, Bıyık Ö (2015) T.T 0001/1 M. Numaralı Fatih Mehmed Devri Aydın İli Mufassal
    Defteri (Metin ve İnceleme). İzmir Araştırma ve Uygulama Merkezi Yayın 11, Ege Üniversitesi,
    İzmir
Ergul F (July 2012) The Ottoman identity: Turkish, Muslim or *Rum*? Middle East Stud 48(4):629–
    645
Fine JVA (1996) The late medieval Balkans. A critical survey from the late twelfth century to the
    Ottoman conquest. The University of Michigan Press, Ann Arbor
Gökbilgin MT (1957) Rumeli'de Yürükler, Tatarlar ve Evlâd-ı Fâtihân. İstanbul Üniversitesi
    Edebiyat Fakültesi Yayınlarından 748, İstanbul
Halaçoğlu Y (1991) XIV-XVII. Yüzyıllarda Osmanlılarda Devlet Teşkilatı ve Sosyal Yapı. Türk
    Tarih Kurumu, Ankara
Hinz W (1955) Islamische Masse und Gewichte umgerechnet ins metrische System. Handbuch der
    Orientalistik 1, E. J. Brill, Leiden
İnalcık H (1954) Ottoman methods of conquest. Stud Islam 2:103–129
İnalcık H (1959) Osmanlılar'da Raiyyet Rüsûmu. Belleten 23(92):575–610
İnalcık H (1982) Rice cultivation and the *çeltükci-reᶜāyā* system in the Ottoman Empire. Turcica,
    Revue d'études turques 14:69–141
İnalcık H (1983) Introduction to Ottoman metrology. Turcica, Revue d'études turques 15:311–348
İnalcık H (2014) 'The Yörüks: Their origins, expansion and economic role. Cedrus 2:467–495
Kasaba R (2009) A movable empire. Ottoman nomads, migrants & refugees. University of
    Washington Press, Seattle and London
Kiel M (1996) Das türkische Thessalien. Etabliertes Geschichtsbild versus osmanische Quellen.
    Ein Beitrag zur Entmythologisierung der Geschichte Griechenlands. In: Lauer R, Schreiner P
    (eds) Die Kultur Griechenlands in Mittelalter und Neuzeit. Bericht über das Kolloquium der
    Südosteuropa-Kommission 28–31 Oktober 1992. Vandenhoeck & Ruprecht, Göttingen, pp 109–
    196

Kiel M (2007) The smaller Aegean islands in the 16th–18th centuries according to Ottoman administrative documents. In: Davies S, Davis JL (eds) Between Venice and Istanbul. Colonial Landscapes in Early Modern Greece. Hesperia Supplement 40, American School of Classical Studies at Athens, Athens, pp 35–54

Kiel M (2013) Yenişehir. In: Türkiye Diyanet Vakfı İslâm Ansiklopedisi. vol 43. Istanbul, pp 473–476

Kostis KP (2020) Στον καιρό της πανώλης. Εικόνες από τις κοινωνίες της ελληνικής χερσονήσου. 14ος-19ος αιώνας. Πανεπιστημιακές Εκδόσεις Κρήτης, Heraklion

Kouli K (2012) Vegetation development and human activities in Attiki (SE Greece) during the last 5,000 Years. Veg Hist Archaeobotany 21:267–278

Kurt Y (2004) Çukurova Tarihinin Kaynakları I. 1525 Tarihli Adana Sancağı Mufassal Tahrir Defteri. Türkiye'nin Sosyal ve Kültürel Tarihi Projesi, Türk Tarih Kurumu, Ankara

Lambros SP (1926) Ἀνωνύμου Πανηγυρικὸς εἰς Μανουὴλ καὶ Ἰωάννην Η΄ τοὺς Παλαιολόγους. In: Παλαιολόγεια καὶ Πελοποννησιακά. vol. 3. Athens, pp 132–199

Liakopoulos GC (2009) Η αμπελοκαλλιέργεια και η οινοπαραγωγή στην πρώιμη οθωμανική Πελοπόννησο, βάσει του κατάστιχου TT10-1/14662. In: Pikoulas YA (ed) Οἶνον ἱστορῶ IX Πολυστάφυλος Πελοπόννησος. ΕΝΟΑΠ Νεμέα, Athens, pp 197–222

Liakopoulos GC (2019) The early Ottoman Peloponnese. A study in the light of an annotated *editio princeps* of the TT10–1/14662 Ottoman taxation cadastre (ca. 1460–1463). Royal Asiatic Society, The Ibrahim Pasha of Egypt Fund Series, The Gingko Library, London

Moutafchieva VP (1988) Agrarian relations in the Ottoman Empire in the 15th and 16th centuries. Boulder, East European Monographs, Columbia University Press, New York

Pamuk Ş (2000) A monetary history of the Ottoman Empire. Cambridge Studies in Islamic Civilization, Cambridge University Press, Cambridge

Panagiotopoulos V (1987) Πληθυσμός και οικισμοί της Πελοποννήσου, 13ος-18ος αιώνας. Μελέτες Νεοελληνικής Ιστορίας, Ιστορικό Αρχείο, Εμπορική Τράπεζα της Ελλάδος, Athens

Savvides A (1995) Τα προβλήματα για την οθωμανική κατάληψη και την εξάπλωση των κατακτητών στο Θεσσαλικό χώρο. Θεσσαλικό Ημερολόγιο 28:33–64

Taşkın Ü (2005) Osmanlı Devleti'nde Kullanılan Ölçü ve Tartı Birimleri. unpublished MA dissertation, T.C. Fırat Üniversitesi, Elazığ

Thiriet F (1959) La Romanie vénitienne au moyen âge. Le développement et l'exploitation du domaine colonial vénitien (XIIe-XVe siècles). De Boccard, Paris

Topping P (1980) Albanian settlements in medieval Greece: some Venetian testimonies. In: Laiou-Thomadakis AE (ed) Charanis studies. Essays in honor of Peter Charanis. Rutgers University Press, New Brunswick, pp 261–271

Triantaphyllou MV, Kouli K, Tsourou T, Koukousioura O, Pavlopoulos K, Dermitzakis MD (2010) Paleoenvironmental changes since 3000 BC in the coastal marsh of Vravron (Attica, SE Greece). Quatern Int 216:14–22

Varlık N (2015) Plague and empire in the early modern Mediterranean world. The Ottoman experience, 1347–1600. Cambridge University Press, New York

Vranoussi E (1998) Deux documents byzantins inédits sur la présence des Albanais dans le Péloponnèse au XVe siècle. In: Gasparis Ch (ed) Οι Αλβανοί στο Μεσαίωνα. Διεθνή Συμπόσια 5, Εθνικό Ίδρυμα Ερευνών / Ινστιτούτο Βυζαντινών Ερευνών, Athens, pp 293–305

Yeni H (2013) The utilization of mobile groups in the Ottoman Balkans: a revision of general perception. Archiv Orientální 81(2):183–205

Yeni, H. Autumn 2017. Osmanlı Rumelisi'nde Yörük Teşkilatı, Kökeni ve Nitelikleri. Osmanlı Tarihi Araştırma ve Uygulama Merkezi Dergisi 42:187–205

Zachariadou EA (1983) Trade and crusade. Venetian Crete and the emirates of Menteshe and Aydin (1300–1415). Hellenic Institute of Byzantine and Post-Byzantine Studies, Venice

Zakythinos DA (1949) La population de la Morée byzantine. L'Hellénisme contemporain 3(1) (1949):7–25

Zakythinos DA (1975) Le Despotat grec de Morée. Histoire politique. Maltezou Ch (ed). vol. 1. Variorum Reprints, London

Eastward Migration in European History: The Interplay of Economic and Environmental Opportunities

Piotr Guzowski

Abstract During the preindustrial era one of the major migration waves headed eastward to Eastern Europe, where scores of migrants, in their pursuit of happiness, hoped to fulfil their dreams, have their own farm or set up a company, achieve a higher social status, and benefit from religious freedom and tolerance. The first wave of migration was connected with German colonization and the establishment of settlements following the German law. The alluringly large expanses of "pristine" land, together with tax privileges and the prospects of relative autonomy, attracted scores of bold, enterprising and hard-working settlers to relocate to the East. Most of them were peasants and townsfolk from the German states and the Netherlands, but there were also Jews escaping discrimination in Western Europe as well as West-European Protestants and Catholics attracted by religious tolerance in the East. Prospects of freedom and economic success encouraged them all to choose Poland and the Polish-Lithuanian Commonwealth as their second homeland.

Keywords German colonization · Migration · The Polish–Lithuanian Commonwealth

Migration forms an important part of the history of Central and Eastern Europe. From the current and past centuries' perspective it might seem surprising that in earlier centuries migration flows actually ran in a different direction to what they do today. Considering the formation of the European Union and the fairly recent migration crisis caused by the political upheaval in Africa and the Middle East, the focus of our attention may have been on the influx of those asylum seekers and economic migrants who have been making their way into Western Europe. It is a little-known fact, however, that during the preindustrial era a major migration flow was headed eastward to Eastern Europe, where scores of migrants, in their pursuit of happiness, hoped to fulfil their dreams, have their own farm or set up a company, achieve a higher social status, and benefit from religious freedom and tolerance. In this article I shall investigate why venturing into the unknown but seemingly bountiful lands of

P. Guzowski (✉)
University of Białystok, Bialystok, Poland
e-mail: guzowski@uwb.edu.pl

what was to become the Polish–Lithuanian Commonwealth seemed to the eastward migrants of the Middle Ages and early modern period to be a risk worth taking. The Polish–Lithuanian state eventually failed, falling prey to the partitioning powers of the late eighteenth century, but it seems to have held considerable appeal among the migrants who decided to relocate within its boundaries. It is also for this reason that its history and extraordinary culture is worth studying.

Until the latter half of the thirteenth century or even the beginning of the four-teenth century (Lopez 1976), most countries of Western Europe had witnessed an era of prosperity that had lasted approximately three hundred years. Throughout that period, those countries had seen an increase in commercialization that had a tremen-dous impact across various social groups. Culturally, one manifestation of such pros-perity was the twelfth-century Renaissance (Benson et al. 1985), while in terms of geography and the history of civilization the era brought about the "Europeaniza-tion" of Europe (Bartlett 1994). The technological advancement in agriculture hardly matched the population growth in Europe's most developed regions, which prompted lords, knights and even common serfs to look for opportunities in the peripheries of European kingdoms. Medieval texts mentioned the instability of agriculture in Flan-ders, Netherlands and Zealand, which, from the perspective of their authors, stood in stark contrast to the fertile soils and attractive landscapes of Central Europe. As for migrants, these were not exclusively penniless peasants on the lookout for a piece of land to claim as their own; relatively well-off landowners would also liquidate their inventories and use the proceeds to start a new life in the lands beyond the Elbe. The institutional framework for the migration was laid out by German, Czech and Polish princes, who accelerated the process by incentivizing new settlers (*hospites*) and proprietors (*locatores*, who recruited the former), as well as through propaganda. Those who took the risk of resettling to the East were thus primarily motivated by the opportunities that the distant lands of Central Europe provided (Piskorski 2017). The anonymous author of an old medieval Flemish song portrayed their enthusiasm in the following terms:

> To the Eastern land we want to venture on horseback.
> To the Eastern land we want to venture together.
> All together towards the woodlands green.
> All together towards the woodlands.
> Where a better life awaits (Zientara 2006, 58).

The alluringly large expanses of land, together with tax privileges and the prospects of relative autonomy, attracted scores of bold, enterprising and hard-working settlers to relocate to the East. The colonization was bolstered by the institu-tional framework provided by Magdeburg law and manifested itself in several distinct forms. Such regions as Brandenburg and Pomerania (among others) saw the prolifer-ation of large villages consisting of several dozen households, which resulted in the establishment of fixed settlement structures. Communities of settlers were granted privileges that allowed the inhabitants to preserve their own national identities and cultures for many decades or even centuries. As regards the settlements where the number of immigrants was relatively low and the majority of the inhabitants were primarily of local (e.g. Slavic) descent, a range of particular legal and organizational models were put in place. While in the case of towns it was more apparent, villages were also affected: in the later Middle Ages, even though most villagers had no experts in written Magdeburg law at hand, they nonetheless adopted the self-government model laid out in the *Sachsenspiegel* and recognized the validity of such written statements as borough rights and court records. Beyond doubt, the settlers from the West had a number of advantages over the locals: better equipped to thrive in an economy of steadily progressing commercialization, they could avail themselves of more advanced agricultural techniques (such as the three-field system) and of more sophisticated appliances (including the iron plough and the harrow).

The limitations of the pre-statistical era make the scale of migration-related colonization difficult to gauge. Historians estimate that between 100,000 and a million German migrants settled in Poland during the Middle Ages and the early modern period (Piskorski 2002). During the early modern period (the 16th–18thc)., numerous immigrants also arrived from the Low Countries, which is why more than 1,000 settlements bearing the name "Olędry" ("from Netherlands") were founded across Poland, indicating either the ethnic origin of the settlers or the settlement's organization following the Dutch model. From an early date, municipal and royal officials enthusiastically installed settlers from the Low Countries in Pomerania and along the Vistula, recognizing their skills at managing wetlands for agriculture. As for the settlers themselves, the prospect of owning their own farm was very promising; most importantly, their chances of obtaining one were much higher in Poland than they were in their homeland, where the capital invested in agriculture had resulted in the majority of villagers losing their property rights and being demoted to the status of employees. Beginning at the end of the sixteenth century, Poland's reputation as a place worth relocating to began to grow steadily, especially as the country was regarded as one that offered attractive economic prospects and guaranteed freedom of religion. The majority of Dutch settlers were Mennonites, a Protestant denomination persecuted in the Low Countries. Mennonite settlers escaped to Poland for safety, being confident that they would be allowed to maintain their cultural identity (Targowski 2016).

Religious tolerance, the right to autonomy and financial prospects also made Poland appealing to Jewish communities. Beginning in the Middle Ages, the safety and various privileges offered by the country's rulers and the great landowners attracted a substantial influx of Jewish migrants, which continued until the fall of the Polish–Lithuanian Commonwealth (Stampfer 2018). In towns large and small, Jewish communities grew, and their businesses thrived. By the end of the eighteenth century Jews constituted approximately 8% of Poland's total population (Mahler 1967).

Another group of migrants were Scots, both Catholics and radical Protestants, who were motivated to relocate to Poland for financial and religious reasons. After arriving in Poland, they would become merchants, soldiers and, at times, even officials at the royal court. At first, they settled primarily in Gdańsk and in other cities; with time, many were granted nobility (Bajer 2012). Similar to German, Dutch and

Jewish settlers, Scottish immigrants would build social networks based on trading relationships and, at times, family ties. While the Polish–Lithuanian Commonwealth is associated with a very strong social and political position of the Catholic Church, it must be emphasized that it incorporated regions with substantial non-Catholic populations, among them Evangelicals (such as Royal Prussia) and members of the Eastern Orthodox Church or the Ruthenian Uniate Church (in the Grand Duchy of Lithuania and in Ukraine). Most of Europe was riven by religious conflicts, while Poland retained its reputation for religious tolerance for many years. In 1573, Polish parliament proclaimed peaceful coexistence between members of different religions (*dissidentes in religione*). This, however, did not mean that religious conflicts in Poland did not occur. Like other countries, Poland, too, saw scenes of rivalry between religions, often intertwined with politics; this rivalry, however, never led to wars of religion.

The Union of Lublin between the Grand Duchy of Lithuania and the Kingdom of Poland incorporated the vast territories of Ukraine, adequately safeguarded against Tartar raids since 1569, into the borders of the latter political entity. As a result, the "effectively limitless stretches of fertile farmland in the Dnieper Ukraine and Bracław (and, since the end of the sixteenth century, also territories to the east of the Dnieper river) acted like a giant vacuum cleaner, absorbing the most active representatives of the peasant class from other regions" (Jakowenko 2000, 184). It also absorbed members of other social groups, including nobles, Jews and German migrants. In the latter half of the sixteenth and first half of the seventeenth centuries no other region in Europe could compete with Ukrainian voivodeships of the Commonwealth when it came to urban and rural population growth. During that period, more than a hundred towns and settlements were established. They enjoyed numerous privileges, partly as a result of the actions of Polish and Ruthenian magnates, who anticipated financial gain, but also of the immigrants themselves, who cherished their freedom and independence.

The steady influx of settlers that continued until the late Middle Ages attests to the appeal of the territories that later formed the Polish–Lithuanian Commonwealth. When, during the early stages of colonization, one of the Polish princes reorganized his city, Cracow, under Magdeburg law, he proclaimed in the document granting new privileges to the city that his goal was to bring together people from "different climates" (Piekosiński 1879, 1). The actions undertaken initially by the central government and its institutions became increasingly more spontaneous, as they came to rely on trade partnerships and social ties that emerged from them. This was further facilitated by the incorporation of Eastern Europe into the world-system in the early modern period. The Polish–Lithuanian Commonwealth, a polity based in a peripheral region of Europe, came to symbolize the attractiveness of peripheral countries. While the modernization of the country and its economy was not as dynamic as in the most developed parts of Western Europe, and the emerging serfdom system progressively disadvantaged its people, the Polish–Lithuanian Commonwealth was still perceived as a place of opportunity. At the height of its territorial expansion, the Commonwealth spanned an area of almost a million square kilometres. Ecologically diverse, it included vast woodlands, fertile chernozem belts and long rivers, which allowed transportation of grain and timber to the port cities of the Baltic Sea, and the vast plain expanses that allowed for herding of oxen from Ruthenia into Central Europe. These territories prospered, even during the challenging climatic conditions of the Little Ice Age; most notably, the recorded consequences of extreme weather and long-lasting climate changes were the same as in Western Europe. Despite the climatic disadvantage and the fact that the Commonwealth was simultaneously at war with Sweden, the Grand Duchy of Moscow, Transylvania and the Cossacks, the Polish–Lithuanian state did not suffer any long-lasting social or economic crises. Neither droughts (such as the drought of 1540) nor the particularly harsh winters at the beginning of the seventeenth century affected the country's economy to such an extent as did the Swedish blockade of the Baltic ports. Peasant immigrants arriving in the Polish–Lithuanian Commonwealth were at an advantage over indigenous peasants. Nonetheless, despite the growing importance of the serfdom system, all subjects (including German and Dutch immigrants) of the estates owned by the nobles, royalty and the church were granted certain privileges, which made it possible for more resourceful individuals to make a profit and boost their social status. One significant privilege was that they could freely trade in grain and also benefit from the removal of internal borders and tariffs that weighed heavily on the vast territories of the Commonwealth, which was composed of large swathes of modern-day Poland, Lithuania, Latvia, Belarus and Ukraine. As a result, the benefits resulting from trade in the Baltic region and from the integration of markets not only lined the pockets of the nobility and clergy, who collected tithes in kind, but also trickled down to peasants, who traded in grain using a network of trading posts (they also formed companies that traded in goods derived from agriculture) and involving local farms in their businesses (Topolski 2000).

Reduced tax rates benefitted the migrants and likewise the locals. Initiated in the late medieval period, the transition from the domain-funded state (sustained primarily by the proceeds from crown lands and monopolies) to the tax-funded state (financed primarily through taxation) was commonly associated with an increase in tax rates. The Kingdom of Poland, and eventually the Grand Duchy of Lithuania as well, also went through this transition; however, they did so with a much greater delay compared to the Western countries and, as evidenced in the later history of the Polish–Lithuanian Commonwealth, to an insufficient degree (Boroda and Guzowski 2016). Nevertheless, from the perspective of both locals and migrants from the West, taxes levied on peasants, which constituted a considerable portion of the state's tax revenue, were still much lower than in the absolutist regimes governed by dukes and monarchs. Compared to the West, the average Polish peasant in the latter half of the sixteenth century had more land at his disposal and had to sell a mere 10–20% of his grain harvest to pay his tax and rent. In Prussia, the Low Countries and France, the tax to be paid in proportion to the income was at least twice as much (Guzowski 2008). The nobility, who had established democratic rule, made sure that taxation was not excessively burdensome. While the nobles perceived the exploitation of peasants as beneficial, the increases in tax resulting from the establishment of the military-fiscal state did not lead to major peasant uprisings.

Nonetheless, the Commonwealth did not enjoy continuous success throughout the late medieval and early modern periods and was ultimately partitioned by its more developed neighbours. Despite its ultimate fate, however, the territories governed by the Polish–Lithuanian Commonwealth attracted rural and urban settlers from more developed areas of Europe. On the one hand, it offered large strips of uncultivated and sparsely inhabited land suitable for farming—in a sense, the migration to European peripheries stood in agreement with the populations' preindustrial Malthusian mechanism, and Eastern territories benefitted to some degree from their own underdevelopment, attracting bold and resourceful individuals—whereas on the other hand, the Kingdom of Poland, which later became part of the Polish–Lithuanian Commonwealth, offered a fair degree of religious freedom and numerous rights that allowed settlers to govern themselves autonomously and maintain their cultural identity. What is equally important is that it also guaranteed various economic privileges, among which free trade, the removal of tariffs and low taxation were the most attractive. In their glory days, such peripheral countries as the Polish–Lithuanian Commonwealth offered considerable business opportunities that were not hindered by the nascent early modern bureaucracy, feudal relationships, the investments in agriculture by towns from the most developed areas of Europe or fierce competition. In addition, entering the global market allowed migrant merchants, particularly those from Western Europe, to maintain and access the social networks they had established in their homeland. States attracted settlers who could exhibit their advanced technical knowledge and organizational skills but did not make them dependent in a colonial-like fashion. Port cities and some large land estates were dependent on Western capital to an extent, but they still benefitted from the opportunities resulting from the integration of the local and regional markets of the Commonwealth with the markets of Europe at large. These processes did not lead to environmental degradation or a substantial transformation of the landscape. To Western immigrants, the natural resources might have seemed boundless; however, as early as in the sixteenth century, during the period of an economic boom, the country put in place its first resource management programmes, including management of woodlands in the crown lands, and of the great land estates owned by the magnates and clergy. Despite the ongoing development of settlements, population density in the Polish–Lithuanian Commonwealth was lower (in some regions even by several times) than Europe's average, which helped preserve the balance between humans and the natural world. Even several hundred years after the ecological revolution that took place in the early modern period, the name of the Polish–Lithuanian border region of "Podlasie", which means "near the forests", is still suitable. The high degree of conservation of forests may have minimized the effects of the most severe historical droughts (such as in the 1530 s, which was by and large the driest decade). The somewhat harsh climate, which the migrants had to grow accustomed to, meant local agriculture had to specialize in less demanding crops. Rye, which constituted 90% of Polish grain exports, was more resistant to snow and sub-zero temperatures and became the primary bread grain for the local populations, contrary to the West, where wheat was the standard. To gain a better understanding of Central and Eastern Europe's history, we may benefit from considering the modern-day context of migration and conclude that, over the course of several centuries, individuals who relocated eastward did so to express their preference and fulfil their need for change: it seems that this was easier when they moved beyond their homelands and ventured east.

References

Bajer PP (2012) Scots in the Polish-Lithuanian commonwealth, 16th to 18th centuries: the formation and disappearance of an ethnic group. Brill, Leiden

Benson RL, Constable G, Lanham CD, Haskins CH (1985) Renaissance and renewal in the twelfth century. Clarendon Press, Oxford

Boroda K, Guzowski P (2016) From King's finance to public finance. Different strategies of fighting financial crisis in the Kingdom of Poland under Jagiellonian Rule (1386–1572). In: The financial crises. Their management, their social implications and their consequences in pre-industrial times. Firenze University Press, Firenze, pp 451–470

Lopez RS (1976) The commercial revolution of the middle ages, 950–1350. Cambridge University Press, Cambridge

Bartlett R (1994) The making of Europe: conquest, colonization, and cultural change, 950–1350. Princeton University Press, Princeton

Guzowski P (2008) Chłopi i pieniądze na przełomie średniowiecza i w początkach epoki wczesnonowożytnej. Wydawnictwo Avalon, Kraków

Jakowenko N (2000) Historia Ukrainy: od czasów najdawniejszych do końca XVIII wieku, Instytut Europy Środkowo-Wschodniej, Lublin

Mahler R (1967) Żydzi w dawnej Polsce w świetle liczb. Struktura demograficzna i społeczno-ekonomiczna Żydów w Koronie w XVIII wieku. Przeszłość Demograficzna Polski 1:131–180

Piekosiński F (1879) Kodeks Dyplomatyczny Miasta Krakowa, vol. 1. Akademia Umiejętności, Kraków

Piskorski JM (2002) Historiographical approaches to medieval colonization of east central Europe: a comparative analysis against the background of other European inter-ethnic colonization processes in the middle ages. Boulder, New York

Piskorski JM (2017) The medieval colonization of central Europe as a problem of world history and historiography. In: Berend (ed) The expansion of central Europe in the Middle Ages. Routledge, London, pp 215–236

Stampfer S (2018) Settling down in Eastern Europe In Grill T (ed) Jews and Germans in eastern Europe. Shared and comparative histories. DeGruyter, München—Wien, pp1–20

Targowski M (2016) in Padian A, Targowski M (eds) Olędrzy: osadnicy znad Wisły: sąsiedzi bliscy i obcy, Fundacja Ośrodek Inicjatyw Społecznych ANRO, Toruń, pp11–26

Topolski J (2000) Przełom gospodarczy w Polsce XVI wieku i jego następstwa. Wydawnictwo Poznańskie, Poznań

Zientara B (2006) Henryk Brodaty i jego czasy. Wydawnictwo Trio, Warszawa

The Environmental Dimension of Migration: The Case of Poland After World War II

Małgorzata Praczyk

Abstract The article deals with the environmental dimension of migrations. It takes as a case study the migration of Poles after 1945 that took place due to the post-war redrawing of political borders in Central Europe. Massive displacement of the populations, who were forced to leave towns and villages that they had inhabited for generations, resulted in disruptions of bonds with the natural environment and the domesticated landscape. The ways of dealing with this rupture and the ways of domesticating the new landscapes and natural environments are key problems discussed here. This analysis helps to better appreciate the environmental dimension of the 21st migrations we increasingly observe today.

Keywords Migration · Poland · World war II · Nature

Migrations are a complex phenomenon that has accompanied humanity since its dawn. As Klaus Bade notes, "they are just as much a part of the human condition as birth, reproduction, illness and death" (Bade 2003: ix). The environmental dimension of migration is, in turn, one of its most salient and immanent components. The increase in global migration in the nineteenth and twentieth centuries, including those motivated by armed conflicts, also affected their scale and importance when it comes to the role of the natural environment. Whilst trying to understand the contemporary, environmental implications of migration caused by natural disasters or wars, it is worth looking at examples from the recent past, which can help us to better prepare for the challenges faced by host societies and migrants moving between different socio-natural systems (Izdebski 2018: 14–15). In doing so, we should not lose sight of the fact that these mass population movements consist of the experiences of individuals who perceive the changes taking place on a personal level, and we must remain mindful of similar issues that make up their migration experience.

In this article, I look at the migrations that took place after the Second World War due to the redrawing of political borders in Central Europe. I am particularly interested in the situation of Poland, which before World War II occupied areas of

M. Praczyk (✉)
Adam Mickiewicz University, Poznań, Poznań, Poland

© The Author(s) 2022
A. Izdebski et al. (eds.), *Perspectives on Public Policy in Societal-Environmental Crises*,
Risk, Systems and Decisions, https://doi.org/10.1007/978-3-030-94137-6_22

what is now western Ukraine, Belarus and Lithuania, and whose access to the sea was limited to the city of Gdynia and its environs, stretching north to the Hel peninsula near Gdańsk. Following the political decisions made at the Potsdam Conference in 1945, for example, it was decided to move the Polish geographical borders westwards and northwards. Thus, the country, having lost territories in the east, gained territories that had previously belonged to Germany, although its area was reduced by more than 70,000 square kilometres. However, this geopolitical decision of the Big Three resulted not only in shifting the borders on the map but also in a massive displacement of the population, who were forced to leave the towns and villages that they had inhabited, often for generations (Halicka 2020). This concerned both the Poles leaving the Polish eastern borderlands ("Kresy") for central Poland and the so-called "Recovered Territories" and the Germans leaving the eastern territories in Germany. As a consequence of these migrations, as well as other movements of the Polish population from other parts of Europe, more than 2.2 million people migrated to new areas. As regards Germans leaving the eastern German territories that were incorporated into Poland after the war, the number of migrants totalled around 2.6 million people. The decision to redraw Poland's borders also resulted in the country gaining a new climate zone in the Baltic region and new hardiness zones in the western parts of Poland, which was particularly significant from the point of view of plant cultivation practices (Praczyk 2018: 54).

On the other hand, when it comes to the condition of the ecosystem that was inhabited by the settlers, we must note two fundamental issues. The first concerns the demographic deficit, as the population that eventually settled the territory abandoned by the Germans was only half of the pre-war population. As a result of this disproportion, very interesting processes of uninhabited parts of villages (or entire villages) and parts of towns being reabsorbed by nature occurred in these areas. These territories witnessed a process of rationalisation, a decomposition of matter, and, in effect, post-anthropogenic landscapes emerged with exceptionally good conditions for the development of synanthropic species. In particular, these processes took place in the border area, due to post-war restrictions limiting settlement there. Consequently, the nature of the riverside areas around the Oder and Lusatian Neisse experienced a favourable transformation.

The second issue concerns the post-war destruction of the settlement area. War damage resulted not only in the destruction of towns and villages, the heritage of material culture, but also in a kind of ecocide. The military operations that were carried out at the end of the war led to the devastation of large swathes of forest, as well as significant areas of arable land. After the war, fallow land accounted for some 3.8 million hectares out of the 5.9 million hectares of agricultural land (Dziurzyński 1983: 183, Łach 1996: 12–16). During the hostilities, animals died en masse, not only mammals but also birds and fish. In turn, other species reproduced on an enormous scale, leading to the emergence of severe plagues, primarily of rodents and insects (Praczyk 2018: 276–283). Subsequently, the area inhabited by migrants, in those places where the fighting had taken place, had a post-apocalyptic quality.

The resettlement of populations, and often of animals migrating with humans, was a major logistical operation and required a series of political and administrative decisions that had environmental implications. In addition to official documents, reports and other source materials confirming the awareness of the environmental consequences of migration, primarily of the PUR (State Repatriation Office), the institution responsible for organising resettlement, I have also used diaries written by the resettled migrants to investigate their experiences.

I have examined more than a thousand memoirs, only some of which have been published (Praczyk 2018). The Polish memoirs about these events are unique on a global scale. The vast number of memoirs written after the war were the result of diary competitions organized by various institutions, encouraging people to send in their written experiences of the war and the post-war period. The idea for these competitions originated in the work of renowned Polish sociologists, dating back to the inter-war period, who tried to reach out to social groups, such as the peasantry, who had no opportunity to share their experiences in any other forum. The most important figure in this sociological milieu was Florian Znaniecki, who worked in the USA for many years (e.g. at the University of Chicago) and was a co-creator of the biographical method and the founder of Polish sociology (Kaźmierska 2015: 96). Some of the memoirs that allowed me to examine the significance of environmental issues in the experience of migration included those produced for the competition for memoirs of settlers in the "Recovered Territories". A particularly valuable feature of this corpus of memoirs is the social cross-section, as the authors of the memoirs were both men and women, representatives of all contemporary generations, with very different social and economic status.

Thanks to the documents from the first group of sources, we can observe the decision-makers' awareness of the importance of environmental conditions in the process of population resettlement. Selection of resettlement sites was to take into account the climate and soil so that the resettled areas resembled as closely as possible the migrants' places of origin. In particular, the aim was to create good conditions for agricultural development and to make use of the environmental and farming skills that the resettlers had at their disposal. In July 1945, therefore, a resettlement plan was drawn up which made provision for the geographical and natural features. This plan was commissioned by the Office for Settlement Studies, which included the Scientific Council for the Recovered Territories, and this council included representatives of the natural sciences, earth sciences and geographical sciences (IV Sesja Rady Naukowej dla Zagadnień Ziem Odzyskanych 1948). According to this plan, the settlement would be carried out in latitudinal zones, and in accordance with the soil and climatic ranges of the sites of origin and settlement. Despite the existing recommendations, difficulties in carrying out the settlement, due to both transport restrictions and traveller activity, for the most part prevented this plan from materialising. The routes of migration, especially in areas destroyed by warfare, were very chaotic. Consequently, the settlers, arriving at an appointed place, often started to journey further on their own to find a suitable settlement. Interestingly, during the meetings of the Scientific Council for the Recovered Territories, attention was paid not only to objective aspects, such as familiarity with the type of farmland, but also to subjective factors. The emotional ties between the population that was forced to migrate and nature were emphasised. It was also noted that in view of the poverty and often very modest living conditions of the peasants in the eastern areas of the Second Polish Republic, their attachment to the land and the natural surroundings resembled emotional attachment to the most precious objects. This was also an effect of identifying life with working the land. For the migrating peasants, the land (soil) was a crucial factor in forging their identity and their sense of belonging to the place they were forced to leave. The natural environment was thus an important identity-forming factor equivalent to, for example, the question of nationality, ethnicity or language community.

Another important factor influencing the shape of local ecosystems was the migration of animals, particularly of horses and cows. These animals were transported to areas previously affected by warfare, and thus also to the Recovered Territories as part of international aid (UNRRA), were purchased by the Polish government or brought with the settlers. The first group, for example, included about 100,000 horses and 16,000 cows mainly from the USA (Łaptos 2018: 197–2013). The animals which were purchased, numbering tens of thousands and coming from Iceland, the Netherlands and Sweden, were used to working in completely different environmental conditions and to living in completely different agroecosystems (Archiwum Akt Nowych, MZO 1444). In addition, about 300,000 animals accompanied the migrants, most of which were terribly exhausted by the journey and traumatised by the experiences of war and the conditions to which they were not accustomed. In addition, there were horses demobilised from the army, which were also distributed among the new hosts. Therefore, the settled areas were marked not only by the cultural diversity of the migrating people but also by the great variety of animals forced to adapt to new natural settings (Praczyk 2018, 189–191).

However, it was not until I looked into the migrants' memoirs that the complexity and importance of the environmental aspect of migration became apparent. The prevalence of memories relating to both pragmatic and emotional environmental conditions far exceeded my initial assumptions. Migration was beyond any doubt an important environmental challenge for the migrants, occupying a central place in their memories. In their recollections, they highlighted a number of elements that comprise the environmental dimension of migration. The most important of these were.

- the grief and fear of severing the emotional bond with the natural environment that they had lost;
- the natural environment perceived as a source of trauma, which emerged from observing the inhabited nature that had been damaged by war;
- the problem of domesticating the new natural space associated with the emotional sense of alienation, strengthened by the memory of being uprooted from their original area;
- ascribing new functions to the environment they were encountering;
- the mutual adjustment of various elements of the natural environment, including the people who inhabited it.

These processes are compounded by two important issues: generational differences, which influenced the perception of these problems; and gender differences, which, as my research has shown, did not play a fundamental role.

The first of these issues has to do with the breakdown of emotional ties with the natural environment caused by forced displacement. Therefore, the despair described by both male and female diarists was not only linked to the loss of their homes (as is usually pointed out) but precisely to the feeling of loss of the entire natural environment. The connection to this environment proved to be of vital importance, and the breaking of this link, traumatic. This applies both to relationships within whole ecosystems and to individual elements of nature, including animals with which people formed emotional bonds; this did not apply only to cats and dogs, but also to cows, horses and other animals which lived on farms with people and sometimes shared their homes with them. Sometimes in country houses, one of the rooms was reserved for livestock.

Memories of the abandoned environment featured descriptions of a whole range of sensory experiences that were not to be found in the new places. Thus, the migrants wrote about the unique smells of plants, the tastes of fruit and tactile impressions, especially in terms of the fertility of the land or the unforgettable microclimate of the lost forests, marshes and lakes. Recalling the soundscape of the places they had abandoned, attention was drawn to the singing of many species of birds, the sounds of insects and the murmur of forests, streams or rivers. These descriptions were in stark contrast to the emptiness and ominous silence of the memories of the areas being settled. One of the strategies for coping with being uprooted from one's former home was to bring at least a symbolic part of the environment along on the journey, such as tree or bush seedlings (mainly fruit trees) and clumps of earth, which were to become a symbolic souvenir in the new place. The same was done with twigs or other dried remains of plants that were important to people. One of the diarists recalled the day of departure as follows: "my father took out of his pocket a handkerchief as white as snow, into which he put a few lumps of earth. He did this with solemnity and great reverence. Tears as big as peas started to fall from his eyes. He did not even try to wipe them off." (Ośrodek "Karta": W.16) Some people wrote about bidding farewell to every tree, every corner of the garden.

Animals also played a very important role. It was sometimes possible to take a certain number of livestock, but only a few, which often meant abandoning the remaining droves of animals. As a rule (although there were exceptions), the opportunity to take animals on the road did not extend to dogs and cats. The memories of migrants are thus filled with heartrending descriptions of abandoned pets. It is worth recalling at least one such memory, written down by Helena Dragan: "A villager with his wife and a small child in nappies ... managed to reach the station. And when the train was about to leave, the villager approached the wagon, where a couple of pretty horses were standing, cried, and kissed the horses as if they were human faces. They both cried and, having left the horses, set off on their journey with their bags" (Instytut Zachodni: P216). The friendly relationships with animals, which the quotation illustrates, points to the subjectivity of the abandoned animals, even if the bonds formed between people and animals were of a professional sort.

This fact was already described by Eric Baratay, who wrote that the roles in which animals are usually inscribed in historiography, portrayed only in a utilitarian and subordinate function to humans, do not reflect the multidimensional relationships that were established between them and humans. The pragmatic rationale behind the presence of animals in people's lives did not cancel out the deep, emotional attachments that often developed in parallel (Baratay 2014: 53). Describing animals only in terms of things or when referring to human cruelty belittles and limits the image of human–non-human relationships, in aristocratic as well as peasant society. Moreover, powerful emotions are shown both by women and men. The stereotypically attributed higher emotionality of women towards nature was not confirmed in the diaries I have analysed.

The moment when the migrants left their homes and natural environment, however, was only the beginning of their migration experience. Gradually, first on the road and then in the process of settling new areas, further facets of environmental influence became evident. Noticing the differences between the abandoned and the new landscapes, vegetation types, landforms and air quality, and the presence of (very limited) animals, the settlers realised how important the nature they had lost was for them. The trauma of settling devastated territories was only partly related to life among the ruins. An equally important component of it was the witnessing of destroyed nature. When people came to the area of the Recovered Territories where warfare had taken place, they were confronted with military equipment, unexploded ordnance and carcasses strewn about the forests and fields. In places that the Germans had already left, forced as they had been by the Poles to leave their animals behind, feral dogs and cats wandered about, as did hungry cows and horses. Sometimes, in farms hurriedly vacated by Germans, settlers would come across the dead bodies of horses and cows tied up or locked inside barns, or dead or starving dogs chained up. For example, one can find such a trail of these traumatic images in the settlers' memories: "In a neighbour's yard across the road there was a cow lying half eaten by dogs. When I walked closer, a large dog jumped out from inside the cow.... Elsewhere, behind the barn, a cow lay calving and dead." (Instytut Zachodni: P145) As another settler remembered, "not far from the main road we entered an estate, or rather livestock buildings. We found over a hundred skeletons of cattle and horses tied to mangers." (OBN: R-822) These and many other similar descriptions dominate reminiscences of the arrival in the Recovered Territories. Interestingly, although the initial period of organising the natural environment usually spanned a relatively short period of time, from a few years to a maximum of a dozen years after settlement, in the settlers' accounts it is envisioned as a different epoch, a liminal situation, a moment of transition between the earlier and the later lives.

In this period, apart from the post-apocalyptic scenes cited above, the overriding emotions are anxiety, fear and disgust. Thus, we can read in the diaries about recurring forest fires caused by abandoned ammunition, about animals and people ripped apart by mines, about recurring infestations of rats and mice, which were very troublesome and revolting for several years: "My parents started cultivating gardens, ploughing and sowing fields. But their efforts were largely in vain. And it was all because of … mice. Because of the uncollected grain from the fields during the last year of the war, the unthreshed wheat stacks, the absence of people and predatory animals such as cats, these rodents multiplied on a scale unprecedented in our latitude" (Ośrodek "Karta": 208). There are even reports in memoirs of rodents eating fruit from trees (Instytut Zachodni: P125). The instability of the post-war ecosystem also haunted migrants in the form of floods and droughts, and fears of epidemics caused by the decaying organic matter of human and animal remains in the areas which they had settled.

Such a picture of the environment into which the settlers had encroached was countered by further contrasting memories of abandoned habitats. In this case, however, they were much more acute for the older generation, as in Zygmunt Sobolewski's recollection: "My father is not particularly enthusiastic about the land. According to him, it dries out too quickly, it is not rich enough. It's a far cry from the soil he left behind, and he's upset that it probably needs a lot of manure. And finally, he says that maybe it is not the worst, but it is not as good as our black soil" (Ośrodek "Karta": AWII/2242/P).

Among younger migrants, besides longing and resentment, there were also descriptions of enthusiasm for work, seeing good opportunities for living in the new place, and sometimes even contentment after leaving behind oppressive living conditions. Satisfaction was also expressed in the recollections of poor peasants, who, if they had settled on only slightly damaged farms, treated the situation as a step up in social hierarchy. They made use of the abandoned farming tools and the locally preserved agricultural infrastructure to improve their quality of life. Sometimes they were also delighted with the new landscape and plant species they had not come across before (e.g. magnolias).

Usually, however, taming the settled space was an arduous process that required sacrifice, and was tainted by a sense of alienation, as in Jan Szozda's memoir: "In the first days of settling in Duchowo it was sad, then slowly we got used to new farmyards, a different land, a different climate .... And so gradually everyone got absorbed in the daily work" (PIN-Instytut Śląski: A3156, Wol. 67O). Sometimes, however, this led to a gradual identification with the new environment, which was treated exceptionally subjectively, as in the memoirs of Józef Pacholak, who wrote that "such roaming in the fields and forests of Kwidzyn lasted two years. I learned about the area by collecting herbs and mushrooms. I developed strong bonds of friendship with the nature of the Recovered Territories" (Instytut Zachodni: P172). Such a culmination of settling and transforming the new ecosystems through one's own work can also be found in memories written years later, in retrospect. At this point I would like to quote one more recollection which captures this well: "Looking at the landscape around me—I compared it in my mind with the landscape of my homeland. And I must admit that I was yearning for the land where I had grown up. How different the local area was from my native land. Today, years later, I must say I did not come to like this land so quickly or get used to it so quickly. The work I put in, the effort and its fruits made me love this land, where one could say I left my sweat, and recognised it as my home" (Książnica Pomorska: 1970).

The examples of the migration of Poles after World War II that I have analysed demonstrate which factors should be considered when assessing the environmental effects of population movements in similar situations. On the one hand, objective factors are important, such as analysis of the degree of destruction, the demographics of the settled/resettled areas, an analysis of risks connected with infestations of various animal species (caused by the disruption of the ecosystem after a temporary or long-lasting disappearance of the population in cultivated areas, or by the devastation or excessive mortality of plant and animal species that had led to the disturbance of ecosystems and their biodiversity) and the possible contamination of part of the environment. On the other hand, one should also factor in aspects which are less frequently discussed, especially those related to the personal perception of nature and the space of the migrants, which influence the way they used the settled environment, as well as their well-being in the new place.

This case study also teaches us that unanticipated population movements and human behaviour that contribute to reshaping existing ecosystems must be accounted for. In effect, as the post-war Polish migrations show, local ecosystems are altered not only by hostilities but also by the entry of new settlers and political decisions (e.g. on the level of settlement of the areas, or new administrative divisions or functions). Due to the large number of farmers settling the Recovered Territories, for example, the character of many smaller towns changed, and they began to perform de facto functions typical of rural centres.

The secondary ecological succession that occurred in these areas as a result of the war led to a new quality of local ecosystems, which were different both from the pre-existing German ecosystems in the region and from those that the settlers had been forced to abandon. However, their habits and skills, as well as the varying levels of devastation of the local ecosystems and the new political circumstances, contributed to the formation of new socio-natural systems, which consisted both of lands reclaimed by nature and those areas of the war-devastated environment that were transformed and re-ordered by the migrants. Therefore, it is not appropriate to speak of an encroachment into a pre-existing "German" ecosystem but of a rupture and discontinuation brought about by a disaster (in this case, war) and a gradual restoration, already a new ecological balance, from which entirely new natural wholes emerged.

The settlers' memories of the environment shows the immense effort required to shape the new natural reality and, most importantly, the fundamental role played by nature that needed to be ordered and tamed. It demonstrates what an enormous challenge it is to restore or, in this case, shape the ecological balance. The settlers' memoirs also reveal that the transition from thinking of a new natural environment as foreign to a point where it becomes personal and "one's own" is usually a long and often psychologically painful process, comprising the experience of migration as much as other factors such as cultural or political adaptation. Ignoring this aspect of the migration experience can have disastrous consequences, not only for the migrating individuals but also for the new ecosystems created by mass human migration.

References

Bade K (2003) Migration in European history. Malden, Oxford, Carlton, s ix
Baratay É (2014) Zwierzęcy punkt widzenia. Inna wersja historii, transl. Paulina Tarasewicz, Gdańsk
Beata H (2020) The Polish Wild West. forced migration and cultural appropriation in the polish-German borderlands, 1945–1948. Routledge
Dziurzyński Patrycy (1983) Osadnictwo rolne na Ziemiach Odzyskanych, Warszawa
Halicka Beata (2020) The Polish Wild West. Forced Migration and Cultural Appropriation in the Polish-German Borderlands, 1945–1948, Routledge
Izdebski Adam (2018) Średniowieczni Rzymianie i przyroda, Kraków
Łach Stanisław (1996) Osadnictwo miejskie na ziemiach odzyskanych w latach 1945–1950, Słupsk
Łaptos Józef (2018) Humanitaryzm i polityka. Pomoc UNRRA dla Polski i polskich uchodźców w latach 1944–1947, Kraków
Poznań Każmierska Kaja (2015) Biobraphical and collective memory: mutual influences in central and eastern European context. In: Pakier M, Wawrzyniak J (eds), Memory and change in Europe. Eastern Perspective, Berghahn Books, pp 96–114
Praczyk Małgorzata (2018) Pamięć środowiskowa we wspomnieniach osadników na "Ziemiach Odzyskanych"
IV Sesja Rady Naukowej dla Zagadnień Ziem Odzyskanych. 18–21 XII
1946 r., z. VI: Potrzeby osadnictwa na ziemiach odzyskanych w zakresie komunikacji, Kraków 1948

Historical Sources

Archiwum Akt Nowych w Warszawie, Ministerstwo Ziem Odzyskanych

Memoirs

Ośrodek "Karta", Wypędzenie ze wschodu (1939–1959) — we wspomnieniach Polaków, Niemców i innych wypędzonych, 1997a, sygn. W.16, Stanisław Jastrzębski
Instytut Zachodni, Pamiętnik Osadnika Ziem Odzyskanych, 1957a, sygn. P216, Helena Dragan
Instytut Zachodni, Pamiętnik Osadnika Ziem Odzyskanych, 1957b, sygn. P145, Franciszek Kurpisz
Ośrodek Badań Naukowych im. J. Kętrzyńskiego w Olsztynie, Moje życie na Warmii i Mazurach, 1984–1985, sygn. R-822, Feliks Murawa
Ośrodek "Karta", Wypędzenie ze wschodu (1939–1959) — we wspomnieniach Polaków, Niemców i innych wypędzonych, 1997b, sygn. 208, Ryszard Bitowt
Instytut Zachodni, Pamiętnik Osadnika Ziem Odzyskanych, 1957c, sygn. P125, Józefa Nogat
Ośrodek "Karta", Wysiedlenie — wspólne doświadczenie narodów, 1994, sygn. AWII/2242/P, Zygmunt Sobolewski
PIN — Instytut Śląski, Pamiętniki trzech pokoleń mieszkańców Ziem Odzyskanych, sygn. A3156. Wol. 67, Jan Szozda
Instytut Zachodni, Pamiętnik Osadnika Ziem Odzyskanych, 1957d, sygn. P172, Józef Pacholak
Książnica Pomorska im. Stanisława Staszica w Szczecinie, Dzieje szczecińskich rodzin, 1970, Andrzej Kostański

Conclusions

Concluding Remarks: Interdisciplinarity and Public Policy

John Haldon

Abstract The main lesson from our discussion and from the combined wisdom of the chapters in this volume is that both critical history and interdisciplinarity are the two essential and inseparable ingredients required to understand the nature of past human responses to environmental and climate challenges. The costs of resilience need to be considered more fully than hitherto, and future planning and policy must take these, and any structural social inequalities underlying unequal cost distribution, into account. The relationship between agency and structure is fundamental to understanding the causal associations behind the outcomes we see in our evidence. To contribute to the achievement of true ecological justice, which might be the only hope for our survival, historians need to challenge environmental and social-economic or cultural determinism while at the same time bearing in mind the realities and struggles of day-to-day lives and the beliefs that are integral to them.

Keywords Environmental challenges · Complexity · Systemic flexibility · Redundancy · Cultural continuity · Costs of resilience · Critical history · Interdisciplinarity

The contributors to this volume have ranged widely across many different themes, but all have been concerned with past societal resilience, how we understand it today, and what this understanding might mean both for the way in which the contemporary world confronts significant environmental and climatic challenge, and for how the history of societal responses to past environmental challenges can be—should be—written. We place our analyses within a broader context, that of the increasing globalization of climate- and environment-related policy-making, and of increasing societal and technological complexity and thus of increasing systemic fragility (or brittleness). From our historical case-studies we have attempted to highlight, where possible, general patterns underlying past responses to environmental challenges, whether failures or successes, and to establish the extent to which systemic mechanisms such as feedback loops, cascades, tipping points and cycles are key underlying

J. Haldon (✉)
Princeton University, Princeton, NJ, USA
e-mail: jhaldon@princeton.edu

© The Author(s) 2022   345
A. Izdebski et al. (eds.), *Perspectives on Public Policy in Societal-Environmental Crises*,
Risk, Systems and Decisions, https://doi.org/10.1007/978-3-030-94137-6_23

characteristics of societies understood through the lens of a complex adaptive systems approach. Where such shared mechanisms can be identified, we suggested that they might provide invaluable help in pinpointing and understanding vulnerabilities not just within historical systems but also in the modern world, and thus contribute to the formation of a more resilient future.

Our various case-studies indicate that how societies in the past were able to respond to stress depended on three key sets of conditions: their complexity (the degree of interdependency across social relationships and structures), their institutional and ideological flexibility, and their systemic redundancy, all of which together determine the resilience of the system. Such conditions did not exist in isolation but combined and recombined in innumerable historical configurations. By examining particular historical case-studies we can detect both shared general patterns, show how each case is subtly different from the next and suggest how that leads into different or diverging developmental pathways.

Several of the chapters addressed these issues and showed that resilience and the potential for a society to maintain cohesion and cultural continuity through periods of system-challenging stress has costs. The distribution of the costs of resilience, and the degree to which this might be built into any system, have varied across time and cultural milieu. In the contributions that follow we present cases where we can observe both top-down and bottom-up responses to significant environmental challenges and the ways in which different sectors of society responded or reacted (and with either positive or negative outcomes); the differential costs of resilience when a state, or a society or a specific sector within a society is faced with substantial economic and political challenges; and the nature of state and society-level responses to stress factors and both planned as well as unintended consequences.

One key element that has become clear is that effective future planning needs to take account not only of a given environmental state, but also that of the dynamics underlying it—what are the various potentials for change or transformation that inhere within a particular context? By the same token, planning needs also to consider the fact that socio-economic inequalities and imbalances need to be addressed as part of any response to an environmental challenge, both at a local/regional as well as at a national and international scale. Resilience consists of multiple contributing factors, including both institutional and systemic frameworks as well as agents' perceptions and their potential to generate socially and culturally viable responses. An idealized response, divorced from the realities and struggles of day-to-day lives, is destined to fail, or to massively underprivilege a substantial sector of a population—with consequent medium- and long-term consequences for the resilience of the society as a whole to respond adequately to existential challenge. In the past, the burden of resilience has often fallen disproportionately upon those least able to bear it, for example, a factor which remains still a significant problem in policy thinking.

Likewise, how we write about past resilience, or its absence, can play a significant role. Historians and archaeologists need to nuance their interpretation of the past in order to understand the whys and wherefores of past human responses and thus contribute to undermine both environmental and social-economic or cultural determinism and bring into focus the range of potential adaptive pathways—an approach that can offer significant important information in thinking about contemporary adaptation strategies in confronting climate. This requires a refusal to engage in an apocalyptic rhetoric of shock and despair or of sensationalist claims about collapse and catastrophe. But it also demands the active intervention of the historian (for example) in the thought-world of her or his audience or readership, whether the general public or a specific target group such as policy-makers and planners, permitting all of these to attribute meaning to the histories they have experienced within the context of their own lives and experiences.

Yet at the same time attempting to make the lessons of the past accessible in the context of planning for the future brings with it the problem of how the future is supposed to look since, as we have now seen, 'anticipatory knowledge' always objectifies the future. Does reflecting on possible futures create greater awareness of the observers' agency, does it chiefly suggest modes of enhancing efficient action, or does it in fact narrow the horizons for alternative ways of building resilience into a system? Given the fact that we are all, always, constrained by the possibilities inhering in our own historically-determined symbolic universe, this is a real issue that needs to be confronted, and some of the contributions to this volume have suggested appropriate strategies for addressing the question.

Our contributors show, collectively, that both critical history and interdisciplinarity are the two essential and inseparable ingredients required to understand the nature of past human responses to environmental and climate challenges. They also show that separating but integrating agency and structure are crucial to understanding the causal relationships underpinning or concealed by the outcomes we can observe in our various types of data, historical, archaeological and palaeoenvironmental. Collaboration between the social or humanistic and the natural sciences can make a key contribution, both in understanding what the primary requirements are for our own survival as a species, as well as in determining how effectively we can respond to the challenges we face and at what level—local, national, hemispheric or global. As one of the editors concluded in his contribution: only a history which is a dialogue of the old and the new can open our minds to move in the direction of true ecological justice, which might be our hope for survival and flourishing.